INTRODUCTION TO GENERAL RELATIVITY, BLACK HOLES, AND COSMOLOGY

Introduction to General Relativity, Black Holes, and Cosmology

Yvonne Choquet-Bruhat

French Academy of Sciences, Paris,
and the American Academy of Arts and Science

With a foreword by

Thibault Damour

Permanent Professor, Institut des Hautes Études Scientifique
Member of the French Academy of Sciences, Paris

OXFORD
UNIVERSITY PRESS

OXFORD

UNIVERSITY PRESS

Great Clarendon Street, Oxford, OX2 6DP,
United Kingdom

Oxford University Press is a department of the University of Oxford.
It furthers the University's objective of excellence in research, scholarship,
and education by publishing worldwide. Oxford is a registered trade mark of
Oxford University Press in the UK and in certain other countries

Published in the United States of America by Oxford University Press
198 Madison Avenue, New York, NY 10016, United States of America

British Library Cataloguing in Publication Data

Data available

Library of Congress Control Number: 2014934913

ISBN 978–0–19–966645–4 (hbk.)
ISBN 978–0–19–966646–1 (pbk.)

Printed and bound by
CPI Group (UK) Ltd, Croydon, CR0 4YY

Foreword

Yvonne Choquet-Bruhat has made many deep and lasting contributions to mathematical and physical aspects of General Relativity, starting with her epoch-making 1952 proof of the well-posedness of the Cauchy problem for Einstein's equations. We are all very fortunate that she has undertaken to present, in terms accessible to all, a comprehensive account of all the aspects of General Relativity. Indeed, this beautiful book is quite unique both in the range of topics it covers and in the way each topic is treated.

First, the range of topics presented by Yvonne Choquet-Bruhat is truly remarkable. She covers successively the basics of Riemannian geometry (Chapter I) and of Special Relativity (Chapter II); the mathematical and physical definition of General Relativity (Chapter III); the main properties and consequences of Einstein's field equations (Chapter IV); the mathematics and physics of the Schwarzschild spacetime (Chapter V); a mathematically rigorous discussion of black holes (Chapter VI); a brief, but up to date, discussion of relativistic cosmology (Chapter VII); a thorough presentation of the Cauchy problem (Chapter VIII); and last, but not least, a detailed discussion of two of the most important phenomenological matter models, namely relativistic fluids (Chapter IX) and the relativistic kinetic theory of N-particle systems (Chapter X).

But, most importantly, the treatment of each one of these broad topics is both very comprehensive and remarkably concise. She has succeeded in reaching a Landau- and Lifshitz-like ideal of covering all the crucial issues in the most concise way, while expounding each topic in a mathematically rigourous way. This rare combination of qualities makes this book particularly valuable. For instance, her discussion of the Schwarzschild spacetime covers, in one go: (i) a derivation of the solution that includes a proof of Birkhoff's theorem; (ii) the form of the solution in five different types of coordinates; (iii) a preliminary discussion of the event horizon (which is developed in the following chapter); (iv) the motion of planets and of light, and their comparison with the most recent observations; (v) the stability of circular orbits; (vi) a presentation of Fermat's principle that includes its little-known generalization to arbitrary spacetimes due to G. Ferrarese; (vii) the redshift and time-delay effects; (viii) a discussion of spherically symmetric interior solutions that includes several theorems about their general properties; (ix) the Reissner-Nordström solution; (x) the Schwarzschild solution in any dimension; and (xi) a precise but concise account of the results of Gu Chao Hao, Christodoulou, and others on spherically symmetric gravitational collapse. The chapter is then nicely capped by some problems, followed by their solutions.

I am sure that this remarkably concise and complete book by Yvonne Choquet-Bruhat will allow many readers to achieve a deep understanding of General Relativity through her unique mathematico-physical knowledge of one of the fundamental pillars of modern physics. Thank you, Yvonne for sharing with us the deep wisdom you have acquired during your lifelong exploration of the beautiful theoretical landscape opened, nearly a century ago, by Albert Einstein.

THIBAULT DAMOUR
Institut des Hautes Études Scientifiques

Preface

Special Relativity (1905) and General Relativity (1915), both due to Einstein's genius, are fundamental pillars of modern physics. They have revolutionized the scientific concepts of space and time, formerly due to everyday experience at a human scale, and also to previous scientific work and experiments made with clocks, very precise and reliable at the considered scale. These concepts of absolute space and absolute time had become ingrained in the minds of common folk as well as scientists, and were difficult to overthrow for a long time.[1] Nowadays, the world explored by humanity, at microscopic as well as cosmological scales, has become very much greater, and old concepts have had to be replaced by more general ones. Spectacular improvements in technology have changed the lives of a large part of humanity, and new information processes have permitted easier access to scientific knowledge and the acceptance of new concepts.

Modern physical theories have a mathematical formulation, often geometrical, with consequences deduced from mathematical theorems. The validity of a physical theory results from the verification of its consequences by observations or experiments.

General Relativity is a beautiful geometric theory, simple in its general mathematical formulation, which has numerous consequences with striking physical interpretations: gravitational waves, black holes, cosmological models. Several of these consequences have been verified with a great accuracy. The Einstein equations present a wide variety of new and interesting mathematical problems with possible physical interpretation.

The aim of this book is to present with precision, but as simply as possible, the foundations and main consequences of General Relativity. It is written for an audience of mathematics students interested in physics and physics students interested in exact mathematical formulations—or indeed for anyone with a scientific mind and curious to know more of the world we live in. The mathematical level of the first seven chapters is that of undergraduates specializing in mathematics or physics; these chapters could be the basis for a course on General Relativity. The next three chapters are more advanced, though not requiring very sophisticated mathematics; they are aimed at graduate students, lecturers, and researchers. No a priori specialized physics knowledge is required. These chapters could serve as the text for a course for graduate students.

I wish to everyone a good trip in this strange but fascinating world.

Yvonne Choquet-Bruhat

[1] For instance, the fact that two twins who live different lives age differently was called the 'twin paradox', although in fact there is no paradox there, except in the human-built definition of time.

Acknowledgements

My gratitude goes to IHES, its scientific director Jean Pierre Bourguignon, and its permanent professor Thibault Damour, for inducing Pierre Cartier, who kindly accepted, to give me a desk in his office. I have thus benefited for the last ten years from a pleasant environment and working facilities together with instructive seminars and conversations that have been intellectually stimulating as well as friendly.

I am very fortunate that I have been able to meet regularly at IHES Thibault Damour, who has always been ready to answer my questions and allow me to benefit from his great and wide knowledge of relativistic physics and cosmology, as well as of the varied elaborate mathematical tools used in this field—from the oldest to the most recent. He was kind enough to read the whole of a preliminary version of this book. The present version has been very much improved through his information and his constructive criticism. Thank you Thibault—this book would never have been completed without your help and encouragement.

I also thank Richard Kerner, who spent time reading my text, pointing out useful corrections.

I thank François Bachelier for his pertinent help with tex.

Cover painting: *Cosmos Birth* by Geneviève Choquet-Kastylevsky.

Notation

The sign $\equiv$ denotes an identity and the sign $:=$ a definition, although sometimes when there is no confusion possible I denote by the simple sign $=$ one or the other of these.

In the text, I usually denote by a single letter a geometric object, for example X or T for a vector or tensor. In equations, I prefer to use indices, since this makes operations simpler to write and understand—without speaking of possible numerical or algebraic computing applications.

The spacetime metric signature is $-++\ldots+$. In the Lorentzian case, the spacetime indices are Greek letter. When specified, a time index is denoted by zero, and the space indices by Latin letters from the middle of the alphabet, $i, j, \ldots$.

Definitions of new notions are in bold characters and are assembled in the index.

Note on references

I have not tried to find for each result an original source. This would have been a difficult and sometimes controversial task. In the side notes, I give references, assembled at the end of the book, to papers I have used, where the interested reader can find relevant details, complements, and references to previous papers about the subjects treated.

- The reference YCB-OUP2009 is to Yvonne Choquet-Bruhat *General Relativity and the Einstein Equations*, published by Oxford University Press in 2009.
- The references CB-DMI and II are to Y. Choquet-Bruhat and C. DeWitt-Morette, *Analysis, Manifolds and Physics*, Parts I and II, published by North-Holland in 1982 and 2000, respectively.

Contents

Part B   Advanced topics

Part A

Fundamentals

This part contains the mathematical definitions and physical interpretations necessary for a basic understanding of Einstein's General Relativity. The basic definitions and properties of Riemannian and Lorentzian geometry are given. Special Relativity is introduced, with discussion of proper time, the equivalence of mass and energy, and physical observations. The spacetimes of General Relativity are defined, and their basic mathematical properties and physical interpretations are described, along with comments on experimental results. The theoretical foundations of the Einstein equations are given and these equations are derived. The Newtonian and Minkowskian approximations of their solutions are described. Gravitational waves and gravitational radiation are introduced. The spherically symmetric Schwarzschild spacetime is deduced from the Einstein equations in vacuum, its mathematical consequences are computed, and their observational confirmation is described. Black holes, a phenomenon unknown to Newtonian mechanics, but predicted by the Einstein equations, are discussed. Both spherically symmetric and axisymmetric (Kerr) black holes are constructed and studied. Some mathematical results and conjectures are introduced, and an introduction to the thermodynamics of black holes and Hawking radiation is provided. Relevant astrophysical observations are described. The construction of the first cosmological models—Einstein, de Sitter, and anti-de Sitter—is given. General Robertson–Walker spacetimes, the cosmological redshift, and the Hubble law are discussed. Friedman–Lemaître universes and the presently accepted ΛCDM model with accelerated expansion are described, as is the current majority view on the content of the universe, including the mysterious dark matter and dark energy. A brief account of primordial cosmology is given.

Riemannian and Lorentzian geometry

I.1 Introduction

We give in this chapter a survey of the basic definitions of Riemannian and Lorentzian differential geometry[1] that are necessary in General Relativity.

I.2 Differentiable manifolds and mappings

I.2.1 Differentiable manifolds

The fundamental arena of differential geometry is a differentiable manifold. For the physicist, the most concrete and useful definition makes apparent the local identification of a manifold with R^n, the space of sets of n real numbers with its usual topology.[2] The definition proceeds as follows.

A **chart** on a set X is a pair (U, ϕ), with U a subset of X, called the domain of the chart, and ϕ a bijection from U onto an open set u of R^n, i.e. a one-to-one invertible mapping $\phi : U \to u$ by $x \mapsto \phi(x) \equiv (x^1, \ldots, x^n)$. The numbers x^i, $i = 1, \ldots, n$, are called **local coordinates** of the point $x \in X$.

An **atlas** on X is a collection of charts (U_I, ϕ_I), with $\{I\}$ an arbitrary set of indices, whose domains cover X.

A mapping f between open sets of R^n, $f : u \to v$, is called a **homeomorphism** if it is bijective and if f and its inverse mapping f^{-1} are continuous. A bijective mapping is a C^k **diffeomorphism** if f and f^{-1} are of class C^k.

Exercise I.2.1 *Prove that a C^1 diffeomorphism with f of class C^k is a C^k diffeomorphism.*

Hint: $\partial(f f^{-1})/\partial x^i \equiv 0$.

An atlas endows X with the structure of a **topological manifold**, of dimension n, if the mappings $\phi_I \circ \phi_J^{-1}$ are homeomorphisms between the open sets of R^n, $\phi_J(U_J \cap U_I)$ and $\phi_I(U_J \cap U_I)$. The mappings $\phi_I \circ \phi_J^{-1}$ define changes of local coordinates in the intersection $U_J \cap U_I$, $(x^1, \ldots, x^n) \mapsto (y^1, \ldots, y^n)$. The manifold is a **differentiable manifold** of class C^k if these mappings are C^k diffeomorphisms; that is, if

---

[1]More complete treatments can be found in many textbooks, in particular in CB-DMI.

[2]A basis of open sets in R comprises the sets determined by the order relation $a < x < b$. The topology of R^n is the direct product of n copies of R.

the functions $y^i(x^1, \ldots, x^n)$ are of class C^k and the Jacobian determinant $D(y)/D(x)$ with elements the partial derivatives $\partial y^i/\partial x^j$ is different from zero. Henceforth, by **smooth** we shall mean of class C^k with k large enough for the statement under consideration to be true; a particular case is $k = \infty$.

Two C^k atlases on X are said to be equivalent if their union is again a C^k atlas. We consider them to define the same C^k manifold. It is known that any C^1 manifold can be endowed with a C^∞ structure. Unless otherwise stated, all manifolds considered here are assumed to be smooth.

The given definition of a manifold does not imply that it is a Hausdorff topological space; that is, it admits the possibility for two points not to have non-intersecting neighbourhoods. In the following, we will, unless otherwise stated, only consider **Hausdorff manifolds** and call them simply manifolds.

Examples of manifolds are the sphere and the torus.

An open set of R^n is oriented by the order of the coordinates $(x^1, \ldots, x^n)$. A differentiable manifold is said to be **orientable** (and oriented by the choice of coordinates) if its defining atlas is such that $D(y)/D(x) > 0$ in all intersections of domains of charts.

Unless otherwise stated, the manifolds considered here will be C^∞, connected and oriented.

I.2.2   Differentiable mappings

A function f on an n-dimensional manifold V^n is a mapping $V^n \to R$ by $x \mapsto f(x)$. Its representative in local coordinates of the chart (U, ϕ) is a function on an open set of R^n, $f_\phi := f \circ \phi^{-1} : (x^i) \mapsto f(\phi^{-1}(x^i))$.

The function f is of class C^k at x if f_ϕ is of class C^k at $\phi^{-1}(x)$. This definition is chart-independent if V^n is smooth. The gradient, also called differential, of f is represented in a chart by the partial derivatives of f_ϕ. If (U, ϕ) and (U', ϕ') are two charts containing x, it holds that at x (calculus relations)

$$\frac{\partial f_\phi}{\partial x^i} = \frac{\partial f_{\phi'}}{\partial x^{j'}} \frac{\partial x^{j'}}{\partial x^i}. \tag{I.2.1}$$

This equivalence relation entitles the differential of f to be called a covariant vector (see below). A covariant vector is a geometric object, independent of a particular of coordinates choice.

A differentiable mapping f between differentiable manifolds and diffeomorphisms can be defined analogously. The differential of f at $x \in W^n$, the source, is represented in a chart around x and a chart at $f(x) \in V^p$, the target, by a linear mapping from R^n into R^p; that is, an $n \times p$ matrix with elements $\partial f^\alpha/\partial x^i$, $\alpha = 1, \ldots, p$, $i = 1, \ldots, n$.

I.2.3   Submanifolds

An **embedded submanifold** of V is the image by an injective differentiable mapping of a smooth manifold W.

I.2.4   Tangent and cotangent spaces

A **tangent vector** v to a differentiable manifold V at a point $x \in V$ is a geometric object. It can be defined through local coordinates as an equivalence class of triplets $(U_I, \phi_I, v_{\phi_I})$, where (U_I, ϕ_I) are charts containing x, while $v_{\phi_I} = (v^i_{\phi_I})$, $i = 1, \ldots, n$, are vectors in R^n. The equivalence relation is given by

$$v^i_{\phi_I} = v^j_{\phi_J} \frac{\partial x^i_I}{\partial x^j_J}, \tag{I.2.2}$$

where x^i_I and x^i_J are respectively local coordinates in the charts (U_I, ϕ_I) and (U_J, ϕ_J). The vector $v_\phi \in R^n$ is the **representative** of the vector v in the chart (U, ϕ). The vector v is attached to the manifold by the assumption, compatible with the equivalence relation, that the numbers v^i_ϕ are the components of v_ϕ in the frame of R^n defined by the tangents to the coordinate curves, curves in R^n where only one coordinate varies.

Tangent vectors at x constitute a vector space, the **tangent space** to V^n at x, which is denoted $T_x V^n$. The set of pairs (x, v_x), $x \in V^n$, $v_x \in T_x V^n$, denoted by TV^n, is called the **tangent bundle** to V^n.

An arbitrary set of n linearly independent tangent vectors at x constitute a **frame** at x. The **natural frame** associated to a chart (U, ϕ) is the set of n vectors $e_{(i)}$, $i = 1, \ldots, n$, such that $e^j_{(i),\phi} = \delta^j_i$; they are represented in local coordinates by tangent vectors to the coordinates curves of the chart. The numbers $v^i_{\phi_I}$ are the components of the vector v in the natural frame. Traditionally, indices of components of vectors are written upstairs.

The **cotangent space** T^*_x to V is the dual of T_x, that is, the space of 1-forms on T_x, which are also called **covariant vectors** at $x \in V$, the tangent vectors being called **contravariant**. The components of a covariant vector at x in a chart (U, ϕ) containing x, are a set of n numbers v_i, $i = 1, \ldots, n$, with indices traditionally written downstairs. Under a change of chart from (U, ϕ) to (U, ϕ'), it holds that

$$v'_i = v_j \frac{\partial x^j}{\partial x'^i}. \tag{I.2.3}$$

Covariant vectors can be defined by this equivalence relation. By (2.1), the differential at x of a differentiable function f is a covariant vector, called the gradient of f. The **natural coframe** is the set of differentials dx^i of the coordinate functions $x \mapsto x^i$.

I.2.5   Vector fields and 1-forms

A **vector field** on V assigns a tangent vector at $x \in V$ to each point x.

The relations (2.1) and (2.2) show that, given a differentiable function f on V, the quantity $v(f)$ defined for points x in the domain U of the chart (U, ϕ) by

$$v(f) := v_\phi^i \frac{\partial f_\phi}{\partial x_\phi^i}, \quad f_\phi := f \circ \phi^{-1} \tag{I.2.4}$$

is chart-independent: v defines a mapping between differentiable functions. It is easy to check that this mapping is additive,[3]

$$v(f + g) = v(f) + v(g), \tag{I.2.5}$$

and satisfies the Leibniz rule,

$$v(fg) = fv(g) + gv(f). \tag{I.2.6}$$

The properties (2.5) and (2.6) characterize a **derivation operator**, and $v(f)$ is called the **derivative of f along v**.

If we take for v a vector of a natural frame $e_{\phi,(i)}$, that is, $v^j = \delta_i^j$, then

$$v(f) \equiv e_{\phi,(i)}(f) = \frac{\partial(f \circ \phi^{-1})}{\partial x_\phi^i}. \tag{I.2.7}$$

Alternatively, a tangent vector at a point x on a differentiable manifold can be defined, without first introducing its representatives and the equivalence relation, as the value at x of a linear first-order derivation operator acting on differentiable functions defined in a neighbourhood of x.[4] The natural frame can thus be defined as the set of operators $\partial/\partial x^i$, $i = 1, \ldots, n$.

A covariant vector field on V assigns a covariant vector at x to each point $x \in V$. It is called a **1-form on V**.

I.2.6   Moving frames

A **moving frame**, often called simply a frame in what follows, in a subset U of a differentiable n-dimensional manifold V is a set of n vector fields on U that are linearly independent in the tangent space $T_x V$ at each point $x \in U$. A manifold that admits global frames (not necessarily global coordinates) is called **parallelizable**.[5]

A coframe on U is a set of n 1-forms θ^i that are linearly independent at each $x \in U$ in the dual space $T_x^* V$. In the domain U of a coordinate chart, a coframe is specified by n linearly independent differential 1-forms

$$\theta^i \equiv a_j^i dx^j, \tag{I.2.8}$$

with a_j^i functions on U.

---

[3]Even linear, that is also such that $v(\lambda f) = \lambda v(f)$, with λ a constant in R, a property included in the Leibniz rule.

[4]See, for instance, CB-DMI, III B 1.

[5]It has been proved that all orientable differentiable 3-manifolds are parallelizable; this is not true in all other dimensions.

The differential of a differentiable function f is a covariant vector field called an **exact 1-form**

$$df := \frac{\partial f}{\partial x^i} dx^i \equiv \partial_i f \theta^i.$$

Its components $\partial_i f$ in the coframe θ^i are called **Pfaff derivatives**.

The dual frame to a coframe θ^i is the set of vector fields such that

$$\theta^i(e_j) = \delta^i_j. \tag{I.2.9}$$

The natural frame $\partial/\partial x^i$ and the natural coframe dx^i are dual.

I.3 Tensors and tensor fields

I.3.1 Tensors, products and contraction

A covariant p-tensor at a point $x \in V$ can be defined as a p-multilinear form on p direct products of the tangent space $T_x V$ with itself. Contravariant tensors can be defined as multilinear forms on direct products of the cotangent space $T_x^* V$.

Tensors, contravariant or covariant, can also be defined, like vectors, through equivalence relations between their components in various charts. For instance a 1-contravariant 2-covariant tensor T at $x \in V$ is an equivalence class of triplets $(U_I, \phi_I, T_{\phi_I, ij}{}^h)$, $i, j, h = 1, \ldots, n$, with the equivalence relation being the law of change of components of T by change of local coordinates from (U, ϕ) to (U, ϕ') given by

$$T_{i'j'}{}^{h'} = \frac{\partial x^k}{\partial x^{i'}} \frac{\partial x^l}{\partial x^{j'}} \frac{\partial x^{h'}}{\partial x^m} T_{kl}{}^m. \tag{I.3.1}$$

The space of tensors of a given type is a vector space, in which summation of elements and their multiplication by scalars are defined in terms of these operations acting on components.

The **tensor product** $S \otimes T$ of a p-tensor S and a q-tensor T is intrinsically defined: it is the $(p + q)$-tensor with components defined by products of components. For example, the tensor product of a covariant vector ω with a contravariant 2-tensor T is the mixed 3-tensor $\omega \otimes T$ with components

$$(\omega \otimes T)_i{}^{jk} = \omega_i T^{jk}. \tag{I.3.2}$$

Although products of components are commutative, tensor products are *non-commutative:* $\omega \otimes T$ and $T \otimes \omega$ of the previous example do not belong to the same vector space. A p-covariant and q-covariant tensor is an element of the tensor product of p copies of T_x and q copies of T_x^*, but different orders of these products give different spaces of tensors: $T_x^* \otimes T_x \neq T_x \otimes T_x^*$.

A basis of the vector space of tensors of a given type is obtained by tensor products of vectors and covectors of bases. For example, a natural basis of the space of covariant 2-tensors at x is $dx^i \otimes dx^j$:

$$T = T_{ij} dx^i \otimes dx^j. \qquad (\text{I.3.3})$$

The **symmetry** [respectively **antisymmetry**] properties, for example $T_{ij} = T_{ji}$ [respectively $T_{ij} = -T_{ji}$], are intrinsic properties (i.e. independent of coordinates).

Exercise I.3.1   *Check these properties.*

The **contracted product** of a p-contravariant tensor and a q-covariant tensor is a tensor of order $p + q - 2$ whose components are obtained by summing over a repeated index appearing once upstairs and once downstairs.

I.3.2   Tensor fields. Pullback and Lie derivative

Tensor fields are assignments of a tensor at x to each point x of the manifold V. Differentiability can be defined in charts; the notion of a C^k tensor field is chart-independent if the manifold is of class C^{k+1}.

The image of a contravariant tensor field on V under a differentiable mapping between differentiable manifolds $f : V \to W$ is not necessarily a tensor field on W unless f is a diffeomorphism.

Exercise I.3.2   *Prove this statement in the case of a vector field.*

The **pullback** f^* on V of a covariant tensor field on W is a covariant tensor field. For instance, for a covariant vector ω and a mapping f given in local charts by the functions $y^\alpha = f^\alpha(x^i)$, with y^α coordinates on W and x^i coordinates on V, the pullback is defined by

$$(f^*\omega)_i(x) = \frac{\partial y^\alpha}{\partial x^i} \omega_\alpha(y(x)). \qquad (\text{I.3.4})$$

Exercise I.3.3   *Extend the definition and property to covariant tensors.*

The **Lie derivative** of a tensor field T with respect to a vector field X is a derivation operator from p-tensors into p-tensors. If f_t is the one-parameter local group of diffeomorphisms[6] generated by X, then the Lie derivative at $x \in V$ of a contravariant tensor T is defined by

$$(\mathcal{L}_X T)(x) := \lim_{t=0}[(f_t^{-1})' T(f_t(x)) - T(x)]. \qquad (\text{I.3.5})$$

For example, for a contravariant 2-tensor, it is computed to be

$$(\mathcal{L}_X T)^{jk} = X^i \frac{\partial T^{jk}}{\partial x^i} - T^{ik} \frac{\partial X^j}{\partial x^i} - T^{ji} \frac{\partial X^k}{\partial x^i}. \qquad (\text{I.3.6})$$

The Lie derivative of a vector field X in the direction of the vector field Y is called the **Lie bracket** of X and Y; it has components

$$(\mathcal{L}_X Y)^j \equiv [X, Y]^j \equiv X^i \partial_i Y^j - Y^i \partial_i X^j.$$

[6]See, for instance, CB-DMI, III C.

An analogous definition gives Lie derivatives of covariant tensors. For instance, for a covariant 2-tensor,

$$(\mathcal{L}_X T)_{jk} = X^i \frac{\partial T_{jk}}{\partial x^i} + T_{ik} \frac{\partial X^i}{\partial x^j} + T_{ji} \frac{\partial X^i}{\partial x^k}. \tag{I.3.7}$$

Exercise I.3.4 *Check that the quantities (3.6) [respectively (3.7)] transform as the components of a 2-contravariant tensor [respectively a 2-covariant tensor] under a change of coordinates.*

I.3.3   Exterior forms

A totally antisymmetric covariant p-tensor field is also called an **exterior p-form**. A natural basis of the space of p-forms is obtained by antisymmetrization. We call the p-**antisymmetrization operator** the mixed tensor with components $\varepsilon^{12\ldots p}_{i_1 \ldots i_p} = 1$ if $i_1 \ldots i_p$ is an even permutation of $12 \ldots p$ and $\varepsilon^{12\ldots p}_{i_1 \ldots i_p} = -1$ if it is an odd permutation The components of a p-form can be written

$$\omega_{i_1 \ldots i_p} = \varepsilon^{12\ldots p}_{i_1 \ldots i_p} \omega_{12 \ldots p}$$

A p-form is determined by the data of its component with indices of increasing order.

We define the **exterior product** of exterior forms by antisymmetrization of tensor products. A basic example is the exterior product of two 1-forms given by[7]

$$dx^i \wedge dx^j = -dx^j \wedge dx^i := \frac{1}{2}(dx^i \otimes dx^j - dx^j \otimes dx^i). \tag{I.3.8}$$

Using exterior products, a p-form reads in a natural coframe

$$\omega = \frac{1}{p!}\omega_{i_1 \ldots i_p} dx^{i_1} \wedge \ldots \wedge dx^{i_p}. \tag{I.3.9}$$

The **exterior derivative** of a p-form is a $(p+1)$-form given in local coordinates by

$$d\omega := \frac{1}{p!}d\omega_{i_1 \ldots i_p} dx^{i_1} \wedge \ldots \wedge dx^{i_p} \equiv \frac{1}{p!}\frac{\partial}{\partial x^j}\omega_{i_1 \ldots i_p} dx^j \wedge dx^{i_1} \wedge \ldots \wedge dx^{i_p}. \tag{I.3.10}$$

A form whose differential is zero is called a **closed form**. A form that is the differential of an exterior form is called an **exact form**. An exact form is a closed form. The reciprocal of this property is true on manifolds diffeomorphic to R^n (**Poincaré lemma**), but not on general manifolds. A necessary and sufficient condition is given by the **de Rham theorem**.[8]

Example I.3.1 *Any n-form on the compact n-manifold S^n is closed, but is an exact differential only if its integral on S^n vanishes.*

[7]Note that there is an alternative definition of the exterior product in which the factor of $\frac{1}{2}$ is omitted: $dx^i \wedge dx^j = dx^i \otimes dx^j - dx^j \otimes dx^i$.

[8]The sufficient condition is that the integral of the p-form is zero on p-cycles.

The **interior product** of a p-form with a vector field X is the $(p-1)$-form given in local coordinates by the formula

$$(i_X\omega)_{i_1\ldots i_{p-1}} := \frac{1}{(p-1)!}X^h\omega_{hi_1\ldots i_{p-1}}.$$

Lemma I.3.1 *The **Lie derivative of an exterior form** ω with respect to a vector field X is given by the formula*

$$\mathcal{L}_X\omega \equiv i_X d\omega + d(i_X\omega).$$

Exercise I.3.5 *Prove this formula for 2-forms by using local coordinates.*

I.4   Structure coefficients of moving frames

For the natural coordinates x^i in a chart of a manifold, which are functions $x \mapsto x^i$, it holds that $\partial x^i/\partial x^j = \delta^i_j$, $d^2x^i \equiv 0$. However, for a moving coframe in a domain U, the differentials of the 1-forms θ^i do not vanish in general: they are given by 2-forms

$$d\theta^i \equiv -\frac{1}{2}C^i_{jh}\theta^j \wedge \theta^h. \tag{I.4.1}$$

The functions C^i_{jh} on U, antisymmetric in j and k, are called the **structure coefficients** of the frame.

The Pfaffian derivatives ∂_i in the coframe θ^i of a function f on U are defined by $df \equiv \partial_i f \, \theta^i$. The identity $d^2 f \equiv 0$ implies that

$$d^2 f \equiv \frac{1}{2}(\partial_i\partial_j f - \partial_j\partial_i f - C^h_{ij}\partial_h f)\theta^i \wedge \theta^j \equiv 0; \tag{I.4.2}$$

hence, Pfaffian derivatives in contrast to ordinary partial derivatives, do not generally commute,

$$(\partial_i\partial_j - \partial_j\partial_i)f \equiv C^h_{ij}\partial_h f. \tag{I.4.3}$$

Exercise I.4.1 *Show that the structure coefficients of a coframe $\theta^i := a^i_j dx^j$ are given by*

$$C^i_{hk} \equiv A^j_k\partial_h a^i_j - A^j_h\partial_h a^i_j$$

where A is the matrix inverse of the matrix a.

We note that the basis e_i dual to θ^i satisfies the commutation conditions

$$[e_i, e_j] = \frac{1}{2}C^h_{ij}e_h,$$

where $[.,]$ denotes the **Lie bracket** of vector fields:

$$[v, w] := \mathcal{L}_v w = -\mathcal{L}_w v.$$

I.5   Pseudo-Riemannian metrics

I.5.1   General properties

A **pseudo-Riemannian metric** on a manifold V is a symmetric covariant 2-tensor field g such that the quadratic form it defines on contravariant vectors, $g(X, X)$, given in local charts by $g_{ij} X^i X^j$, is non-degenerate; that is, the determinant $\mathrm{Det}(g)$ with elements g_{ij} does not vanish in any chart; this property is independent of the choice of the charts because, under a change of local coordinates $(x'^i) \to (x^i)$, it holds that

$$\mathrm{Det}(g) = \mathrm{Det}(g') \left(\frac{D(x')}{D(x)} \right)^2. \tag{I.5.1}$$

The inverse of the matrix (g_{ij}) is denoted by (g^{ij}).

Exercise I.5.1   *Show that the Kronecker symbol δ_i^j, $\delta_i^j = 1$ for $i = j$, $\delta_i^j = 0$ for $i \neq j$, is invariant under change of frame.*
   Show that the g^{ij} are the components of a contravariant symmetric 2-tensor.

A metric is traditionally written in a moving frame:

$$g = g_{ij} \theta^i \theta^j, \qquad \text{i.e., in a natural frame,} \quad g = g_{ij} dx^i dx^j. \tag{I.5.2}$$

It is known from algebra that any non-degenerate quadratic form over the reals can be written as a sum

$$g_{ij} dx^i dx^j \equiv \sum_i \varepsilon_i (\theta^i)^2 \quad \text{with} \quad \varepsilon_i = \pm 1, \tag{I.5.3}$$

where the θ^i are independent real linear forms of the dx^i, that is, a moving frame. Given g, this decomposition can be done in many ways, but the number of ε_i that are equal to $+1$ and the number that are equal to -1 are independent of the decomposition; this is called the **signature** of the quadratic form. A moving frame where a metric takes the form (5.3) is called an **orthonormal frame**, whatever its signature.
   A pseudo-Riemannian metric g on V defines at each point $x \in V$ a **scalar product**, that is, a bilinear non-degenerate function on the tangent space $T_x V$:

$$(v, w) := g_x(v, w) \equiv g_{ij}(x) v^i w^j, \quad v, w \in T_x V.$$

Two vectors of a tangent space $T_x V$ are said to be orthogonal if their scalar product vanishes.
   Through contracted products with the metric or its contravariant counterpart, one associates canonically contravariant and covariant tensors, for example

$$T_{ij} = g_{ih} g_{jk} T^{hk}. \tag{I.5.4}$$

An **isometry** of the pseudo-Riemannian manifold (V, g) is a diffeomorphism f that leaves g invariant; that is,

$$(f^* g)(x) = g(x) \quad \text{at each point } x \in V. \tag{I.5.5}$$

If a metric g is invariant under the action of a one-parameter group generated by the vector field X, then its Lie derivative with respect to X vanishes; the converse is locally true. The vector field X is then called a **Killing vector** of g.

Exercise I.5.2   *Show that the Lie bracket of two Killing vectors is also a Killing vector.*

A pseudo-Riemannian manifold of dimension n admits at most $n(n+1)/2$ Killing vectors.

Two pseudo-Riemannian manifolds (V, g) and (V', g') are called **locally isometric** if there exists a differentiable mapping f such that any point $x \in V$ admits a neighbourhood U, and $f(x)$ a neighbourhood U', with (U, g) and (U', g') isometric. Locally isometric manifolds have the same dimension but can have different topologies (for example, a plane and a cylinder).

A pseudo-Riemannian manifold is called a **flat** space if it is isometric with a pseudo-Euclidean space, that is, R^n with metric

$$g_{ij} dx^i dx^j \equiv \sum_i \varepsilon_i (dx^i)^2 \quad \text{with} \quad \varepsilon_i = \pm 1.$$

It is called **locally flat** if it is locally isometric to a flat space.

The **volume form** of the metric g is the exterior n-form that reads in local coordinates

$$\omega_g = |\text{Det} g|^{\frac{1}{2}} dx^1 \wedge \ldots \wedge dx^n. \tag{I.5.6}$$

This exterior form induces, on domains oriented by the order $x^1, \ldots, x^n$, a **volume element** on a pseudo-Riemannian manifold, often denoted by the same symbol ω_g, although the volume element concerns only oriented manifolds. It reads,

$$\mu_g = |\text{Det} g|^{\frac{1}{2}} dx^1 \ldots dx^n, \tag{I.5.7}$$

the order of the differential dx^i is irrelevant in integration (Fubini's theorem).[9]

[9]For integration of forms on manifolds, see for instance CB-DMI, IV B 1.

I.5.2   Riemannian metrics

A metric is called **Riemannian** (or, for emphasis, properly Riemannian) if the quadratic form defined by g is positive-definite.[10]

[10]That is, $\varepsilon_i = +1$ for all i in the formula (5.3).

If v is a tangent vector at x to a Riemannian manifold (V, g), the non-negative scalar

$$g_x(v, v)^{\frac{1}{2}} := [g_{ij}(x) v^i v^j]^{\frac{1}{2}} \tag{I.5.8}$$

is the **norm** (in g) of v; it vanishes only if v is the zero vector, whose components are $v^i = 0$.

The **length** ℓ of a parametrized curve $\lambda \mapsto x(\lambda)$ joining two points of V with parameters $\lambda = a$ and $\lambda = b$ is

$$\ell := \int_a^b \left[g_{ij}(x(\lambda)) \frac{dx^i}{d\lambda} \frac{dx^j}{d\lambda} \right]^{\frac{1}{2}} d\lambda. \tag{I.5.9}$$

Elementary analysis shows that the definition is independent of the parametrization. The **distance** between these two points is the lower bound of the length of arcs joining them. A **geodesic** of a properly Riemannian manifold is a curve for which this bound is attained.

Distance endows a properly Riemannian manifold with the structure of a metric space with a topology that coincides with its topology as a manifold. Riemannian manifolds and their topologies have been extensively studied by differential geometers, and their metrics are closely related to their topological properties. They are of interest to General Relativists as possible properties of space (see Chapter VII).

I.5.3   Lorentzian metrics

A pseudo-Riemannian metric g is called a **Lorentzian metric** if the signature of the quadratic form defined by g is $(-++\cdots+)$. In the case of a manifold with a Lorentzian metric, we denote its dimension by $n+1$ and use Greek indices $\alpha, \beta, \ldots = 0, 1, 2, \ldots, n$ for local coordinates and local frames. A Lorentzian metric is a quadratic form

$$g \equiv g_{\alpha\beta} dx^\alpha dx^\beta \tag{I.5.10}$$

that admits a decomposition as a sum of squares of 1-forms θ^α with the following signs, independent of the choice of these 1-forms:

$$g \equiv -(\theta^0)^2 + \sum_{i=1,\ldots n} (\theta^i)^2, \quad \theta^\alpha = a^\alpha{}_\beta dx^\beta. \tag{I.5.11}$$

Remark I.5.1   *We adopt the 'mostly plus' MTW convention.*[11] *Some authors prefer the opposite convention, giving to Lorentzian metrics the signature $(+--\ldots-)$. Each of these conventions has its advantages and disavantages, but of course they give equivalent geometrical results (note, however, that, surprisingly, this is not always true on non-orientable manifolds*[12]*).*

[11] Misner–Thorne–Wheeler.

[12] See CB-DMII Chapter I Section 7, Pin and spin groups.

A metric of fundamental physical importance in the case $n = 3$ is the **Minkowski metric**. For a general n, it is the flat metric on R^{n+1} that reads

$$g \equiv -(dx^0)^2 + \sum_{i=1,2,\ldots,n} (dx^i)^2. \tag{I.5.12}$$

The group of linear maps of R^{n+1},

$$\theta^\alpha = L^\alpha{}_{\alpha'} \theta^{\alpha'}, \tag{I.5.13}$$

that preserves orthonormal frames (i.e. frames θ^α where the metric takes the form (5.12)) is called the **Lorentz group** L^{n+1}.

I.6   Causality

I.6.1   Causal and null cones

At each point x of a Lorentzian manifold V, one defines in $T_x V$, the tangent space to V at x, a double cone C_x called the **causal cone**, by the inequality

$$g_x(v, v) \leq 0, \quad v \in T_x V. \tag{I.6.1}$$

The boundary of C_x in $T_x V$ is the double cone Γ_x, called the **null cone** or **light cone**,

$$g_x(v, v) = 0. \tag{I.6.2}$$

A vector $v \in T_x v$ is called **causal** if

$$g_x(v, v) \leq 0. \tag{I.6.3}$$

It is called **timelike** [respectively **spacelike**] if

$$g_x(v, v) < 0 \quad [\text{respectively } g_x(v, v) > 0]. \tag{I.6.4}$$

It is called a **null** vector, not to be confused with the zero vector, if

$$g_x(v, v) = 0 \tag{I.6.5}$$

I.6.2   Future and past

The causal cone C_x splits into two convex cones, C_x^+ and C_x^-, characterized in an orthonormal Lorentzian frame by the properties $v \in C_x$ and

$$C_x^+ : \quad \theta^0(v) \equiv v^0 > 0, \quad C_x^- : \quad v^0 < 0. \tag{I.6.6}$$

If it is possible to choose this splitting continuously on V, the manifold is said to be **time-orientable**. It is time-oriented by the choice. Unless otherwise stated, the manifolds called Lorentzian manifolds are oriented manifolds with time-oriented Lorentzian metrics.

A causal vector is **future-** [respectively **past-**] oriented if it is such that $v \in C_x^+$ [respectively $v \in C_x^-$].

A curve γ joining x_a to x_b in a manifold V is the image in V of a segment of R, $\lambda \mapsto \gamma(\lambda) \in V$, $a \leq \lambda \leq b$. The tangent to γ at a point $\gamma(\lambda)$ is the derivative $d\gamma/d\lambda$, a vector in $T_{\gamma(\lambda)}$ with components $d\gamma^\alpha/d\lambda$ in local coordinates.

In a Lorentzian manifold (V, g), a future **causal curve** is a curve with future causal tangent vectors, and a future **timelike curve** has future timelike tangent vectors. The future of a point x is the set of points $y \in V$ that can be reached from x by a future timelike curve. The future of a subset is the union of the future of its points.

I.6.3   Spacelike submanifolds

Let M be an n-dimensional differentiable submanifold of the $(n+1)$-dimensional differentiable manifold V. Assume that f is defined by an equation $f = 0$, with f a differentiable function with non-vanishing gradient. Then, we denote by ν the contravariant vector associated in the metric g to the gradient of f at $x \in V$.

Exercise I.6.1   *Show that ν is orthogonal in g to vectors v in $T_x M$.*

The submanifold M is called **spacelike** [respectively, timelike, null] if its normal ν is timelike [respectively, spacelike, null].

Exercise I.6.2   *Show that in a small enough neighbourhood, local coordinates of a Lorentzian manifold can always be chosen such that the time lines are orthogonal to spacelike submanifolds; that is, the metric can be written*

$$ds^2 := -N^2(dx^0)^2 + g_{ij} dx^i dx^j.$$

I.6.4   Length and geodesics

The **length** of a causal curve γ joining x_a to x_b is

$$\ell := \int_a^b \left(-g_{\alpha\beta} \frac{d\gamma^\alpha}{d\lambda} \frac{d\gamma^\beta}{d\lambda} \right)^{\frac{1}{2}} d\lambda. \tag{I.6.7}$$

Null curves have zero length.

To define geodesics on a Lorentzian manifold, we replace the integral (6.7), which causes problem at points where the curve has a null tangent vector, by the integral, which is no longer independent of the parameter λ,

$$\int_a^b \mathcal{L} \, d\lambda, \quad \text{with} \quad \mathcal{L} := g_{\alpha\beta}(x(\lambda)) \dot{x}^\alpha \dot{x}^\beta, \quad \dot{x}^\alpha := \frac{dx^\alpha}{d\lambda}. \tag{I.6.8}$$

A geodesic joining $x(a)$ and $x(b)$ is defined as a critical point of this Lagrangian, that is, a solution of the Euler equations

$$\frac{d}{d\lambda} \frac{\partial \mathcal{L}}{\partial \dot{x}^\alpha} - \frac{\partial \mathcal{L}}{\partial x^\alpha} = 0. \tag{I.6.9}$$

These equations read explicitly

$$2 g_{\alpha\beta} \frac{d^2 x^\beta}{d\lambda^2} + \left(2 \frac{\partial g_{\alpha\beta}}{\partial x^\gamma} - \frac{\partial g_{\beta\gamma}}{\partial x^\alpha} \right) \frac{dx^\beta}{d\lambda} \frac{dx^\gamma}{d\lambda} = 0;$$

that is, because of the symmetry in the indices β and γ,

$$\frac{d^2 x^\rho}{d\lambda^2} + \Gamma^\rho_{\beta\gamma} \frac{dx^\beta}{d\lambda} \frac{dx^\gamma}{d\lambda} = 0, \quad \Gamma^\rho_{\beta\gamma} := \frac{1}{2} g^{\alpha\rho} \left(\frac{\partial g_{\alpha\beta}}{\partial x^\gamma} + \frac{\partial g_{\alpha\gamma}}{\partial x^\beta} - \frac{\partial g_{\beta\gamma}}{\partial x^\alpha} \right). \tag{I.6.10}$$

Lemma I.6.1   *On a curve solution of the equations (6.10), it holds that*

$$g_{\alpha\beta}(x(\lambda)) \dot{x}^\alpha \dot{x}^\beta = \text{constant} \tag{I.6.11}$$

Proof. Equation (6.11), as can easily be checked directly, is a special case of the energy equality satisfied by solutions of the Euler equations of a Lagrangian:

$$\mathcal{L} - \frac{\partial \mathcal{L}}{\partial \dot{x}^\alpha} \dot{x}^\alpha = \text{constant}. \tag{I.6.12}$$

∎

The above lemma shows that the definition (6.10) of geodesics defines not only geometric curves but parametrized ones. It implies that the critical points of the functional (6.10) are also critical points of (6.7), but the converse is not true.

Exercise I.6.3 *Prove this statement.*

It can be proved that a geodesic between two points that are **causally related** realizes, in the Lorentzian case, a local (i.e. among nearby such curves) **maximum of distance**.

Exercise I.6.4 *Check this statement by drawing in a plane with Cartesian coordinates (x, t) a broken causal line joining the origin $(0,0)$ to a point $(0,a)$ and comparing the Minkowskian lengths of this line and the straight line joining the same points.*

I.7   Connections

I.7.1   Linear connection

Partial derivatives of components of tensor fields are not components of tensor fields – they do not transform as such under a change of coordinates. Geometric derivation operators are defined by endowing the manifold with a new structure called a **connection**. These operators are called **covariant derivatives** and are usually denoted by ∇. They map differentiable vector fields into 2-tensor fields and obey the following laws:

$$\nabla(v + w) = \nabla v + \nabla w \qquad \text{additivity}, \tag{I.7.1}$$

$$\nabla(fv) = f\nabla v + df \otimes v \quad \text{Leibnitz rule for a product with a function.} \tag{I.7.2}$$

When acting on a scalar function, they coincide with the differentiation operator; that is,

$$\nabla f := df, \quad \text{a covariant vector.}$$

Consider in the domain U of a chart a vector field $v = v^i e_i$ in an arbitrary frame. By the previous rules, one has

$$\nabla v = v^i \nabla e_i + dv^i \otimes e_i, \tag{I.7.3}$$

and hence ∇v is determined in U by the covariant derivatives of the basis vectors e_i, 2-tensor fields that we write in the chosen frame as

$$\nabla e_i = \omega_{ji}^h \theta^j \otimes e_h. \tag{I.7.4}$$

The ω_{ij}^h are functions of the local frame on U. They are called **connection coefficients**. The **covariant derivative** of a vector reads, by (7.3) and (7.4),

$$\nabla v = (\nabla_j v^i)\theta^j \otimes e_i, \quad \text{with} \quad \nabla_j v^i - \partial_j v^i + \omega_{jh}^i v^h. \tag{I.7.5}$$

The connection coefficients ω_{ji}^h are not components of a tensor;[13] their transformation law under a change of local frame results from the tensorial character of ∇v and is found to be

$$\omega_{ji}^h = A_{h'}^h \partial_j A_i^{h'} + A_{h'}^h A_j^{j'} A_i^{i'} \omega_{j'i'}^{h'}, \tag{I.7.6}$$

where A is the change-of-frame matrix. In the case of a change of natural frame, it holds that

$$A_{i'}^i = \frac{\partial x^i}{\partial x^{i'}}.$$

Exercise I.7.1 *Prove the connection frame-change formula (7.6). Show that the difference of two connections on a manifold V is a tensor on V.*

The covariant derivative of a covariant vector is defined so that the covariant derivative of the scalar $v^i u_i$ is the ordinary derivative and obeys the Leibniz rule:

$$\nabla_j(v^i u_i) = \partial_j(v^i u_i) \equiv (\partial_j v^i)u_i + v^i(\partial_j u_i) \quad \text{and}$$
$$\nabla_j(v^i u_i) = (\nabla_j v^i)u_i + v^i(\nabla_j u_i).$$

It then follows from (7.5) that

$$\nabla_j u_i := \partial_j u_i - \omega_{ji}^h u_h. \tag{I.7.7}$$

The covariant derivatives of a (contravariant) vector v or a covariant vector u in the direction of another vector w are vectors of the same type $\nabla_w v$ and $\nabla_w u$ with components

$$w^i \nabla_i v^j \quad \text{and} \quad w^i \nabla_i u_j.$$

Remark I.7.1 *The set of numbers $\nabla_i v^j$ or $\nabla_i u_j$ defined as components of 2-tensors can also be interpreted for any given i as the components of the vector or covector that is the derivative of v or u in the direction of the frame vector $e_{(i)}$.*

The property that the covariant derivative in the direction of a vector w of a contravariant or covariant vector is also a contravariant or covariant vector, respectively, allows us to define the covariant derivatives of tensor fields by assuming that, as is usual for derivation operators, they satisfy the additivity condition and also obey the Leibniz rule, but here

[13]They define a matrix valued 1-form ω_h^i given by $\omega_i^h := \omega_{ji}^h \theta^j$.

only for derivation in the direction of a given vector,[14] equivalently on representatives (i.e. components). We have, for two arbitrary tensors,

$$\nabla_v(S + T) = \nabla_v S + \nabla_v T \qquad \text{linearity,}$$
$$\nabla_v(S \otimes T) = \nabla_v S \otimes T + S \otimes \nabla_v T \qquad \text{Leibniz rule for tensor product}$$
and the derivation ∇_v.

These general definitions give, for example for the 2-covariant, 2-contravariant tensor

$$T := T_j{}^{hk}{}_l \theta^j \otimes e_h \otimes e_k \otimes \theta^l, \quad \nabla_i T := \nabla_{e_i} T$$

the components of its covariant derivative, a 3-covariant, 2-contravariant tensor

$$\nabla_i T_j{}^{hk}{}_l = \partial_i T_j{}^{hk}{}_l - \omega_{ij}^m T_m{}^{hk}{}_l + \omega_{im}^h T_j{}^{mk}{}_l + \omega_{im}^k T_j{}^{hm}{}_l - \omega_{il}^m T_j{}^{hk}{}_m.$$

Remark I.7.2   *Lie derivatives obey the Leibniz rule*

$$\mathcal{L}_X(S \otimes T) \equiv \mathcal{L}_X S \otimes T + S \otimes \mathcal{L}_X T.$$

I.7.2   Riemannian connection

The **Riemannian connection** of the pseudo-Riemannian metric g is the linear connection ω such that

- The covariant derivative of the metric is zero; that is,

$$\partial_\alpha g_{\beta\gamma} - \omega_{\alpha\gamma}^\lambda g_{\beta\lambda} - \omega_{\alpha\beta}^\lambda g_{\lambda\gamma} = 0. \tag{I.7.8}$$

- The second covariant derivatives of scalar functions commute:

$$\nabla_\alpha \partial_\beta f - \nabla_\beta \partial_\alpha f \equiv 0. \tag{I.7.9}$$

One says that the connection has **vanishing torsion**.

Theorem I.7.1   *The conditions (7.8) and (7.9) determine one and only one Riemannian connection, given by*

$$\omega_{\alpha\gamma}^\beta \equiv \Gamma_{\alpha\gamma}^\beta + g^{\beta\mu} \tilde{\omega}_{\alpha\gamma,\mu}, \tag{I.7.10}$$

with

$$\tilde{\omega}_{\alpha\gamma,\mu} \equiv \frac{1}{2}(g_{\mu\lambda} C_{\alpha\gamma}^\lambda - g_{\lambda\gamma} C_{\alpha\mu}^\lambda - g_{\alpha\lambda} C_{\gamma\mu}^\lambda), \tag{I.7.11}$$

where the $C_{\alpha\gamma}^\lambda$ are the structure coefficients of the moving frame, vanishing in a natural frame, while

$$\Gamma_{\alpha\gamma}^\beta \equiv \frac{1}{2} g^{\beta\mu}(\partial_\alpha g_{\gamma\mu} + \partial_\gamma g_{\alpha\mu} - \partial_\mu g_{\alpha\gamma}). \tag{I.7.12}$$

*The quantities Γ are called the **Christoffel symbols** of the metric g; they are zero for an orthonormal frame and, more generally, when the*

$g_{\alpha\beta}$ *are constant. In a natural frame, the connection coefficients reduce to the Christoffel symbols, which then read*

$$\Gamma^{\beta}_{\alpha\gamma} \equiv \frac{1}{2} g^{\beta\mu} \left(\frac{\partial g_{\gamma\mu}}{\partial x^{\alpha}} + \frac{\partial g_{\alpha\mu}}{\partial x^{\gamma}} - \frac{\partial g_{\alpha\gamma}}{\partial x^{\mu}} \right). \tag{I.7.13}$$

Applying the general formula (7.6) to a change $(x^{\alpha}) \mapsto (x'^{\alpha'})$ of local coordinates in a Riemannian connection gives

$$1'^{\beta'}_{\alpha'\gamma'} = \frac{\partial x'^{\beta'}}{\partial x^{\beta}} \frac{\partial x^{\alpha}}{\partial x'^{\alpha'}} \frac{\partial x^{\gamma}}{\partial x'^{\gamma'}} \Gamma^{\beta}_{\alpha\gamma} + \frac{\partial x'^{\beta'}}{\partial x^{\delta}} \frac{\partial}{\partial x'^{\alpha'}} \left(\frac{\partial x^{\delta}}{\partial x'^{\gamma'}} \right). \tag{I.7.14}$$

Proposition I.7.1   *By a suitable choice of local coordinates, the connection coefficients can be made to vanish at any one given point, and even along any one given curve.*

Exercise I.7.2   *Show that these properties hold.*

This proposition is very useful to remember when computing tensorial expressions containing covariant derivatives.

Exercise I.7.3   *Prove that a Killing vector field satisfies the equations*

$$\nabla_i X_j + \nabla_j X_i = 0.$$

I.8   Geodesics, another definition

I.8.1   Pseudo-Riemannian manifolds

Parallel transport: a vector field v is said to be parallely transported along a curve $\lambda \mapsto x(\lambda)$ with tangent vector u if it satisfies along this curve the differential equation

$$u^{\alpha} \nabla_{\alpha} v^{\beta} = 0, \quad u^{\alpha} = \frac{dx^{\alpha}}{d\lambda}. \tag{I.8.1}$$

A differentiable curve is called a **geodesic** if its tangent vector is parallely transported:

$$u^{\alpha} \nabla_{\alpha} u^{\beta} = 0; \tag{I.8.2}$$

that is, in local coordinates, if

$$\frac{dx^{\alpha}}{d\lambda} \left(\frac{\partial u^{\beta}}{\partial x^{\alpha}} + \Gamma^{\beta}_{\alpha\lambda} u^{\beta} \right) = 0 \tag{I.8.3}$$

or, equivalently,

$$\frac{d^2 x^{\beta}}{d\lambda^2} + \Gamma^{\beta}_{\alpha\gamma} \frac{dx^{\alpha}}{d\lambda} \frac{dx^{\gamma}}{d\lambda} = 0. \tag{I.8.4}$$

Lemma I.8.1   *The scalar*

$$u^{\alpha} u_{\alpha} \equiv g_{\alpha\beta} \frac{dx^{\alpha}}{d\lambda} \frac{dx^{\beta}}{d\lambda}$$

is constant along a geodesic with parallely transported tangent u^{α}.

Proof. Along a geodesic, it holds that

$$u^\alpha u_\beta \nabla_\alpha u^\beta \equiv u^\alpha \nabla_\alpha (u_\beta u^\beta) \equiv \frac{d}{d\lambda}(u_\beta u^\beta) = 0. \qquad \blacksquare$$

It should be noted that the given definition of geodesics concerns parametrized curves—the property (8.2) is not conserved under a change of parameter; indeed, if $\lambda' = f(\lambda)$, $u^\alpha := dx^\alpha/d\lambda$ and $u'^\alpha := dx^\alpha/d\lambda'$, then

$$\frac{du^\alpha}{d\lambda} \equiv \frac{d^2 x^\alpha}{d\lambda^2} = \frac{d^2 x^\alpha}{(d\lambda')^2}\left(\frac{d\lambda'}{d\lambda}\right)^2 + \frac{dx^\alpha}{d\lambda'}\frac{d^2\lambda'}{(d\lambda)^2},$$

and (8.4) becomes

$$\left(\frac{d\lambda'}{d\lambda}\right)^2 \left[\frac{d^2 x^\alpha}{(d\lambda')^2} + \Gamma^\beta_{\alpha\gamma}\frac{dx^\alpha}{d\lambda}\frac{dx^\gamma}{d\lambda}\right] + \frac{dx^\alpha}{d\lambda'}\frac{d^2\lambda'}{(d\lambda)^2} = 0, \qquad (\text{I.8.5})$$

or, equivalently,

$$u'^\alpha \nabla_\alpha u'^\beta + c(\lambda)u'^\alpha = 0 \quad \text{with} \quad c(\lambda) := \frac{d^2\lambda'}{(d\lambda)^2}\left(\frac{d\lambda}{d\lambda'}\right)^2.$$

Therefore, u'^α is parallely transported only if $d^2\lambda'/(d\lambda)^2 = 0$, i.e. $\lambda' = a\lambda + b$; that is, λ and λ' are linked by an affine transformation. The parameter λ on a geodesic, with $u^\alpha \equiv dx^\alpha/d\lambda$ parallely transported, is called a **canonical affine parameter**. The geodesics considered in Section I.6.4 are canonically parametrized.

I.8.2　Riemannian manifolds

In a Riemannian manifold, a geodesic joining two points attains the minimum length among neighbouring curves joining these points.

I.8.3　Lorentzian manifolds

Since $u^\alpha u_\alpha$ is constant along a geodesic, the timelike, null or spacelike character of a geodesic is the same along the whole curve. The canonical parameter λ is proportionnal to arc length if the curve is timelike.

In contrast to the Riemannian case, we have the following theorem.

Theorem I.8.1 *In a Lorentzian manifold, a timelike geodesic joining two points attains the maximum length among neighbouring curves joining these points.*

I.9　Curvature

I.9.1　Definitions

The non-commutativity of covariant derivatives is a geometrical property of the metric, manifesting itself via its Riemannian connection.

The commutation $(\nabla_\alpha \nabla_\beta - \nabla_\alpha \nabla_\beta) v^\lambda$ of two covariant derivatives of a vector v is a 2-covariant, 1-contravariant tensor that is found by calculus to depend linearly on v, and not on its derivatives; that is, there exist coefficients $R_{\alpha\beta}{}^\lambda{}_\mu$ such that

$$(\nabla_\alpha \nabla_\beta - \nabla_\beta \nabla_\alpha) v^\lambda \equiv R_{\alpha\beta}{}^\lambda{}_\mu v^\mu. \tag{I.9.1}$$

These coefficients are the components of a tensor,[15] $Riem(g)$, antisymmetric in its first two indices, called the **Riemann curvature tensor**.[16] The identity (9.1) is called the **Ricci identity**.

It can be proved[17] that a manifold with vanishing Riemann curvature is locally flat.

Straightforward calculus using the expressions for the components in natural coordinates of the covariant derivative of a vector shows that the components of the Riemann tensor in natural coordinates are

$$R_{\alpha\beta}{}^\lambda{}_\mu \equiv \frac{\partial}{\partial x^\alpha} \Gamma^\lambda_{\beta\mu} - \frac{\partial}{\partial x^\beta} \Gamma^\lambda_{\alpha\mu} + \Gamma^\lambda_{\alpha\rho} \Gamma^\rho_{\beta\mu} - \Gamma^\lambda_{\beta\rho} \Gamma^\rho_{\alpha\mu}. \tag{I.9.2}$$

The **Ricci tensor** $Ricci(g)$ is defined by the contraction $\alpha = \lambda$ of the Riemann tensor and has components

$$R_{\beta\mu} \equiv \frac{\partial}{\partial x^\alpha} \Gamma^\alpha_{\beta\mu} - \frac{\partial}{\partial x^\beta} \Gamma^\alpha_{\alpha\mu} + \Gamma^\rho_{\beta\mu} \Gamma^\alpha_{\alpha\rho} - \Gamma^\lambda_{\beta\rho} \Gamma^\rho_{\mu\lambda}. \tag{I.9.3}$$

The **scalar curvature** is

$$R := g^{\alpha\beta} R_{\alpha\beta}. \tag{I.9.4}$$

The **Einstein tensor**[18] is

$$S_{\alpha\beta} := R_{\alpha\beta} - \frac{1}{2} g_{\alpha\beta} R. \tag{I.9.5}$$

Exercise I.9.1 *Show that in an arbitrary moving frame,*

$$R_{\lambda\mu}{}^\alpha{}_\beta \equiv \partial_\lambda \omega^\alpha_{\mu\beta} - \partial_\mu \omega^\alpha_{\lambda\beta} + \omega^\alpha_{\lambda\rho} \omega^\rho_{\mu\beta} - \omega^\alpha_{\mu\rho} \omega^\rho_{\lambda\beta} - \omega^\alpha_{\rho\beta} C^\rho_{\lambda\mu}. \tag{I.9.6}$$

Remark I.9.1 *The Riemann curvature tensor is an exterior 2-form taking values in the set of linear maps from the tangent plane to itself.*

I.9.2 Symmetries and antisymmetries

The Riemann tensor is obviously antisymmetric in its first pair of indices. It is straightforward to show that when written in full covariant form

$$R_{\alpha\beta,\gamma\mu} := g_{\gamma\lambda} R_{\alpha\beta}{}^\lambda{}_\mu,$$

it is also antisymmetric in its last pair of indices and satisfies the so-called **algebraic Bianchi identity**

$$R_{\alpha\beta,\lambda\mu} + R_{\beta\lambda,\alpha\mu} + R_{\lambda\alpha,\beta\mu} \equiv 0. \tag{I.9.7}$$

This identity can be used to show that the Riemann tensor is invariant under the interchange of these two pairs. One says that it is a symmetric double 2-form.

The Ricci and Einstein tensors are symmetric 2-tensors.

[15] Because the left-hand side is a tensor and v is an arbitrary vector.

[16] Note that the sign convention for the Riemannian curvature tensor in (I.9.1) is that used in CB-DMII; an alterative convention is used in CB-DMI.

[17] See CB-DMI, V B 2.

[18] Denoted $G_{\alpha\beta}$ by some authors.

I.9.3   Differential Bianchi identity and contractions

The definition of the Riemann tensor implies the differential **Bianchi identity**

$$\nabla_\alpha R_{\beta\gamma,\lambda\mu} + \nabla_\beta R_{\gamma\alpha,\lambda\mu} + \nabla_\gamma R_{\alpha\beta,\lambda\mu} \equiv 0. \tag{I.9.8}$$

Contracting this identity gives the identity

$$\nabla_\alpha R_{\beta\gamma}{}^\alpha{}_\mu - \nabla_\beta R_{\gamma\mu} + \nabla_\gamma R_{\beta\mu} \equiv 0, \tag{I.9.9}$$

and a further contraction gives the following identity satisfied by the Einstein tensor:

$$\nabla_\alpha S^{\alpha\beta} \equiv 0. \tag{I.9.10}$$

This identity, called the **conservation identity,** plays a fundamental role in the choice of the Einstein equations.

I.10   Geodesic deviation

An important phenomenon that signals to observers the curvature of spacetime is its influence on the distance between nearby geodesics.

Let C_σ be a one-parameter family of geodesics with canonical parameter denoted by s, $C_\sigma(s) := \psi(\sigma, s)$. Denote by $v = \partial\psi/\partial s$ the tangent vector to C_σ and by $X = \partial\psi/\partial\sigma$ the vector that characterizes the infinitesimal displacement of C_σ.

Lemma I.10.1   *These two vector fields commute; that is,*

$$v^\alpha \nabla_\alpha X^\beta - X^\alpha \nabla_\alpha v^\beta = 0 \tag{I.10.1}$$

Proof. The relation is tensorial and pointlike. It is sufficient to prove it at an arbitrary point in particular coordinates. Choose coordinates such that at that point the Christoffel symbols vanish. Then

$$v^\alpha \nabla_\alpha X^\beta - X^\alpha \nabla_\alpha v^\beta = \frac{\partial^2 \psi^\beta}{\partial s \partial\sigma} - \frac{\partial^2 \psi^\beta}{\partial\sigma \partial s} = 0. \qquad \blacksquare$$

Theorem I.10.1   *The rate of acceleration of distance between nearby geodesics is linked with the curvature by the equation*

$$\nabla^2_{v^2} X^\beta \equiv \frac{D^2}{Ds^2} X^\beta = X^\alpha v^\lambda v^\mu R_{\lambda\alpha}{}^\beta{}_\mu, \tag{I.10.2}$$

*called the equation of **geodesic deviation**.*[19]

Proof. Differentiating the relation (10.1) in the direction of v gives, using the parallel transport of v,

$$v^\lambda v^\alpha \nabla_\lambda \nabla_\alpha X^\beta - v^\lambda \nabla_\lambda X^\alpha \nabla_\alpha v^\beta - X^\alpha v^\lambda \nabla_\lambda \nabla_\alpha v^\beta = 0. \tag{I.10.3}$$

[19]One also says that the right-hand side is the relativistic **tidal force**. This terminology is inspired by the fact that tides on Earth are due to the differential of the gravitational attraction (the Hessian of the gravitational potential in the Newtonian approximation) mainly of the Moon, on the seas.

Hence, using the Ricci identity,

$$v^\lambda v^\alpha \nabla_\lambda \nabla_\alpha X^\beta - v^\lambda \nabla_\lambda X^\alpha \nabla_\alpha v^\beta - X^\alpha v^\lambda (\nabla_\alpha \nabla_\lambda v^\beta + R_{\lambda\alpha}{}^\beta{}_\mu v^\mu) = 0. \tag{I.10.4}$$

and hence, by the parallel transport of v,

$$v^\lambda v^\alpha \nabla_\lambda \nabla_\alpha X^\beta - v^\lambda \nabla_\lambda X^\alpha \nabla_\alpha v^\beta + X^\alpha \nabla_\alpha v^\lambda \nabla_\lambda v^\beta - X^\alpha v^\lambda R_{\lambda\alpha}{}^\beta{}_\mu v^\mu = 0. \tag{I.10.5}$$

The commutation (10.1) of v and X shows that this equation simplifies to

$$v^\lambda v^\alpha \nabla_\lambda \nabla_\alpha X^\beta = X^\alpha v^\lambda v^\mu R_{\lambda\alpha}{}^\beta{}_\mu, \tag{I.10.6}$$

which can also be written as

$$\nabla_{v^2}^2 X^\beta \equiv \frac{D^2}{Ds^2} X^\beta = X^\alpha v^\lambda v^\mu R_{\lambda\alpha}{}^\beta{}_\mu. \tag{I.10.7}$$

∎

I.11   Linearized Ricci tensor

The **linearization**,[20] also called **first variation**, of an operator $P : u \mapsto P(u)$ between open sets of normed vector spaces E_1 and E_2, at a point $u \in E_1$, is a linear operator acting on vectors $\delta u \in E_2$ given by the (Fréchet) derivative P'_u of P at u, that is, such that

[20]See, for instance, CB-DMI, II A, in particular Problem 1.

$$\delta P := P'_u \delta u, \quad \text{with} \quad P(u + \delta u) - P(u) = P'_u(u)\delta u + o(|\delta u|). \tag{I.11.1}$$

We consider a pseudo-Riemannian metric, with components $g_{\alpha\beta}$ in local coordinates. The relation between $g_{\alpha\beta}$ and the inverse matrix $g^{\alpha\beta}$,

$$g^{\alpha\lambda} g_{\alpha\beta} = \delta^\lambda_\beta, \quad \text{the Kronecker symbol,} \tag{I.11.2}$$

implies that

$$\delta g^{\alpha\beta} = -h^{\lambda\mu} := -g^{\alpha\lambda} g^{\beta\mu} h_{\lambda\mu}, \quad \text{where we have set} \quad h_{\lambda\mu} := \delta g_{\lambda\mu}. \tag{I.11.3}$$

The definition and the above relation applied to the Christoffel symbols and the Ricci tensor gives by straightforward computation that $\delta\Gamma^\lambda_{\alpha\beta}$ is the following 3-tensor (indices raised with $g^{\lambda\mu}$):

$$\delta\Gamma^\lambda_{\alpha\beta} \equiv \frac{1}{2} \left(\nabla_\alpha h^\lambda_\beta + \nabla_\beta h^\lambda_\alpha - \nabla^\lambda h_{\alpha\beta} \right). \tag{I.11.4}$$

From this formula and the expression for the Ricci tensor, it follows that $\delta Ricci(g)$ is the linear operator on $h := \delta g$ given in local coordinates by

$$\delta R_{\alpha\beta} \equiv -\frac{1}{2}\nabla^\lambda \nabla_\lambda h_{\alpha\beta} + \frac{1}{2}\left(\nabla_\lambda \nabla_\alpha h^\lambda_\beta + \nabla_\lambda \nabla_\beta h^\lambda_\alpha - \nabla_\alpha \nabla_\beta h^\lambda_\lambda \right). \tag{I.11.5}$$

From this identity, there results the linearization of the scalar curvature:

$$\delta R \equiv g^{\alpha\beta} \delta R_{\alpha\beta} + R_{\alpha\beta} \delta g^{\alpha\beta},$$

where $g^{\alpha\beta}\delta R_{\alpha\beta}$ is the divergence of a vector, namely,

$$g^{\alpha\beta}\delta R_{\alpha\beta} \equiv -\nabla_\lambda \left(\nabla^\lambda h_\alpha^\alpha - \nabla_\alpha h^{\lambda\alpha}\right). \tag{I.11.6}$$

I.11.1   Linearized Bianchi identities

The contracted Bianchi identities

$$\nabla_\alpha S^{\alpha\beta} \equiv 0$$

imply

$$\delta(\nabla_\alpha S^{\alpha\beta}) \equiv 0;$$

that is,

$$\nabla_\alpha(\delta S^{\alpha\beta}) + S^{\alpha\lambda}\delta\Gamma_{\alpha\lambda}^\beta + S^{\lambda\beta}\delta\Gamma_{\alpha\lambda}^\alpha \equiv 0,$$

with

$$\delta S^{\lambda\mu} \equiv \delta R^{\lambda\mu} - \frac{1}{2}(g^{\lambda\mu}\delta R + h^{\lambda\mu}R) \equiv 0.$$

I.12   Physical comment

We shall see in the following chapters that in Relativity (and already in Newton's theory) a model for reality at a macroscopic scale is based on differentiable manifolds, geometric objects whose elements are called points. Each point of, let us say, a 3-manifold is represented by a family of sets of three numbers, each set being the coordinates of the point in a reference frame; different elements of the family are linked by an equivalence relation between reference frames. The physically realistic problem is to link the abstract reference frames with a concrete observable one. We will return to this subject in the following chapters.

I.13   Solutions of selected exercises

Exercise I.3.2 Image of a vector field

Let $f : V \to W$ by $x \mapsto y := f(x)$, and denote by (y^α) coordinates on W and by (x^i) coordinates on V. The differential of f at x is a linear map between T_xV and T_yW, denoted by f', such that

$$(f'v)^\alpha(y) \equiv \frac{\partial f^i}{\partial x^j}v^i(x(y)).$$

It is not a vector field on W if the inverse mapping of f, $f^{-1} : y \mapsto x(y)$, is not defined on W.

Exercise I.3.4 Components of Lie derivatives

For a covariant 2-tensor, the components of the Lie derivative are

$$(\mathcal{L}_X T)_{jh} = X^i \frac{\partial T_{jh}}{\partial x^i} + T_{ih} \frac{\partial X^i}{\partial x^j} + T_{ji} \frac{\partial X^i}{\partial x^h}. \qquad (\text{I.13.1})$$

Under the change of coordinates $x^i = \phi^i(y^{j'})$, we have $X^i = A^i_{j'} X^{j'}$, with

$$A^i_{j'} := \frac{\partial x^i}{\partial y^{j'}} \text{ and } \frac{\partial}{\partial x^i} = A^{j'}_i \frac{\partial}{\partial y^{j'}},$$

with $(A^{j'}_i)$ the inverse matrix of $(A^i_{j'})$. Straightforward computation gives

$$(\mathcal{L}_X T)_{jh} = A^i_{m'} X^{m'} A^{n'}_i \frac{\partial (A^{p'}_j A^{q'}_h T_{p'q'})}{\partial y^{n'}} + A^{n'}_i A^{q'}_h T_{n'q'} A^{p'}_j \frac{\partial (A^i_{m'} X^{m'})}{\partial y^{p'}}$$

$$+ A^{p'}_j A^{n'}_i T_{p'n'} A^h_q \frac{\partial (A^i_{m'} X^{m'})}{\partial y^{q'}}.$$

Since $A^{j'}_i A^i_{h'} = \delta^{j'}_{h'}$, the terms containing no derivatives of the A are easily seen to reduce to

$$A^{p'}_j A^{q'}_h \left(X^{m'} \frac{\partial T_{p'q'}}{\partial y^{m'}} + T_{m'q'} \frac{\partial X^{m'}}{\partial y^{p'}} + T_{p'm'} \frac{\partial X^{m'}}{\partial y^{q'}} \right) \equiv A^{p'}_j A^{q'}_h (\mathcal{L}_X T)_{p'q'}.$$

$$(\text{I.13.2})$$

The terms containing derivatives of the A are

$$X^{m'} T_{p'q'} \frac{\partial (A^{p'}_j A^{q'}_h)}{\partial y^{m'}} + A^{n'}_i A^{q'}_h T_{n'q'} A^{p'}_j X^{m'} \frac{\partial A^i_{m'}}{\partial y^{p'}}$$

$$+ A^{p'}_j A^{n'}_i T_{p'n'} A^{q'}_h X^{m'} \frac{\partial A^i_{m'}}{\partial y^{q'}},$$

which can be written

$$X^{m'} T_{p'q'} \frac{\partial (A^{p'}_j A^{q'}_h)}{\partial y^{m'}} + A^{n'}_i A^{q'}_h T_{n'q'} A^{p'}_j X^{m'} \frac{\partial A^i_{m'}}{\partial y^{p'}}$$

$$+ A^{p'}_j A^{n'}_i T_{p'n'} A^{q'}_h X^{m'} \frac{\partial A^i_{m'}}{\partial y^{q'}}.$$

Renaming indices then gives

$$T_{p'q'} X^{m'} \left[\frac{\partial (A^{p'}_j A^{q'}_h)}{\partial y^{m'}} + A^{p'}_i A^{q'}_h A^{n'}_j \frac{\partial A^i_{m'}}{\partial y^{n'}} + A^{p'}_j A^{q'}_i A^{n'}_h \frac{\partial A^i_{m'}}{\partial y^{n'}} \right] \equiv 0.$$

Since partial derivatives commute, it holds that

$$\frac{\partial A^i_{m'}}{\partial y^{n'}} \equiv \frac{\partial^2 x^i}{\partial y^{n'} \partial y^{m'}} = \frac{\partial A^i_{n'}}{\partial y^{m'}},$$

and we can write

$$T_{p'q'}X^{m'}\left[\frac{\partial(A_j^{p'}A_h^{q'})}{\partial y^{m'}} + A_i^{p'}A_h^{q'}A_j^{n'}\frac{\partial A_{n'}^i}{\partial y^{m'}} + A_j^{p'}A_i^{q'}A_h^{n'}\frac{\partial A_{n'}^i}{\partial y^{m'}}\right] \equiv 0.$$

But, since $A_{n'}^i$ and $A_j^{n'}$ are elements, of inverse matrices,

$$A_i^{p'}A_j^{n'}\frac{\partial A_{n'}^i}{\partial y^{m'}} = -A_i^{p'}A_{n'}^i\frac{\partial A_j^{n'}}{\partial y^{m'}} = -\frac{\partial A_j^{p'}}{\partial y^{m'}}.$$

An analogous computation replacing j by h completes the proof that

$$(\mathcal{L}_X T)_{jh} \equiv A_j^{p'}A_h^{q'}(\mathcal{L}_X T)_{p'q'}.$$

Exercise I.3.5 Lie derivative of exterior forms

On a two-dimensional manifold, the maximal order of an exterior form is 2:

$$\omega := \frac{1}{2}\omega_{ij}dx^i \wedge dx^j \equiv \omega_{12}dx^1 \wedge dx^2$$

The general formula for the Lie derivative of a covariant tensor gives for $\mathcal{L}_X\omega$ the 2-form

$$(\mathcal{L}_X\omega)_{12} = X^h\frac{\partial}{\partial x^h}\omega_{12} + \omega_{12}\left(\frac{\partial X^h}{\partial x^h}\right) \equiv \frac{\partial}{\partial x^h}(X^h\omega_{12}), \qquad \text{(I.13.3)}$$

while

$$i_X\omega = \omega_{12}(X^1 dx^2 - X^2 dx^1), \qquad d\omega \equiv 0.$$

Hence,

$$d(i_X\omega) = \frac{\partial(\omega_{12}X^1)}{\partial x^1} + \frac{\partial(\omega_{12}X^2)}{\partial x^2}dx^1 \wedge dx^2$$

and

$$\mathcal{L}_X\omega \equiv d(i_X\omega) \equiv d(i_X\omega) + i_X d\omega.$$

Exercise I.5.1 Kronecker symbol and contravariant components of g

A mixed tensor X with components $X_j^i := \delta_j^i$ in coordinates x^i has the following components in coordinates x'^i:

$$X_{j'}^{i'} \equiv \frac{\partial x^{i'}}{\partial x^i}\frac{\partial x^j}{\partial x^{i'}}\delta_j^i.$$

Hence,

$$X_{j'}^{i'} \equiv \frac{\partial x^{i'}}{\partial x^i}\frac{\partial x^i}{\partial x^{i'}} = \delta_{j'}^{i'},$$

because $\partial x^{i'}/\partial x^i$ and $\partial x^i/\partial x^{i'}$ are elements of inverse matrices.

The elements g^{ih} of the inverse of the metric g_{ij} are such that

$$g^{ih} g_{jh} = \delta^i_j,$$

and in another coordinate system they are such that

$$g^{i'h'} g_{j'h'} = \delta^i_j;$$

that is,

$$g^{i'h'} \frac{\partial x^j}{\partial x^{j'}} \frac{\partial x^h}{\partial x^{h'}} g_{jh} = \delta^i_j.$$

Hence,

$$g^{ih} = g^{i'h'} \frac{\partial x^j}{\partial x^{j'}} \frac{\partial x^h}{\partial x^{h'}},$$

and the g^{ih} are the components of a contravariant tensor.

Exercise I.7.1 Connection frame-change formula

Let

$$v^i = A^i_{i'} v^{i'}, \qquad \nabla_j v^i = \partial_j v^i + \omega^i_{jh} v^h, \qquad \nabla_{j'} v^{i'} = \partial_{j'} v^{i'} + \omega^{i'}_{j'h'} v^{h'}.$$

We have $\partial_j = A^{j'}_j \partial_{j'}$, with $A^{j'}_j$ and $A^j_{j'}$ inverse matrices, and hence

$$\nabla_j v^i = A^{j'}_j \partial_{j'} (A^i_{i'} v^{i'}) + \omega^i_{jh} A^h_{h'} v^{h'}$$
$$= A^{j'}_j A^i_{i'} \partial_{j'} v^{i'} + (A^{j'}_j \partial_{j'} A^i_{i'}) v^{i'} + \omega^i_{jh} A^h_{h'} v^{h'}.$$

But, since ∇v is a mixed tensor,

$$\nabla_j v^i = A^i_{i'} A^{j'}_j \nabla_{j'} v^{i'} = A^i_{i'} A^{j'}_j (\partial_{j'} v^{i'} + \omega^{i'}_{j'h'} v^{h'}).$$

Identifying these two expressions for $\nabla_j v^i$ gives the connection change-of-frame formula and the tensorial law for the difference of two connections, because the $v^{h'}$ are arbitrary numbers.

Exercise I.7.3 Killing equations

A Killing vector X of a metric g is such that

$$(\mathcal{L}_X g)_{hk} \equiv X^i \partial_i g_{hk} + g_{ih} \partial_k X^i + g_{ki} \partial_h X^i = 0.$$

Using $\nabla_i g_{hk} \equiv 0$ gives the result by straightforward computation.

I.14   Problems

I.14.1   Liouville theorem

1. Let (V, g) be a pseudo-Riemannian manifold with tangent space TV. Prove (Liouville theorem) that the volume form θ in TV given by

$$\theta := |\mathrm{Det} g| dx^0 \wedge dx^1 \wedge \ldots \wedge dx^n \wedge dp^0 \wedge dp^1 \wedge \ldots \wedge dp^n$$

is invariant under the geodesic flow with tangent vector X; that is, with $\mathcal{L}_X$ denoting the Lie derivative with respect to X, it holds that

$$\mathcal{L}_X \theta = 0. \tag{I.14.1}$$

2. Show that $i_X \theta$ is a closed form.

Solution

In local coordinates $(x^A) := (x^\alpha, p^\beta)$ on TV, the Lie derivative of the exterior $2(n+1)$-form θ with respect to the vector field $X = (p, G)$, that is, $(X^A) = (p^\alpha, G^{\tilde{\alpha}})$, reads, using the fact that the components of θ do not depend on p,

$$(\mathcal{L}_X \theta)_{01\ldots n, n+1\ldots 2(n+1)} = p^\alpha \frac{\partial}{\partial x^\alpha} \theta_{01\ldots n, n+1\ldots 2(n+1)} + \theta_{A1\ldots 2(n+1)} \frac{\partial X^A}{\partial x^0}$$

$$+ \theta_{0A\ldots 2(n+1)} \frac{\partial X^A}{\partial x^1} + \ldots + \theta_{01\ldots (2n+1)A} \frac{\partial X^A}{\partial x^{2(n+1)}}. \tag{I.14.2}$$

The expression for θ gives that

$$\frac{\partial}{\partial x^\alpha} \theta_{01\ldots nn+1\ldots 2(n+1)} = g^{\lambda \mu} \left(\frac{\partial}{\partial x^\alpha} g_{\lambda \mu} \right) \theta_{01\ldots nn+1\ldots 2(n+1)}; \tag{I.14.3}$$

on the other hand, since θ is antisymmetric, the same index cannot appear twice in its components; therefore, the second line of (I.14.2) is equal to (recall that the components p^α of X do not depend on x^α)

$$\theta_{01\ldots 2(n+1)} \frac{\partial X^A}{\partial x^A} = \theta_{01\ldots 2(n+1)} \frac{\partial G^\alpha}{\partial p^\alpha}. \tag{I.14.4}$$

The expression for G gives that

$$\frac{\partial G^\alpha}{\partial p^\alpha} = -2\Gamma^\lambda_{\lambda \alpha} p^\alpha. \tag{I.14.5}$$

The result $\mathcal{L}_X \theta = 0$ follows from the expression for the Christoffel symbols.

2. The $2(n+1)$-exterior form θ is closed ($d\theta \equiv 0$) on TV, a $2(n+1)$-dimensional manifold. The equality $\mathcal{L}_X \theta \equiv i_X d\theta + d i_X \theta = 0$ implies that $d i_X \theta = 0$.

I.14.2   Codifferential δ and Laplacian of an exterior form

The **codifferential**, denoted by δ, of an exterior p-form

$$\omega := \frac{1}{p!}\omega_{i_1\ldots i_p}dx^{i_1} \wedge \ldots \wedge dx^{i_p}$$

on a pseudo-Riemannian manifold is defined to be the exterior $(p-1)$-form

$$\delta\omega := \frac{-1}{(p-1)!}\nabla^i\omega_{ij_1\ldots j_{p-1}}dx^{j_1} \wedge \ldots \wedge dx^{j_{p-1}}.$$

1. Show that δ and the exterior differential d are adjoint operators in the sense that for forms with compact support in an open set of R^n, θ of degree $p+1$ and ω of degree p, it holds that

$$(\omega, \delta\theta) = (\delta\omega, d\theta),$$

where the scalar product of p-forms in the metric g is defined by

$$(\omega, \theta) = \int_{R^n} \omega_{i_1\ldots i_p}\psi^{i_1\ldots i_p}\mu_g,$$

with μ_g the volume element of g.

2. Show that $\Delta := -(\delta d + d\delta)$, called the **Laplacian**,[21] is an operator from p-forms into p-forms given by (a hat means that the index is absent from the sequence)

$$(\Delta\omega)_{i_1\ldots i_p} \equiv g^{jh}\nabla_j\nabla_h\omega_{i_1\ldots i_p} - \sum_{1 \leq q \leq p} R^h_{i_q}\omega_{hi_1\ldots\hat{i}_q\ldots i_p}$$

$$- 2\sum_{r<q\leq p} R^h{}_{i_q}{}^j{}_{i_r}\omega_{jhi_1\ldots;\hat{i}_r\ldots\hat{i}_q\ldots i_p}.$$

If g is Lorentzian, then $\delta d + d\delta$ is usually denoted by $\square_g$ and called the **wave operator** of g (the **d'Alembertian** if g is the Minkowski metric η).

[21]Note that the Laplacian is sometimes (for instance in CB-DMI) defined without the minus sign: $\triangle := \delta d + d\delta$.

Solution

Take a 1 form $\omega = \omega_\alpha dx^\alpha$; then

$$d\omega = \frac{1}{2}(\nabla_\beta\omega_\alpha - \nabla_\alpha\omega_\beta)dx^\beta \wedge dx^\alpha, \quad \delta\omega = -\nabla_\alpha\omega^\alpha \equiv -\nabla^\alpha\omega_\alpha$$

$$(\delta d\omega)_\beta := -\nabla^\alpha(\nabla_\alpha\omega_\beta - \nabla_\beta\omega_\alpha), \quad (d\delta\omega)_\beta = \nabla_\beta\nabla^\alpha\omega_\alpha.$$

Therefore, the components of the Laplacian of ω, a 1-form like ω, are

$$(\Delta\omega)_\beta \equiv \nabla^\alpha\nabla_\alpha\omega_\beta - \nabla^\alpha\nabla_\beta\omega_\alpha + \nabla_\beta\nabla^\alpha\omega_\alpha;$$

that is, using the Ricci formula for commutation of covariant derivatives,

$$(\Delta\omega)_\beta \equiv \nabla^\alpha\nabla_\alpha\omega_\beta - R^\alpha{}_{\beta\alpha}{}^\lambda\omega_\lambda \equiv \nabla^\alpha\nabla_\alpha\omega_\beta - R_\beta{}^\lambda\omega_\lambda.$$

The general result is obtained similarly by straightforward computation.

I.14.3 Geodesic normal coordinates

The geodesics arcs on a pseudo-Riemannian manifolds (V, g) are, in the tangent bundle TV, solutions of a first-order differential system that reads in local coordinates (x^α, v^α)

$$\frac{dx^\alpha}{d\lambda} = v^\alpha, \quad \frac{dv^\alpha}{d\lambda} = -\Gamma^\alpha_{\beta\gamma} v^\beta v^\gamma. \tag{I.14.6}$$

Classical theorems on such differential systems say that if the coefficients are Lipschitzian, i.e. if the metric is $C^{1,1}$, then this system has one and only one solution, defined in a neighbourhood of $\lambda = 0$; taking given values for $\lambda = 0$, $x^\alpha(0) = x_0^\alpha$, $v^\alpha(0) = a^\alpha$.

1. Show that if the functions

$$x^\alpha := \phi^\alpha(\lambda, a^\alpha)$$

satisfy the system (14.6) with $\phi^\alpha(0, a^\alpha) = 0$, $\partial\phi^\alpha/\partial\lambda(0, a^\alpha) = a^\alpha$, then

$$x^\alpha = x_k^\alpha := \phi^\alpha\left(k\lambda, \frac{a^\alpha}{k}\right), \quad k \text{ a real number.}$$

Hint: Show that x^α and x_k^α satisfy the same differential system and take the same initial value.

2. Deduce from the result of Part 1 that there exists a neighbourhood of x_0^α such that the functions

$$x^\alpha = \phi^\alpha(1, y^\alpha)$$

define an admissible change of coordinates $(x^\alpha) \to (y^\alpha)$. Show that in the coordinates y^α, called **normal geodesic coordinates**, the geodesic arcs issuing from x_0 are represented by straight lines, $\lambda \mapsto y^\alpha = a^\alpha\lambda$. Show in particular that in these coordinates the Christoffel symbols vanish at x_0.

I.14.4 Cases $d = 1, 2,$ and 3

One-dimensional manifolds are locally isometric with straight lines.

For two-dimensional manifolds, the Riemann tensor has only (up to sign) one non-zero component, $R_{12,12}$. What are the Ricci tensor and the scalar curvature?

Show that in dimension $d = 3$, the Riemann tensor is linear in the Ricci tensor (see Problem I.14.7 on the Weyl tensor).

I.14.5 Wave equation satisfied by the Riemann tensor

Deduce from the Bianchi identities a system of semilinear wave equations satisfied by the Riemann tensor[22] when the Ricci tensor is equal to a given tensor ρ.

[22]Bel (1958).

Solution

The Riemann tensor satisfies the Bianchi identities

$$\nabla_\alpha R_{\beta\gamma,\lambda\mu} + \nabla_\gamma R_{\alpha\beta,\lambda\mu} + \nabla_\beta R_{\gamma\alpha,\lambda\mu} \equiv 0. \qquad (I.14.7)$$

These identities together with the Ricci identity imply that

$$\nabla^\alpha \nabla_\alpha R_{\beta\gamma,\lambda\mu} + \nabla_\gamma \nabla^\alpha R_{\alpha\beta,\lambda\mu} + \nabla_\beta \nabla^\alpha R_{\gamma\alpha,\lambda\mu} + S_{\beta\gamma,\lambda\mu} \equiv 0, \quad (I.14.8)$$

where $S_{\beta\gamma,\lambda\mu}$ is a homogeneous quadratic form in the Riemann tensor:

$$\begin{aligned}
S_{\beta\gamma,\lambda\mu} &\equiv \{R_\gamma{}^\rho R_{\rho\beta,\lambda\mu} + R^\alpha{}_{\gamma\beta}{}^\rho R_{\alpha\rho\lambda\mu} + [(R^\alpha{}_{\gamma\lambda}{}^\rho R_{\alpha\beta,\rho\mu}) \\
&\quad - (\lambda \to \mu)]\} - \{\beta \to \gamma\}.
\end{aligned} \qquad (I.14.9)$$

On the other hand, the Bianchi identities imply by contraction

$$\nabla_\alpha R_{\beta\gamma,}{}^\alpha{}_\mu + \nabla_\gamma R_{\alpha\beta,}{}^\alpha{}_\mu + \nabla_\beta R_{\gamma\alpha,}{}^\alpha{}_\mu \equiv 0, \qquad (I.14.10)$$

which can be written, using the symmetry $R_{\alpha\beta,\lambda\mu} \equiv R_{\lambda\mu,\alpha\beta}$, as

$$\nabla_\alpha R^\alpha{}_{\beta,\lambda\mu} + \nabla_\mu R_{\lambda\beta} - \nabla_\lambda R_{\mu\beta} \equiv 0, \qquad (I.14.11)$$

and the identities become

$$\nabla^\alpha \nabla_\alpha R_{\beta\gamma,\lambda\mu} + S_{\beta\gamma,\lambda\mu} + \{\nabla_\gamma(\nabla_\mu R_{\lambda\beta} - \nabla_\lambda R_{\mu\beta}) - (\beta \to \gamma)\} \equiv 0, \qquad (I.14.12)$$

If the Ricci tensor $R_{\alpha\beta}$ satisfies the Einstein equations

$$R_{\alpha\beta} = \rho_{\alpha\beta}, \qquad (I.14.13)$$

then the previous identities become the following quasidiagonal system of semilinear wave equations[23] for the Riemann tensor:

$$\nabla^\alpha \nabla_\alpha R_{\beta\gamma,\lambda\mu} + S_{\beta\gamma,\lambda\mu} + J_{\beta\gamma,\lambda\mu} = 0, \qquad (I.14.14)$$

[23] These equations are analogous to the Maxwell equations for the electromagnetic 2-form F.

where $J_{\beta\gamma,\lambda\mu}$ depends on the sources $\rho_{\alpha\beta}$ and is zero in vacuum:

$$J_{\beta\gamma,\lambda\mu} \equiv \nabla_\gamma(\nabla_\mu \rho_{\lambda\beta} - \nabla_\lambda \rho_{\mu\beta}) - (\beta \to \gamma). \qquad (I.14.15)$$

I.14.6   The Bel–Robinson tensor

The **Bel–Robinson tensor** associated with the Riemann tensor on a four-dimensional pseudo-Riemannian manifold was defined by Bel[24] through the use of the left and right adjoints of the Riemann tensor defined by

[24] Bel (1959).

$$(^*Ri)_{\alpha\beta,\lambda\mu} \equiv \frac{1}{2}\eta_{\alpha\beta\rho\sigma} R^{\rho\sigma}_{\lambda\mu}, \qquad (R^*)_{\alpha\beta,\lambda\mu} \equiv \frac{1}{2}\eta_{\rho\sigma\rho\sigma} R^{\rho\sigma}_{\alpha\beta}, \qquad (I.14.16)$$

with $\eta_{\alpha\beta\lambda\mu}$ the volume form of the spacetime metric.

1. Prove that the left and right adjoints of the Riemann tensor are equal.

2. Prove the **Lanczos identity**

$$(Ri +^* Ri^*)_{\alpha\beta,\lambda\mu} \equiv S_{\alpha\lambda}g_{\beta\mu} + S_{\beta\mu}g_{\alpha\lambda} - S_{\alpha\mu}g_{\beta\lambda} - S_{\beta\lambda}g_{\alpha\mu}, \quad (I.14.17)$$

with S the Einstein tensor

$$S_{\alpha\lambda} \equiv R_{\alpha\lambda} - \frac{1}{2}g_{\alpha\lambda}R. \quad (I.14.18)$$

3. The **Bel–Robinson tensor** Q is the fourth-order tensor given by

$$Q^{\alpha\beta\lambda\mu} \equiv \frac{1}{4}\left[R^{\alpha\rho,\lambda\sigma}R^{\beta\ \mu}_{\rho\ \sigma} + 2(^*R)^{\alpha\rho,\lambda\sigma}(^*R)^{\beta\ \mu}_{\rho\ \sigma}\right.$$
$$\left. + (^*R^*)^{\alpha\rho,\lambda\sigma}(^*R^*)^{\beta\ \mu}_{\rho\ \sigma}\right]$$

Show that it is symmetric in its two first and two last indices, and under commutation of these pairs of indices.

4. Show that the Bel–Robinson tensor of a vacuum Einstein spacetime, possibly with cosmological constant $(S_{\alpha\beta} = \Lambda g_{\alpha\beta})$, is conservative; namely, that it satisfies the equation

$$\nabla_\alpha Q^{\alpha\beta\lambda\mu} = 0. \quad (I.14.19)$$

It can be proved that the contraction of the Bel–Robinson tensor with timelike vectors is positive, thus giving the definition of a positive energy density for the Riemann tensor. This is used in some existence proofs (see Chapter VIII).

Solution

Parts 1–3 follow by straightforward computation (easier to do in an orthonormal frame).

Part 4 uses the Bianchi identities.

I.14.7   The Weyl tensor

Two metrics g and $\tilde{g}$ on a manifold V are called conformal if there is a positive scalar function $\Omega \equiv e^{2\Phi}$ on V such that

$$\tilde{g} \equiv e^{2\Phi}g, \quad \text{i.e.} \quad \tilde{g}_{\alpha\beta} = e^{2\Phi}g_{\alpha\beta}, \quad \tilde{g}^{\alpha\beta} = e^{-2\Phi}g^{\alpha\beta}. \quad (I.14.20)$$

Two conformal Lorentzian metrics have the same light cones and hence define the same causal structures.

1. Show that the Christoffel symbols of two conformal metrics are linked by the relations

$$\tilde{\Gamma}^\alpha_{\beta\gamma} = \Gamma^\alpha_{\beta\gamma} + S^\alpha_{\beta\gamma}, \quad (I.14.21)$$

where S is the tensor

$$\tilde{\Gamma}^\lambda_{\beta\mu} - \Gamma^\lambda_{\beta\mu} := S^\lambda_{\beta\mu} \equiv \delta^\lambda_\mu\partial_\beta\Phi + \delta^\lambda_\beta\partial_\mu\Phi - g_{\beta\mu}\partial^\lambda\Phi. \quad (I.14.22)$$

2. Show that the Riemann tensors of g and $\tilde{g}$ are linked by

$$\tilde{R}_{\alpha\beta}{}^{\lambda}{}_{\mu} - R_{\alpha\beta}{}^{\lambda}{}_{\mu} \equiv \delta_{\beta}^{\lambda}\Phi_{\alpha\mu} - \delta_{\alpha}^{\lambda}\Phi_{\beta\mu} + g_{\alpha\mu}\Phi_{\beta}{}^{\lambda} - g_{\beta\mu}\Phi_{\alpha}{}^{\lambda}$$
$$+ (\delta_{\beta}^{\lambda}g_{\alpha\mu} - \delta_{\alpha}^{\lambda}g_{\beta\mu})\partial_{\rho}\Phi\partial^{\rho}\Phi,$$

with

$$\Phi_{\alpha\beta} := \nabla_{\alpha}\partial_{\beta}\Phi - \partial_{\alpha}\Phi\partial_{\beta}\Phi.$$

3. Show that the Ricci tensors $R_{\beta\mu}$ and $\tilde{R}_{\beta\mu}$ on a d dimensional manifold are linked by

$$\tilde{R}_{\beta\mu} - R_{\beta\mu} = - g_{\beta\mu}\nabla^{\alpha}\partial_{\alpha}\Phi - (d-2)\nabla_{\mu}\partial_{\beta}\Phi + (d-2)$$
$$\times \left(\partial_{\mu}\Phi\partial_{\beta}\Phi - g_{\beta\mu}\partial^{\lambda}\Phi\partial_{\lambda}\Phi\right) \qquad \text{(I.14.23)}$$

and the scalar curvatures by

$$e^{2\Phi}\tilde{R} - R = -2(d-1)\nabla^{\alpha}\partial_{\alpha}\Phi - (d-2)(d-1)\partial^{\lambda}\Phi\partial_{\lambda}\Phi. \qquad \text{(I.14.24)}$$

4. Define the **Weyl tensor**[25] by the relation

$$W_{\alpha\beta,\gamma\delta} = R_{\alpha\beta,\gamma\delta} - \frac{1}{d-2}(g_{\beta\delta}R_{\alpha\gamma} - g_{\alpha\gamma}R_{\beta\delta} + g_{\beta\gamma}R_{\alpha\delta} - g_{\alpha\delta}R_{\beta\gamma})$$
$$- (g_{\beta\gamma}g_{\alpha\delta} - g_{\alpha\gamma}g_{\beta\delta l})R.$$

Check that the Weyl tensor has the same symmetries as the Riemann tensor and has zero trace.

Show that the components $W_{\alpha\beta}{}^{\lambda}{}_{\mu}$ are the same for two conformal metrics.

Since W is obviously zero for a flat metric, it is also zero if the metric is conformal to a flat metric.[26]

5. Show that the Weyl tensor of a three-dimensional pseudo-Riemannian manifold (M, g) is identically zero. Show that a three-dimensional pseudo-Riemannian manifold with Ricci tensor identically zero is locally flat.

[25] It is equal to the Riemann tensor in a Ricci-flat space ($R_{\alpha\beta} \equiv 0$). It is considered that the Weyl tensor embodies in some sense the non-Newtonian properties of the gravitational field, in particular its radiation properties. This point of view is supported by the fact that the equations for massless fields, at least in four spacetime dimensions, are conformally invariant.

[26] It can be proved that if $d > 3$, then the identical vanishing of the Weyl tensor implies that the metric is locally conformally flat.

I.14.8   The Cotton–York tensor

It can be proved that a 3-manifold is locally conformally flat if and only if its **Cotton tensor** vanishes. This tensor is a 3-tensor with components given in terms of the Ricci tensor and the scalar curvature of a metric g_{ij} by

$$C_{kij:} := \nabla_i\left(R_{jk} - \frac{1}{4}g_{jk}R\right) - \nabla_j\left(R_{ik} - \frac{1}{4}g_{ik}R\right). \qquad \text{(I.14.25)}$$

Define a 2-tensor, called the **Cotton–York** tensor, by

$$Y^{ij} := -\varepsilon^{ikl}g^{mj}C_{mkl}.$$

Show that Y is symmetric, traceless and transverse, that is, such that $\nabla_i Y^{ij} = 0$.

I.14.9   Linearization of the Riemann tensor

Write the linearization of the Riemann tensor as a geometric second-order operator on $h_{\alpha\beta} := \delta g_{\alpha\beta}$.

Solution

$$\delta R_{\alpha\beta}{}^{\lambda}{}_{\mu} \equiv \frac{\partial}{\partial x^{\alpha}}\Gamma^{\lambda}_{\beta\mu} - \frac{\partial}{\partial x^{\beta}}\Gamma^{\lambda}_{\alpha\mu} + \Gamma^{\lambda}_{\alpha\rho}\Gamma^{\rho}_{\beta\mu} - \Gamma^{\lambda}_{\beta\rho}\Gamma^{\rho}_{\alpha\mu}, \qquad (I.14.26)$$

$$\delta R_{\alpha\beta}{}^{\lambda}{}_{\mu} \equiv \frac{\partial}{\partial x^{\alpha}}\delta\Gamma^{\lambda}_{\beta\mu} - \frac{\partial}{\partial x^{\beta}}\delta\Gamma^{\lambda}_{\alpha\mu} + \Gamma^{\lambda}_{\alpha\rho}\delta\Gamma^{\rho}_{\beta\mu} + (\delta\Gamma^{\lambda}_{\alpha\rho})\Gamma^{\rho}_{\beta\mu}$$
$$- \delta\Gamma^{\lambda}_{\beta\rho}\Gamma^{\rho}_{\alpha\mu} - \Gamma^{\lambda}_{\beta\rho}(\delta\Gamma^{\rho}_{\alpha\mu}), \qquad (I.14.27)$$

where $\delta\Gamma^{\lambda}_{\alpha\beta}$ is the tensor

$$X^{\lambda}_{\alpha\beta} := \delta\Gamma^{\lambda}_{\alpha\beta} \equiv \frac{1}{2}\left(\nabla_{\alpha}h^{\lambda}_{\beta} + \nabla_{\beta}h^{\lambda}_{\alpha} - \nabla^{\lambda}h_{\alpha\beta}\right). \qquad (I.14.28)$$

Therefore, $\delta R_{\alpha\beta}{}^{\lambda}{}_{\mu}$ is the tensor

$$\delta R_{\alpha\beta}{}^{\lambda}{}_{\mu} \equiv \nabla_{\alpha}X^{\lambda}_{\beta\mu} - \nabla_{\beta}X^{\lambda}_{\alpha\mu}.$$

I.14.10   Second derivative of the Ricci tensor

The second (Fréchet) derivative, also called the **second variation,** of an operator P at $u \in E_1$ is, if E_1 is a Banach algebra, a quadratic form P''_{u^2} in δu such that

$$\delta^2 P := P''_{u^2}(\delta u, \delta u), \qquad (I.14.29)$$

with

$$P(u + \delta u) - P(u) = P'_u(u)\delta u + \frac{1}{2}P''_{u^2}(u)(\delta u, \delta u) + o(|\delta u|^2). \quad (I.14.30)$$

Compute the **second variation** of the Ricci tensor at a metric g as a quadratic form in $h := \delta g$:

$$\delta^2 R_{\alpha\beta} := R''_{\alpha\beta}(g)(h, h). \qquad (I.14.31)$$

Solution

[27]Choquet-Bruhat (2000).

Straightforward though lengthy computation gives[27]

$$\delta^2 R_{\alpha\beta} \equiv -h^{\lambda\mu}\left[\nabla_{\lambda}(\nabla_{\alpha}h_{\beta\mu} + \nabla_{\beta}h_{\alpha\mu} - \nabla_{\mu}h_{\alpha\beta}) - \nabla_{\alpha}\nabla_{\beta}h_{\lambda\mu}\right]$$
$$-\nabla_{\lambda}h^{\lambda\mu}(\nabla_{\alpha}h_{\beta\mu} + \nabla_{\beta}h_{\alpha\mu} - \nabla_{\mu}h_{\alpha\beta}) + \frac{1}{2}\nabla_{\beta}h^{\lambda\mu}\nabla_{\alpha}h_{\lambda\mu}$$
$$+\frac{1}{2}\nabla^{\lambda}h^{\rho}_{\rho}(\nabla_{\alpha}h_{\beta\lambda} + \nabla_{\beta}h_{\alpha\lambda} - \nabla_{\lambda}h_{\alpha\beta})$$
$$+\nabla_{\lambda}h^{\mu}_{\alpha}\nabla^{\lambda}h_{\beta\mu} - \nabla_{\lambda}h^{\mu}_{\alpha}\nabla_{\mu}h^{\lambda}_{\beta}. \qquad (I.14.32)$$

Special relativity

II.1 Introduction

Special Relativity, formulated by Enstein in 1905, revolutionized our conceptions of time and space.[1]

II.2 Newtonian mechanics

II.2.1 The Galileo–Newton Spacetime

In the spacetime of Galileo and Newton, the simultaneity of two events is a notion independent of observers. Space and time are absolute objects,[2] which exist independently of matter and events that may happen in them. The mathematical model of Newton spacetime is the direct product of R, where time varies, and space, which is assumed to be a Euclidean space E^3, that is, R^3 with the Euclidean metric, which reads, in natural frames that mathematicians call 'Cartesian' and physicists 'inertial',

$$ds^2 = \sum_{i=1,2,3} (dx^i)^2. \tag{II.2.1}$$

To link this mathematical model with observations, one has to identify an inertial frame of this Euclidean space with observed objects (see Section II.2.3).

The isometry group of the Euclidean model of space permits the existence of solid bodies remaining isometric to themselves under motions. In Galileo–Newton spacetime, lengths can be measured by comparison with a standard metre, a piece of metal deposited in Sèvres. For distances that are too large to be compared directly with a copy of the standard metre, one can use properties of Euclidean geometry, such as triangulation. The existence of clocks measuring absolute time is predicted by the periodic phenomena resulting from Newtonian dynamics.[3]

II.2.2 Newtonian dynamics. Galileo group

The fundamental law of Newtonian dynamics for a particle assumed to be pointlike is

$$F = m\gamma, \tag{II.2.2}$$

[1]See e.g. the historical discussion in Damour (2006).

[2]Newton's contemporary Leibniz was in disagreement with this postulate.

[3]The small oscillations of a pendulum are approximately such a phenomenon.

where m is a constant, the mass of the particle (called inertial mass), γ is its acceleration in absolute time with respect to absolute space, and F is a phenomenological vector, the applied force. The components of γ in Cartesian coordinates x^i for the particle in the absolute space E^3 are the second partial derivatives $\gamma^i = d^2 x^i / dt^2$.

Newton's law (2.2) is invariant under the following time-dependent change of coordinates in Newton's absolute space E^3:

$$x'^i = x^i + v^i t + x_0^i, \qquad (\text{II.2.3})$$

where the v^i and x_0^i are constants. The corresponding t-dependent Cartesian coordinates system in E^3 is in uniform translation with respect to the original absolute space. All such reference frames are inertial frames. The set of transformations (2.3) forms a group, called the **Galileo group**. It had already been remarked by Galileo that a uniform-in-time translation of a boat cannot be detected by observers in the hold. More generally, all the physical laws of Newtonian mechanics are assumed to be invariant under the Galileo group, in the sense that they admit the same formulation in all inertial frames.

II.2.3   Physical comment

Inertial frames are a physical reality. Someone seating on a disk turning with respect to the Earth feels a 'centrifugal force' due the fact that axes fixed on the disk are not inertial for the Galileo–Newton spacetime. More generally, non-inertial coordinates manifest themselves through non-vanishing Christoffel symbols of the Euclidean metric of the physical space E^3 in these axes. These non-zero Christoffel symbols cause the appearance of what in Newtonian mechanics are called Coriolis or inertial forces. So the problem is posed, what is the physical origin of the inertial frames? The answer appeared simple for Newton, namely the a priori given absolute space and time, but the problem of identification with observed reality was there all the same. The answer for Mach was that 'absolute' space was determined by the matter content of the universe.[4] In practice, Mach and Newton agree, in an approximate way. Experiments on Earth have shown that Newton's laws of dynamics written in his absolute spacetime are satisfied to a good approximation if one assumes an inertial frame linked with the Earth, and to a better approximation if the inertial frame, called a Copernican frame, has its origin at the center of the Sun and with axes directed to specific stars. This was observed by Foucault by recording the oscillations of a pendulum on Earth in a vertical plane that remains approximately fixed for a short time, but rotates through $360°$ in a time that depends on the latitude λ, namely[5] 1 day/$\sin\lambda$ because of the centrifugal force due to the rotation of the Earth. When the Solar System was known to be rotating in the Milky Way, another inertial frame was considered, which led to verification of the existence of a further very small correction for experiments made in the Solar System. Astronomers now know that the

[4] Mach's considerations partly inspired Einstein's General Relativity.

[5] In Paris, 31 h 50 m, as can be verified by a visit to the pendulum exhibited at the Musée des arts et métiers or probably soon to one suspended beneath the dome of the Panthéon (its original location). This dependence results from the fact that the rotation of axes linked with the Earth's rotation with respect to the Copernican inertial system leads to what is called in Newtonian mechanics a complementary inertial force. The appearance of $\sin\lambda$, equal to 1 at the North Pole, is due to the value of the vertical (i.e. orthogonal to the Earth's surface) component of the rotation vector of the Earth, parallel to the line between the poles. The horizontal component does not contribute to the complementary inertial force, because it is parallel to the velocity of the pendulum (see e.g. Bruhat, 1934, p. 134).

Milky Way itself does not stay still in the universe, but it is impossible to determine an inertial frame for the universe if such a thing exists. Finally, the physicists of the nineteenth century introduced a mysterious medium, called aether, invisible and intangible, though with all the properties of a solid.[6]

[6]Perhaps physics is coming back to the idea of an absolute spacetime determined by its content, everywhere-present pairs of virtual particles, but perhaps spacetime is not a differentiable manifold, either below the Planck length or at cosmological scales.

II.2.4 The Maxwell equations in Galileo–Newton spacetime

The Maxwell equations unify the various classical physical laws that govern the electric and magnetic fields (E, H). These equations in non-inductive media read, in 3-vector notation on R^3,

$$\operatorname{div} E = q, \quad \text{Coulomb's law,} \tag{II.2.4}$$

$$\operatorname{div} H = 0, \quad \text{Gauss's law, non-existence of magnetic charges.} \tag{II.2.5}$$

The scalar q is the charge density. These equations are 'constraints', i.e. they do not contain time derivatives. The following are evolution equations, with c, the speed of light, a dimensionless constant that depends on the units of space and time (and can be made equal to 1 by choice of a relation between these units[7]):

[7]See Section III.5 in Chapter III.

$$\operatorname{curl} E = -\frac{1}{c} \frac{\partial H}{\partial t}, \quad \text{Faraday's law,} \tag{II.2.6}$$

$$\operatorname{curl} H = j + \frac{1}{c} \frac{\partial E}{\partial t}, \quad \text{Ampère–Maxwell law.} \tag{II.2.7}$$

The vector j is the electric current. The term $\partial E/\partial t$, called the 'displacement current', was introduced by Maxwell and led to the consideration of the electric and magnetic fields as the splitting, in an observer-dependent fashion, of one entity called the electromagnetic field, into these electric and magnetic fields. The set of equations (2.4)–(2.7), still valid today, are the shining success of nineteenth-century theoretical physics.

The various laws written above were interpreted before Einstein with t the absolute time and E^3 the absolute space determined by the mysterious medium of the aether. The Maxwell equations imply that electric and magnetic fields—and also light,[8] which is an electromagnetic phenomenon—propagate in vacuum with the constant velocity c, independent of the time t and the location in this absolute space.[9] The number c can be made equal to 1 by appropriate choice of space and time units.

The problem, for Newton's mechanics, is that the Maxwell equations are not invariant under the Galileo group.

It was the mathematical genius of Lorentz and Poincaré to discover the group that leaves invariant the Maxwell equations, but neither of them discarded Newton's absolute time, nor the aether filling the absolute space.

[8]It was already known that the speed of light is finite, and it had been measured by Römer in 1675 from observations of the eclipses of Jupiter's Moons.

[9]In the absence of electric current and charge, this set of equations implies that E and H satisfy the wave equation in the Minkowski metric

$$-\frac{1}{c^2} \frac{\partial^2 H}{\partial t^2} + \Delta H = 0,$$

$$-\frac{1}{c^2} \frac{\partial^2 E}{\partial t^2} + \Delta E = 0,$$

$$\Delta := \sum_{\hat{\imath}=1,2,,3} \frac{\partial^2}{(\partial t^i)^2}.$$

II.3   The Lorentz and Poincaré groups

The $(n+1)$-dimensional **Lorentz group** is the group of linear maps of R^{n+1},

$$X^\alpha = L^\alpha_{\alpha'} X^{\alpha'},$$

that preserves the quadratic form

$$-(X^0)^2 + \sum_{i=1}^n (X^i)^2;$$

that is, the elements $L^\alpha_{\alpha'}$ of matrices L that represent the Lorentz group on R^{n+1} are such that for each set $\{X^{\alpha'}\} \in R^{n+1}$, it holds that

$$- (L^0_{\alpha'} X^{\alpha'})^2 + \sum_{i=1,\dots,n} (L^i_{\alpha'} X^{\alpha'})^2 \equiv -(X'^0)^2 + \sum_{i=1,\dots,n} (X'^i)^2. \quad \text{(II.3.1)}$$

This identity with $X^{0'} = 1$, $X^{i'} = 0$ implies that

$$(L^0_{0'})^2 = 1 + \sum_{i=1,\dots n} (L^i_{0'})^2 \geq 1. \quad \text{(II.3.2)}$$

The **orthochronous Lorentz group** is the subgroup that preserves the orientation of R defined by the element $L^0_{0'}$, i.e. such that

$$L^0_{0'} > 0. \quad \text{(II.3.3)}$$

The elements with $L^0_{0'} < 0$ reverse the orientation of R. They do not constitute a group.

Equation (3.1) implies that the determinant of L is equal to $+1$ or -1. **Proper Lorentz transformations** are those that preserve the spacetime orientation; i.e. they are such that

$$\text{Det}(L) = 1. \quad \text{(II.3.4)}$$

Particular Lorentz transformations are the space rotations

$$L^0_{i'} = L^i_{0'} = 0, \qquad \sum_{i=1,\dots,n} (L^i_{j'} X^{j'})^2 \equiv \sum_{i=1,\dots,n} (X'^i)^2.$$

Standard geometrical considerations show that every proper Lorentz transformation can be written

$$L = R_1 L_S R_2, \quad \text{(II.3.5)}$$

where R_1 and R_2 are space rotations and L_S is a so-called **special Lorentz transformation**, acting only in the time and one space directions, namely such that (3.1) reduce to

$$(L^0_{0'})^2 - (L^1_{0'})^2 = 1, \quad \text{(II.3.6)}$$

$$(L^1_{1'})^2 - (L^0_{1'})^2 = 1, \quad \text{(II.3.7)}$$

$$L^0_{0'} L^0_{1'} - L^1_{1'} L^1_{0'} = 0. \quad \text{(II.3.8)}$$

Exercise II.3.1 *Show that the general solution of these equations is, for an orthochronous transformation,*

$$L^0_{0'} = L^1_{1'} = \cosh \varphi, \quad L^1_{0'} = L^0_{1'} = \sinh \varphi. \tag{II.3.9}$$

The $(n + 1)$-**Poincaré group** is the group of isometries of a flat Lorentzian $(n + 1)$-manifold, the Minkowski spacetime M^{n+1}. It is the semidirect product of the translation group R^{n+1} of R^{n+1} and the Lorentz group L^{n+1}.

See Problem II.1 for the statement and proof of the invariance of the Maxwell equations written for a pair of vectors E, H.

II.4   Lorentz contraction and dilation

Before the new physics introduced by Einstein—that is, redefinition of time and space as observable reality instead of abstract a priori concepts—the results obtained by Lorentz and Poincaré led to controversial studies on the dynamics of charged bodies. We give some of their results, obtained by considering a special Poincaré transformation. We set $x^0 = t$, $\xi^0 = \tau$, and $V := \tanh \varphi$, and hence $\sinh \varphi = V/\sqrt{1 - V^2}$, and $\cosh \varphi = 1/\sqrt{1 - V^2}$ (note that $V \leq 1$). We obtain the transformation law

$$t - \tau = \frac{t' - \tau' + V(x'^1 - \xi'^1)}{\sqrt{1 - V^2}}, \quad x^1 - \xi^1 = \frac{x'^1 - \xi'^1 + V(t' - \tau')}{\sqrt{1 - V^2}}, \tag{II.4.1}$$

where (x^1, t) and (ξ^1, τ) are the coordinates of two points in the (x^1, t) plane in an inertial system, and the primed quantities are the coordinates of these points in another inertial system in uniform translation along the x^1 axis with respect to the first one. With classical interpretation of time and space, the relative velocity of this translation is dx'^1/dt for fixed x^1, while the relative velocity of the unprimed frame with respect to the primed one is dx^1/dt' for fixed x'^1. These velocities resulting from the above formulas are found to be

$$\frac{dx^1}{dt'}|_{x'^1 = \text{const}} = \frac{V}{\sqrt{1 - V^2}} \quad \text{and} \quad \frac{dx'^1}{dt}|_{x^1 = \text{const}} = \frac{-V}{\sqrt{1 - V^2}}. \tag{II.4.2}$$

- Consider two events simultaneous in the primed frame: $t' = \tau'$. The formulas (4.2) give that

$$x^1 - \xi^1 = \frac{x'^1 - \xi'^1}{\sqrt{1 - V^2}}, \tag{II.4.3}$$

and hence the spatial distance observed in the primed frame is smaller than that observed in the unprimed one. This is the **Lorentz contraction**.

- Consider two events with the same spatial location in the primed frame: $x'^1 = \xi'^1$. Then

$$t - \tau \geq t' - \tau'. \tag{II.4.4}$$

This is the **Lorentz dilation.**

The Lorentz contraction and dilation are not intrinsic phenomena. They are relative to the observers, linked with their reference frames, as is obvious from the fact that they are reversed by exchanging the roles of these frames when defining simultaneity or spatial coincidence.

II.5   Electromagnetic field and Maxwell equations in Minkowski spacetime M^4

[10]It was first introduced as a technical tool (in the Euclideanized version with $x^4 = ix^0$) by Poincaré in 1905. Later, in 1908, Minkowski realized the deeper mathematical–physical importance of the four-dimensional structures η and F.

[11]As in Chapter I, we have adopted the MTW convention, although the signature $(+---)$, which is more convenient for some problems, is used by many authors.

Recall that the Poincaré–Minkowski spacetime[10] M^4 is the $(3 + 1)$-dimensional manifold R^4 endowed with a Lorentzian, flat metric that reads,[11] in so-called **inertial coordinates,**

$$\eta = -(dx^0)^2 + \sum_{i=1,2,3} (dx^i)^2. \tag{II.5.1}$$

We define the **electromagnetic 2-form** F on M^4,

$$F \equiv \frac{1}{2} F_{\alpha\beta} dx^\alpha \wedge dx^\beta, \tag{II.5.2}$$

by its components in inertial coordinates deduced from the electric and magnetic vectors E and H previously considered on E^3. We set

$$F_{i0} = E_i, \quad F_{23} = H_1, \; F_{31} = H_2, \; F_{12} = H_3.$$

The electric current is defined as the vector J^α on M^4 with components $J^i = -j^i, \; J^0 = q$.

Theorem II.5.1 *Giving to the velocity of light its geometric value $c = 1$, the Maxwell equations (2.4)–(2.7) can be written as a pair of equations for the electromagnetic 2-form F on Minkowski spacetime M^4. In arbitrary coordinates, the equations are, with ∇ the covariant derivative in the Minkowski metric in these coordinates,*

$$dF = 0, \text{ i.e. } \nabla_\alpha F_{\beta\gamma} + \nabla_\gamma F_{\alpha\beta} + \nabla_\beta F_{\gamma\alpha} = 0, \tag{II.5.3}$$

and

$$\nabla \cdot F = J, \quad \text{i.e.} \quad \nabla_\alpha F^{\alpha\beta} = J^\beta, \tag{II.5.4}$$

where indices are raised with the Minkowski metric.

Equation (5.4) implies the conservation of the electric current J:

$$\nabla_\alpha J^\alpha = 0. \tag{II.5.5}$$

Proof. It is straightforward to check in inertial coordinates that equations (2.4) and (2.5) are equivalent[12] to $dF = 0$ and that equations (2.6) and (2.7) are equivalent to $\nabla \cdot F = J$. This equation implies $\nabla \cdot J = 0$ (see Chapter I), which could have been deduced directly from equations (2.4) and (2.5) in inertial coordinates. Since d and $\nabla\cdot$ are geometric operators, the Maxwell equations for F considered as an exterior 2-form are valid in any frame of M^4.

Since the operator d does not depend on the spacetime metric, and $\nabla \cdot F$ is invariant under the group leaving the metric invariant, the Maxwell equations are invariant under the isometry group of Minkowski spacetime. ∎

Conversely, given on spacetime an exterior 2-form, we can define for a given observer a corresponding electric field as follows. We consider a local coordinate system for which the observer is at rest, i.e. the observer describes a time line with unit tangent vector u with $u^\alpha = \delta_0^\alpha$. Then, the observer's electric vector field, orthogonal to u, is $E^\beta = F^{\beta\alpha}u_\alpha$, i.e., in his **proper frame**,[13] $E^i = F^{0i}$.

The magnetic 2-form of an observer is the trace in his proper space of the electromagnetic 2-form. A magnetic space vector field is defined only if this space is three-dimensional. In this case the magnetic field is the adjoint of the magnetic 2-form, given by the contracted product with the volume form of the three-dimensional Euclidean metric:

$$H^h := \frac{1}{2}\eta^{ijh}F_{ij}.$$

Note that H so defined (for $n = 3$) is a vector on the space manifold, but it is not a spacetime vector.

Under a change of observer, i.e. of proper frame, the space and time components of the electromagnetic field are mixed. This fact has been checked over and over in laboratories.

Exercise II.5.1   *Write transformation of the electric and magnetic fields under a change of Lorentzian frame with only non-zero components $L_\alpha^{\alpha'}$ with α and α' equal to zero or one.*

Hint: Use $F_{\alpha\beta} \equiv \dfrac{\partial x^{\alpha'}}{\partial x^\alpha} \dfrac{\partial x^{\beta'}}{\partial x^\beta} F_{\alpha'\beta'} = L_\alpha^{\alpha'} L_\beta^{\beta'} F_{\alpha'\beta'}.$

Maxwell tensor

The Maxwell tensor τ is the symmetric 2-tensor given on Minkowski spacetime by

$$\tau_{\alpha\beta} := F_\alpha{}^\lambda F_{\beta\lambda} - \frac{1}{4}\eta_{\alpha\beta}F^{\lambda\mu}F_{\lambda\mu}. \tag{II.5.6}$$

Exercise II.5.2   *Show that the Maxwell tensor is traceless in spacetime dimension 4.*

The definitions of E and H in space dimension 3 show that the components $\tau_{0\alpha}$ of τ in inertial coordinates read as follows, with ε_{ijl}

---

[12] Recall that in inertial coordinates on M^4 with $c = 1$, it holds that $F^{ij} = F_{ij}$, $F_{i0} = -F^{i0}$, and $\nabla_\alpha = \partial/\partial x^\alpha$.

[13] The proper frame of an observer is the orthonormal frame with timelike vector the unit tangent vector u to its trajectory. Its proper space is the hyperplane orthogonal to u.

totally antisymmetric and $\varepsilon_{123} = 1$:

$$\tau_{00} = \frac{1}{2}(E^2 + H^2), \quad \tau_{0i} = -\varepsilon_{ijl}E^j H^l. \tag{II.5.7}$$

The component $\tau_{00} = \tau^{00}$ is the energy density of the electromagnetic field, and the $\tau^{0i} = -\tau_{0i}$, $i = 1, 2, 3$, are the components of the Poynting energy flux vector P, given in vector product notation on R^3 by

$$P = E \wedge H. \tag{II.5.8}$$

Lemma II.5.1   *Modulo the Maxwell equations, it holds that*

$$\nabla_\alpha \tau^{\alpha\beta} = J^\lambda F_{\beta\lambda}. \tag{II.5.9}$$

Proof. Straightforward calculation gives

$$\nabla_\alpha \tau^\alpha_\beta \equiv (\nabla_\alpha F^{\alpha\lambda})F_{\beta\lambda} + F^{\alpha\lambda}\nabla_\alpha F_{\beta\lambda} - \frac{1}{2}F^{\lambda\mu}\nabla_\beta F_{\lambda\mu}.$$

Hence, changing names of indices and using antisymmetries,

$$\nabla_\alpha \tau^\alpha_\beta \equiv (\nabla_\alpha F^{\alpha\lambda})F_{\beta\lambda} + \frac{1}{2}F^{\alpha\lambda}(\nabla_\alpha F_{\beta\lambda} + \nabla_\lambda F_{\alpha\beta} + \nabla_\beta F_{\lambda\alpha}). \tag{II.5.10}$$

The Maxwell equations (5.3) and (5.5) imply the result.   ∎

The vector $J^\lambda F_{\beta\lambda}$ is called the **Lorentz force.**

When the Lorentz force is identically zero, the equations $\nabla_\alpha \tau^{\alpha\beta} = 0$ written in inertial coordinates are the usual equations of conservation of energy and momentum.

II.6   Maxwell equations in arbitrary dimensions

The Minkowski spacetime M^{n+1} is the manifold R^{n+1} endowed with a flat metric that reads in inertial coordinates

$$\eta = -(dx^0)^2 + \sum_{i=1,\dots,n} (dx^i)^2. \tag{II.6.1}$$

In a Minkowski spacetime of dimension $n + 1$, it is natural to define an electromagnetic field as an exterior 2-form F that satisfies the **Maxwell equations**

$$dF = 0, \quad \nabla \cdot F = J, \quad \text{with} \quad \nabla \cdot J = 0.$$

Let u be a unit timelike vector field. The electric field relative to u is the vector field $u^\alpha F_{\alpha\beta}$; it is orthogonal to u. Choose for u the time vector $\partial/\partial x^0$ of a system of inertial coordinates. The electric field is then a space vector with components $E_i := F_{i0}$, while the components F_{ij} define on each submanifold $M_0 := \{ x^0 = \text{const}\}$ an exterior 2-form, which can be called the **magnetic 2-form.** As mentioned previously the magnetic vector field is defined only in space dimension 3.

The definition (5.6) of the Maxwell stress energy tensor, as well as its conservation law (5.9), extend to arbitrary dimension.

II.7   Special Relativity

II.7.1   Proper time

Lorentz and Poincaré had made mathematical studies, but it was Einstein who made the conceptual jump to discard Newton's absolute time and space as physical realities.[14] Einstein, in Special Relativity, replaces the direct product $E^3 \times R$ by the Minkowski spacetime M^4. An event is now a point in spacetime. Its history is a timelike trajectory in M^4. An Einsteinian revolution was to discover that the parameter t appearing in the Minkowski metric has no physical meaning: the quantity measurable by well-defined clocks (mechanical, atomic, or biological) is the length of their timelike trajectories in spacetime; it is called the **proper time**. More precisely, if $C : t \mapsto C(t)$, $t \in [t_1, t_2]$ is a future causal curve parametrized by t joining two points of M_4, then the proper time along that curve is the parameter-independent quantity

$$\int_{t_1}^{t_2} \left[-\eta \left(\frac{dC}{dt}, \frac{dC}{dt} \right) \right]^{\frac{1}{2}} dt. \tag{II.7.1}$$

Relativity postulates the existence of universal clocks defined by specific physical phenomenon that measure the proper time. Such clocks are nowadays obtained by using the frequency of radiation emitted by specific atomic transitions, predicted by quantum theory to have a constant universal value. The actually adopted standard clock is the caesium atom, which exhibits a particularly stable (of the order of 10^{-16}) microwave transition between two particular energy levels. The second is now defined through the time measured by the caesium clock.[15]

Einstein's concept of proper time has been proved valid in all experiments performed to test it.

In a Lorentzian manifold, the length of timelike curves joining two points has a local maximum for a timelike geodesic. Geodesics of Minkowski spacetime are represented by straight lines in inertial coordinates. In particular, the line where only the parameter t of inertial coordinates varies is a timelike geodesic. Therefore, the time measured by an observer at rest in some inertial coordinate system is greater than the time measured by a traveller which does not follow such a straight line between their separation and their reunion. This so-called '**twin paradox**' (see Fig. II.1)[16] has been long verified for elementary particles in modern accelerators. The reality of proper time has also been checked in 1971 in an airplane flight of a caesium clock which was late with respect to a similar clock that remained on ground.[17] In long space travels it could be verified with the human biological clock.

The case of spatial distances is more delicate to treat because of the lack of an absolute notion of simultaneity.

[14]See Damour (2013a).

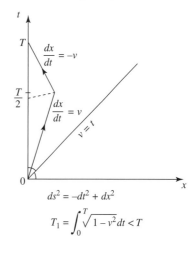

$$ds^2 = -dt^2 + dx^2$$

$$T_1 = \int_0^T \sqrt{1 - v^2}\, dt < T$$

Fig. II.1 The twin paradox.

[15]It is likely, however, that in the future, the caesium clock will be replaced as a standard by an optical clock.

[16]There is actually no paradox, because the two twins do not have the same history: one describes a geodesic in Minkowski spacetime, and the other does not—in fact, he has to use a motor to follow his trajectory. This physical effect is not to be confused with the apparent time dilation, which is a reciprocal effect.

[17]It is pleasant to know that the traveller ages less than the one who stays put.

II.7.2   Proper frame and relative velocities

A velocity, even in Galileo–Newton mechanics, is always defined with respect to some observer. The problem is more complex in Relativity, since there is neither absolute time nor absolute simultaneity. In Relativity, an observer is defined by its world line, a timelike curve. The norm (in the Minkowski metric) of its timelike tangent vector is dependent on the choice of the parameter on the curve; this norm has no physical meaning if the parameter is not specified. We label as a 'unit tangent vector' u this tangent vector normalized by

$$\eta(u, u) = -1. \tag{II.7.2}$$

We call the **proper frame** of the observer (not necessarily following a geodesic) at some point of spacetime an orthonormal Lorentzian frame with u as its timelike axis. Consider an object following a causal line with tangent v at the considered point. If in the proper frame the vector v has time component v^0 and space components v^i, we say[18] that the object has velocity V with respect to this observer, where V is a space vector, orthogonal to u, with components

$$V^i = \frac{v^i}{v^0}. \tag{II.7.3}$$

[18]This definition comes from associating with an observer at some point of spacetime the inertial system in which he is momentarily at rest.

Since v is causal, it holds that

$$|V| \equiv \left[\sum_i (V^i)^2 \right]^{\frac{1}{2}} \leq 1, \tag{II.7.4}$$

with $|V| = 1$, the speed of light, if and only if v is a null vector.

In particular, consider two observers at a point of spacetime with unit velocities u and u'. Choose their proper frames such that they are linked by a special Lorentz transformation, i.e. $e_2 = e'_2$, $e_3 = e'_3$. The velocity of the primed observer with respect to the unprimed observer is then along the axis e_1 and is given by, with $\tilde{u}^\alpha$ the components of u' in the unprimed frame,

$$V = \frac{\tilde{u}^1}{\tilde{u}^0} = \frac{L^1_{0'}}{L^0_{0'}}, \tag{II.7.5}$$

since the components of u' in the primed frame are $u'^1 = 0$, $u'^0 = 1$.

Addition of velocities

As foreseen from the overturning of the notion of absolute time and simultaneity, relativistic relative velocities do not add simply as in Newtonian mechanics.

Consider two systems of inertial coordinates (t, x^1, x^a) and (t', x'^1, x^a). They define at each point of spacetime two Lorentz frames, linked by a special Lorentz transformation, in relative motion with velocity V. The velocity of a given point particle is dx^1/dt with respect to the unprimed frame and dx'^1/dt' with respect to the primed frame. The formula (4.1) implies that

$$\frac{dx^1}{dt} = \frac{dx'^1 + V dt'}{dt' + V dx'^1}, \qquad \text{(II.7.6)}$$

and, therefore, setting $dx^1/dt = U$, $dx'^1/dt' = U'$,

$$U = \frac{U' + V}{1 + V U'}. \qquad \text{(II.7.7)}$$

This coincides in a first approximation with the classical formula for addition of velocities when $U'V$ is small with respect to 1, the velocity of light.

The formula (7.7) implies that $U = 1$ when $U' = 1$: *the speed of light is independent of the relative velocity V of different observers.* This property, very surprising in the framework of Galilean kinematics, but already implicit in the Maxwell equations, was found experimentally by Michelson[19] and verified more accurately by **Michelson and Morley**. They compared the speed of light in the direction of the orbital velocity of the Earth and in a transverse direction. To the surprise of the scientific world, they found the same value in both cases to an accuracy higher than possible experimental error. Since then, it has been verified with greater and greater accuracy, in a medium that does not interfere with this velocity, the vacuum. The most recent experiments give, with a 10^{-9} accuracy, 299 792.458 km/s. The Michelson–Morley experiment was in part[20] a source of inspiration to Einstein to discover Special Relativity.

Since the speed of light in vacuum is a universal constant, it can be used to define the standards of length and time from one another. Physics has shown that time can be measured more accurately, with a caesium clock, than length, for which the previous standard was related to the wavelength of a krypton transition.[21] It has been decided by scientific authorities that the metre is now officially defined to be the distance covered by light in $(299\,792\,458)^{-1}$ seconds. The speed of light in vacuum is therefore fixed to be a universal constant. Unless otherwise specified, we follow the mathematical usage of choosing units of length and time such that this constant is equal[22] to 1 (its 'geometric' value). Of course, in making comparison with observations or experiments, other units may be more appropriate.

[19]See footnote 8.

[20]It seems that Einstein was more inspired by Faraday's law of induction than by the Michelson–Morley experiment.

[21]The caesium clock defines the second as 9 192 631 770 periods of the radiation from a specific caesium-133 atomic transition. Optical clocks defined by electromagnetic radiation are even more precise. See, for instance, Salomon (2013).

[22]If the second is physically defined (for instance through the caesium clock), then unit of length for which the velocity of light is 1 is the distance covered by light in 1 second, that is, 299 792 458 metres.

II.8   Some physical comments

Like all mathematical models, Special Relativity aims at providing as accurate as possible a picture of physical reality. No mathematical model can replace reality, but the first problem is to be able to compare the results given by equations with observed facts. In Special Relativity, one should physically identify spacetime inertial reference frames for Minkowski spacetime. The choice posed a puzzle for Einstein, since Minkowski spacetime considered as a global model is empty.

II.9   Dynamics of a pointlike mass

II.9.1   Newtonian law

The Newtonian equation of motion (2.5) of a particle with rest[23] mass m_0 a constant, subjected to a force f, can be written

$$\frac{d(m_0 v)}{dt} = f, \qquad (II.9.1)$$

with v its velocity with respect to a Galilean inertial frame and t the Newtonian absolute time. When $f = 0$, the particle is in uniform rectilinear motion in all Galilean inertial frames. In a non-Galilean frame, a term, called the inertial force, must be added to f.

From (9.1), we obtain the Newtonian energy equation

$$\frac{d}{dt}\left(\frac{1}{2}m_0 v^2\right) = f \cdot v, \qquad (II.9.2)$$

where the raised dot denotes the Euclidean scalar product.

If we want to write Newton's law (9.1) in a non-Galilean frame, we must add to f the so-called inertial forces due to the motion of the considered frame with respect to a Galilean one.

II.9.2   Relativistic law

The trajectory of a pointlike massive particle in Minkowski spacetime is a timelike curve. Since there is no absolute time to define its velocity, we consider its **unit velocity** u, the tangent vector to its trajectory parametrized by the proper time (the Minkowskian arc length s). The components of u are, in arbitrary coordinates,[24]

$$u^\alpha = \frac{dx^\alpha}{ds}, \quad \text{satisfying} \quad u^\alpha u_\alpha = -1. \qquad (II.9.3)$$

The **acceleration** of the particle is the derivative of u in the direction of itself, $u^\alpha \nabla_\alpha u^\beta$. The time-dependent, coordinate-dependent, Newtonian equations in space, (9.1) and (9.2), are replaced in relativistic dynamics by a spacetime, coordinate-independent, equation

$$u^\alpha \nabla_\alpha (m_0 u^\beta) = F^\beta, \qquad (II.9.4)$$

where F is now a spacetime vector. Using $u_\beta \nabla_\alpha u^\beta = 0$, a consequence of $u_\beta u^\beta = \text{constant}$, (9.4) gives

$$u^\alpha \partial_\alpha m_0 = u_\beta F^\beta. \qquad (II.9.5)$$

This equation holds with m_0 a constant[25] on the trajectory if and only if $u_\beta F^\beta = 0$, i.e. if F is orthogonal (in the Minkowski metric) to the trajectory,[26] which we shall assume in what follows.

Remark II.9.1 *Equation (9.4) reads (see Chapter I) in an arbitrary frame*

$$m_0 \left(\frac{d^2 x^\beta}{ds^2} + \Gamma^\beta_{\alpha\lambda} \frac{dx^\alpha}{ds} \frac{dx^\lambda}{ds} \right) = F^\beta. \tag{II.9.6}$$

In a Minkowskian inertial reference frame, the connection coefficients vanish, and (9.6) with index i reduce to an analogue of Newton's equation in a Galilean frame. This remark leads to interpretation of the term involving Γ as a kind of inertial force.

II.9.3 Newtonian approximation of the relativistic equation

We denote by V as in Section II.7.2 the relative velocity of a massive particle with respect to some Minkowskian inertial coordinates x^i, $x^{s0} = t$, where its unit velocity is $u^\alpha = dx^\alpha/ds$; that is, we set

$$V^i := \frac{dx^i}{dt} = u^i \frac{ds}{dt} = \frac{u^i}{u^0}. \tag{II.9.7}$$

Using $(u^0)^2 - \sum_i (u^i)^2 = (u^0)^2 (1 - |V|^2) = 1$ with $|V|^2 = \sum_i (V^i)^2$ gives

$$u^0 = \frac{dt}{ds} = \frac{1}{\sqrt{1 - |V|^2}} \quad \text{and} \quad u^i = \frac{dx^i}{ds} = \frac{V^i}{\sqrt{1 - |V|^2}}. \tag{II.9.8}$$

These relations imply that the relativistic equations (9.4) can be written in the inertial frame as

$$u^\alpha \frac{\partial}{\partial x^\alpha} \left(\frac{m_0 V^i}{\sqrt{1 - |V|^2}} \right) = F^i, \quad u^\alpha \frac{\partial}{\partial x^\alpha} \left(\frac{m_0}{\sqrt{1 - |V|^2}} \right) = F^0.$$

On the trajectory, it holds that

$$u^\alpha \frac{\partial}{\partial x^\alpha} = \frac{d}{ds} = \frac{dt}{ds} \frac{d}{dt},$$

and hence the above equations read

$$\frac{d}{dt} \frac{m_0 V^i}{\sqrt{1 - |V|^2}} = F^i \sqrt{1 - |V|^2}, \quad \frac{d}{dt} \frac{m_0}{\sqrt{1 - |V|^2}} = F^0 \sqrt{1 - |V|^2}. \tag{II.9.9}$$

The formulas with indices i look like the Newtonian ones if we replace m_0 by

$$m = \frac{m_0}{\sqrt{1 - |V|^2}}, \quad \text{and} \quad f^i = F^i \sqrt{1 - |V|^2}. \tag{II.9.10}$$

[25] The rest mass of a molecule is modified by chemical reactions. The rest mass of an atom is modified by its absorption or emission of photons.

[26] This is the case for the electromagnetic Lorentz force.

For small velocities, i.e. for $|U|$ small with respect to 1, we have $\sqrt{1 - |U|^2} \tilde{=} 1 - \frac{1}{2}|U|^2$. Then,

$$m \tilde{=} m_0 \left(1 + \frac{1}{2}|U|^2\right). \tag{II.9.11}$$

Therefore, in this approximation, the mass m appearing in the relativistic equation is the sum of the rest mass of the particle and its kinetic energy.

Remark II.9.2 *The force f^i is replaced in the new notation by $F^i\sqrt{1 - |V|^2}$. The condition for the constancy of m_0 reads*

$$u_\beta F^\beta \equiv u_i F^i + u_0 F^0 \equiv \frac{V_i F^i - F^0}{\sqrt{1 - |V|^2}} = 0$$

and hence is satisfied by the condition

$$F^0 = \frac{V_i f^i}{\sqrt{1 - |V|^2}}$$

where $V_i f^i$ is in Newtonian mechanics the power of the force f.

II.9.4 Equivalence of mass and energy

The addition of the 'kinetic energy' and the energy of applied forces to the rest mass m_0 in the relativistic equation led Einstein to his famous postulate of the equivalence of mass and energy,[27] with a conversion factor of the order of c^2, where the speed of light c is taken to be 1 in the expression used for the Minkowski metric (geometrized units). The equivalence of mass and energy has been verified in nuclear reactions in a spectacular fashion. The energy can be created by the fission of a uranium atom induced by a collision with a neutron. The sum of the rest masses of the incoming particles is greater than that of the post-fission particles, the difference being seen as kinetic (or radiated) energy. Nuclear fusion of particles into a single particle with rest mass smaller than the sum of the rest masses of the incoming particles also produces energy in accord with $E = mc^2$. Complex fusion reactions involving hydrogen and its isotopes deuterium and tritium, leading in particular to helium nuclei, is the main source of the heat produced by the Sun. Researchers are trying actively to reproduce it on Earth. Constant rest masses are assigned to elementary particles, the photon (rest mass zero), electron, proton, neutrino,[28] etc. The proton has now been experimentally found to be composed of quarks and gluons.

Remark II.9.3 *A useful physical quantity is the **energy–momentum** P, a vector tangent to the trajectory and defined for massive particles by*

$$P^\alpha = mu^\alpha, \quad \text{hence} \quad P^\alpha P_\alpha = -m^2.$$

[27]It is possible that all fundamental elementary particles have zero rest mass and that the positive rest mass of the particles that appear to us as elementary is only an interaction energy.

[28]After many years during which the rest mass of the neutrino was believed to be zero, it has now been established that this rest mass is very small, but non-zero.

In an arbitrary Lorentzian frame, the component P^0 is considered as the energy of the particle with respect to that frame, while the components P^i define its momentum. This splitting is frame-dependent.

II.9.5   Particles with zero rest mass

An energy–momentum is also defined for particles with zero rest mass. Indeed, it was recognized by Einstein in 1905 that light can be described both by waves and by particles, the latter being called photons. A particle with zero rest mass moves on a null geodesic, and the **energy–momentum of a photon is a null vector** P tangent to this geodesic, satisfying the equation

$$(P^0)^2 - \sum_i (P^i)^2 = 0.$$

Its energy with respect to some Minkowskian reference frame is the component P^0.

To determine P, we consider that the wave associated to the considered photon is given by a scalar function $e^{i\varphi}$ with hypersurfaces of constant φ normal to the trajectory of the null particle and hence with $\operatorname{grad}\varphi$ a null vector proportional to P. The wave has a frequency ν in some inertial coordinates if $\partial\varphi/\partial x^0 = 2\pi\nu$.

Einstein introduced in 1905 the postulate that the energy of a photon with respect to some Minkowskian reference frame, i.e. the component P^0, is proportional to the frequency observed in that frame of the associated wave. The energy–momentum of a photon in some inertial coordinates is, with h Planck's constant (introduced by Planck in 1900 with a different meaning),

$$P = \frac{h}{2\pi}\operatorname{grad}\varphi, \quad P^0 = h\nu.$$

The relation $P^0 = h\nu$ introduced by Einstein to explain the photoelectric effect is one of the key discoveries that led to quantum mechanics.

II.10   Continuous matter

Our never-ending improvement in the exploration of reality has made us see that matter is discontinuous at all scales: galaxies in the cosmos, stars in galaxies, molecules in stars, atoms in molecules, 'elementary' particles in atoms, quarks in protons, strings, . . . But, at some scales, which we call 'macroscopic', we are not interested in the impossible task[29] of following the individual motions of constituents. We wish to describe the behaviour of volumes, small at the observed scale, but large at the scale of the constituents, which we call particles, of the model that we are studying. Particles eventually go in and out of such a volume. If it is possible to

---

[29]Even without speaking of the intrusion of quantum mechanics!

define, pointwise on spacetime, measurable macroscopic quantities that characterize the behaviour of matter at the scale of interest, we say that we are dealing with continuous matter.

Real matter is much too complicated to be represented by a single model. In fact, all models are only an approximation of a type of matter. Roughly speaking, fluids are matter models where, in the absence of external forces, the only action on an elementary volume is an interfacial action with neighbouring elements that does not prevent them from slipping against each other. Perfect fluids are those for which the antislipping force has minimal action, that is, is orthogonal to the interface.

In *Newtonian mechanics* with absolute time and simultaneity, the state of a fluid is characterized, if one leaves aside thermodynamical considerations, by its density function μ and its flow vector field v. These are absolute time-dependent quantities on Euclidean space E_3. They satisfy on the one hand the conservation-of-matter equation

$$\frac{\partial \mu}{\partial t} + \partial_i(\mu v^i) = 0 \qquad (\text{II}.10.1)$$

and on the other hand the equations of motion obtained by passing to the pointlike limit of the Newtonian dynamical laws applied to a small volume in E_3. These equations are

$$\frac{d(\mu v^i)}{dt} + \partial_k t^{ik} = 0, \quad \frac{d}{dt} := \frac{\partial}{\partial t} + v^i \frac{\partial x^i}{\partial x}, \qquad (\text{II}.10.2)$$

where t^{ik} is the stress tensor, deduced in the case of fluids from the force f that the surrounding matter exerts at a point on an element of surface with normal n by the linear relation

$$f^i = t^{ij} n_j. \qquad (\text{II}.10.3)$$

The tensor t depends on the nature of the matter. It can be proved to be symmetric when the matter has no intrinsic momentum density.

In *relativistic dynamics*, the fundamental macroscopic, spacetime quantities characterizing a fluid are the **flow vector field** u, a timelike unit vector,[30] i.e. $u^\alpha u_\alpha = -1$, and the **energy density** function μ. A Lorentzian frame with timelike vector u is called a **comoving** or **proper frame** of the fluid. The function μ on spacetime is the time component in the proper frame of the energy–momentum vector field $P = \mu u$.

If there are no interactions (chemical, nuclear, inelastic shocks, etc.) modifying the nature of the underlying particles, we can define for the fluid a rest mass density r, proportional to the particle number density if all particles have the same rest mass. This scalar function on spacetime satisfies a conservation equation that reads in arbitrary coordinates,

$$\nabla_\alpha(r u^\alpha) = 0. \qquad (\text{II}.10.4)$$

[30] Except for null fluids, where $u^\alpha u_\alpha = 0$.

II.10.1   Case of dust (incoherent matter), massive particles

The dust model of matter is a good approximation to other models, owing to the high conversion factor in usual units between rest mass and other types of energy. A fluid is called dust if neighbouring volume elements exert no action on each other. It is then supposed that the flow lines are geodesics,

$$u^\alpha \nabla_\alpha u^\beta = 0. \tag{II.10.5}$$

The following lemma will be of fundamental importance in General Relativity.

Lemma II.10.1   *Equations (10.4) and (10.5) imply that the tensor*

$$T^{\alpha\beta} = r u^\alpha u^\beta \tag{II.10.6}$$

satisfies the conservation law

$$\nabla_\alpha T^{\alpha\beta} = 0. \tag{II.10.7}$$

Proof. From $u^\alpha u_\alpha =$ constant, it results that $u^\beta \nabla_\alpha u_\beta = 0$. The proof is then obtained by computing first $u_\beta \nabla_\alpha T^{\alpha\beta}$.   ∎

Remark II.10.1   *In a proper frame, the components of the tensor (10.6) are*

$$T^{00} = r, \quad T^{0i} = T^{i0} = 0, \quad T^{ij} = 0. \tag{II.10.8}$$

II.10.2   Perfect fluids

The equations of motion for Newtonian fluids are generalized to relativistic fluids by introducing a spacetime energy–momentum stress tensor T, or stress–energy tensor for short, such that its components T^{00}, T^{0i} and T^{i0} in a proper frame of the fluid represent energy, energy flux and momentum per unit volume[31] relative to that frame, while the space part of this tensor is, in a first consideration, identified with the stress tensor of Newtonian mechanics.

A relativistic fluid is called a **perfect fluid** if the components of T in a *proper frame* reduce to

$$T^{00} = \mu, \quad \text{energy density,} \tag{II.10.9}$$

$$T^{0i} = 0, \quad \text{no energy flux (no heat flow),} \tag{II.10.10}$$

$$T^{i0} = 0, \quad \text{zero momentum.} \tag{II.10.11}$$

and, with p a scalar function called the pressure and e the Euclidean metric,

$$T^{ij} = p e^{ij}. \tag{II.10.12}$$

[31]We do not give an axiomatic formulation of these physical notions: physics cannot be axiomatized (nor, by the way, can mathematics at its very foundations). Eventually, energy and momentum are defined as mathematical objects in elaborate theories. For instance, energy appears as the conserved quantity associated to a time-independent Hamiltonian.

The formula (10.12) coincides with the stress tensor of a Newtonian perfect fluid.[32]

Proposition II.10.1 *The stress–energy tensor of a perfect fluid in Minkowski spacetime is, in arbitrary coordinates,*

$$T^{\alpha\beta} = \mu u^\alpha u^\beta + p(\eta^{\alpha\beta} + u^\alpha u^\beta) \qquad \text{(II.10.13)}$$

Proof. One checks that in the proper frame the above tensor has components given by (10.9)–(10.12). ∎

Remark II.10.2 *In the usual pressure and mass units, p is of the order of c^{-2} with respect to μ; hence, at ordinary scales, p is very small compared with μ.*

The proper frame varies from point to point. It is not in general the natural frame of an inertial coordinate system. It is therefore not legitimate to write the usual equations for separate conservation of energy and momentum by using ordinary partial derivatives and the components (10.9)–(10.12) of the stress–energy tensor in a proper frame of the fluid. The dynamical equations of a perfect fluid are postulated to be spacetime tensorial equations that read, in arbitrary coordinates,

$$\nabla_\alpha T^{\alpha\beta} = 0. \qquad \text{(II.10.14)}$$

These equations can be written in an arbitrary Lorentzian metric. We will return to them after introducing General Relativity.

II.10.3   Yang–Mills fields

The electromagnetic field, and the basics of Special Relativity have been treated in Sections II.5–II.7. The definition of the electromagnetic potential has been omitted so far. It is a locally defined 1-form A such that

$$dA = F.$$

Its existence, non-unique, results (Poincaré lemma) from the Maxwell equation $dF = 0$.[33] The non-uniqueness of the electromagnetic potential A, the local character of its definition on a multiply connected manifold, and the physical properties of spinor fields in the presence of an electromagnetic field have led to the interpretation of A as the representative on spacetime of a $U(1)$ connection on its tangent bundle, the electromagnetic field being the curvature of this connection.

After the discovery of other fundamental interactions, strong and weak, it was found by Yang and Mills that these interactions could also be mathematically modelled by curvatures of connections on the tangent bundle of spacetime, but with higher-dimensional and non-Abelian groups. The electromagnetic and weak interactions have been unified by Weinberg and Salam as the curvature of an $SU(2) \times U(1)$ connection, with the addition of a scalar field, the Higgs field, made necessary by the physical fact of the distinction (called symmetry breaking) of these interactions at lower energy scales. The strong interactions have now been identified as a Yang–Mills field, namely the curvature of a $SU(3)$ connection. These connections together with associate spinor fields constitute what is now called the standard model.[34] We recall briefly[35] the descriptions of a classical (non-quantized) Yang–Mills field, the Yang–Mills equations and the conserved stress–energy tensor, which can be taken as a source[36] for the Einstein equations.

Let G be a Lie group and $\mathcal{G}$ its Lie algebra. A representative on spacetime of a G-connection is a locally defined 1-form with values in $\mathcal{G}$; it is called a Yang–Mills potential. The curvature of this connection is represented by a $\mathcal{G}$-valued 2-form, called a Yang–Mills field, given in terms of A by (with $[\,,]$ the Lie bracket in $\mathcal{G}$)

$$F = dA + [A, A], \quad \text{i.e.} \quad F_{\alpha\beta} = \partial_\alpha A_\beta - \partial_\beta A_\alpha + [A_\alpha, A_\beta]. \quad \text{(II.10.15)}$$

The 2-form F satisfies the identity

$$\hat{d}F \equiv 0, \quad \text{i.e} \quad \hat{\nabla}_\alpha F_{\beta\gamma} + \hat{\nabla}_\gamma F_{\alpha\beta} + \hat{\nabla}_\beta F_{\gamma\alpha} = 0, \quad \text{(II.10.16)}$$

where $\hat{d}$ is the gauge-covariant exterior differential and $\hat{\nabla}$ is the metric- and gauge-covariant derivative:

$$\hat{\nabla}_\alpha F_{\beta\gamma} := \nabla_\alpha F_{\beta\gamma} + [A_\alpha, F_{\beta\gamma}], \quad \text{(II.10.17)}$$

where ∇ is the covariant derivative associated with the spacetime metric. The Yang–Mills equations, which can be written in an arbitrary Lorentzian metric, consist of the identity (10.16) and the following equations that generalize the second set of Maxwell equations:

$$\hat{\nabla}_\alpha F^{\alpha\beta} \equiv \nabla_\alpha F^{\alpha\beta} + [A_\alpha, F^{\alpha\beta}] = 0. \quad \text{(II.10.18)}$$

Note that (except for Abelian groups G) the Yang–Mills potential does not disappear from these equations.

The Yang–Mills stress–energy tensor, a generalization of the Maxwell tensor, is an ordinary tensor given by

$$\tau_{\alpha\beta} := F_\alpha{}^\lambda \cdot F_{\beta\lambda} - \frac{1}{4} g_{\alpha\beta} F^{\lambda\mu} \cdot F_{\lambda\mu}. \quad \text{(II.10.19)}$$

where a raised dot denotes the scalar product with respect to the Killing form[37] of $\mathcal{G}$. It is divergence-free if the Yang–Mills equations are satisfied:

$$\nabla_\alpha \tau^{\alpha\beta} = 0.$$

Remark II.10.3 *On a d-dimensional spacetime, the Einstein equations with Yang–Mills source are equivalent to*

$$R_{\alpha\beta} = \rho_{\alpha\beta} \equiv F_\alpha{}^\lambda \cdot F_{\beta\lambda} - \frac{1}{2(d-2)} g_{\alpha\beta} F^{\lambda\mu} \cdot F_{\lambda\mu}. \quad \text{(II.10.20)}$$

The Yang–Mills stress–energy tensor has zero trace in four-dimensional spacetimes, as the Maxwell tensor. In this case, $\rho_{\alpha\beta} = \tau_{\alpha\beta}$.

[34]These Yang–Mills and spinor fields appear in physics, up to now, only as quantum fields.

[35]See, for instance, CB-DMI Vbis Problem 1.

[36]Though, at a macroscopic scale, the Yang–Mills fields are not directly observed.

[37]See, for instance, CB-DMI III D 6.

II.11   Problems

II.11.1   Lorentz transformation of the Maxwell equations

The Maxwell equations on $R^3 \times R$ in non-inductive media read, with $c = 1$ the velocity of light, as follows:

1. The evolution equations are

$$\frac{\partial E^1}{\partial t} = \frac{\partial H^3}{\partial x^2} - \frac{\partial H^2}{\partial x^3} - j, \quad \frac{\partial H^1}{\partial t} = \frac{\partial E^2}{\partial x^3} - \frac{\partial E^3}{\partial x^2}, \quad \text{(II.11.1)}$$

and equations obtained by circular permutation of the indices 1, 2, 3.

2. The constraint equations are

$$\frac{\partial E^1}{\partial x^1} + \frac{\partial E^2}{\partial x^2} + \frac{\partial E^3}{\partial x^3} = q, \quad \frac{\partial H^1}{\partial x^1} + \frac{\partial H^2}{\partial x^2} + \frac{\partial H^3}{\partial x^3} = 0. \quad \text{(II.11.2)}$$

The separation into evolution and constraints is coordinate-dependent, as is the splitting of the electromagnetic field in electric and magnetic parts. Consider the Lorentz transformation

$$t = \frac{t' + Vx'^1}{\sqrt{1 - V^2}}, \quad x^1 = \frac{x'^1 + Vt'}{\sqrt{1 - V^2}}, \quad x^2 = x'^2, \quad x^3 = x'^3 \quad \text{(II.11.3)}$$

and determine E', H', j', q' depending on E, H, j, q, V and satisfying the Maxwell equations.

II.11.2   The relativistic Doppler–Fizeau effect

It has been well known since Doppler (in 1842) and experimentally verified with great accuracy (by Fizeau in 1848) that light appears reddened when the source recedes from the observer, with the opposite occuring when the source approaches the observer. This fact is easily explained by the longer time between two successive pulses needed to reach the observer when the source recedes from him (or her). The classical formula results trivially from addition of velocities in the Newton–Galileo kinematics. Give an expression for the relativistic Doppler–Fizeau effect using the addition-of-velocities formula in Special Relativity.

General Relativity

III.1   Introduction

Besides his desire to reconcile Special Relativity, valid when gravitational effects are negligible, and Newton's law of gravitation, Einstein was led to his theory by physical facts and new ideas. We will not try in this book to follow Einstein's long path (1907–1915) of discovery of General Relativity and the Einstein equations. Their mathematics rely on Lorentzian differential geometry, now a well-understood subject,[1] with which we started this book.

The physical facts that inspired Einstein's genius are the principle of general covariance and the Galileo–Newton equivalence principle.

III.2   Principle of general covariance

The **principle of general covariance** is an extension of the principle called 'material indifference' in Newtonian mechanics, which essentially says that physical phenomena do not depend intrinsically on the reference frame in which we express their laws.[2] The laws may look very different for different observers, and hence in different frames, but it should be possible to find frame-independent formulas for them. Tensor fields are good candidates as objects for physical laws, being intrinsic geometric objects on a manifold. Tensors can be represented by their components, specific numbers attached to them by the choice of a particular reference frame but with general laws for passage from one frame to another.

III.3   The Galileo–Newton equivalence principle

The fundamental law of Newtonian dynamics,

$$F = m_I \gamma, \qquad\qquad\qquad (\text{III.3.1})$$

relates the acceleration γ of a test particle[3] in Galileo–Newton absolute spacetime $E^3 \times R$ to the force F acting on it and to its **inertial mass** m_I. In Newtonian mechanics, this inertial mass is a constant, depending only on the nature of the particle.

[1] In fact, General Relativity has given a great impulse to the study of Riemannian and Lorentzian differential geometry.

[2] A principle that seems trivial, if we believe there exists a reality, although it becomes debatable in quantum mechanics.

[3] Which is considered to be pointlike.

[4]The story, perhaps not history, is that Galileo verified this universality of fall by dropping objects from the Leaning Tower of Pisa. This fact has been observed repeatedly down the years, perhaps since Philiponos Eramnatikos in Alexandria around AD 300. Astronauts on the Moon verified it again by dropping at the same time a piece of metal and a feather (which on Earth would have been slowed down by the atmosphere).

[5]Traditionnally, one introduces a passive and an active gravitational mass—we bypass this distinction. By analogy with the electric field, we could have called the function ρ_G a gravitational charge density, but since it will be identified with the inertial mass density, a variable quantity in Special Relativity, the analogy would be misleading.

[6]The gravitational acceleration of a point particle of inertial mass m_I due to ρ_G is the gradient of the potential U, independent of m_I. On the Earth's surface, the acceleration due to gravity is approximately $9.81\,\mathrm{cm\,s^{-2}}$.

[7]Newton himself had some doubts about this instantaneous "action at a distance", which contradicts our human experience.

[8]With proper units of mass relative to the units of time and length, one can set $G_N = 1$, which is done by many authors. Others prefer, as we do in this book because it is more convenient when working with the Einstein equations, to choose units such that the proportionality factor of the Einstein tensor and the stress–energy tensor of sources is equal to 1.

In its primitive, fundamental form, the equivalence principle is the expression of the fact that the acceleration γ due to gravity of a massive body is independent of its mass.[4]

In **Newtonian gravitation theory**, the acceleration of a test particle in a gravitational field depends only on its location in space (*universality of free fall*); it satisfies a differential equation that reads, in inertial coordinates,

$$\frac{d^2 x^i}{dt^2} = \frac{\partial U}{\partial x^i}. \tag{III.3.2}$$

The function U, called the **gravitational potential**, satisfies the **Poisson equation**

$$\Delta U = -4\pi \rho_G, \ \Delta := \sum_{i=1,2,3} \frac{\partial^2}{(\partial x^i)^2}, \tag{III.3.3}$$

where Δ is the Laplace operator on Euclidean space E^3 while the source ρ_G is a, possibly time-dependent, positive scalar function on R^3 equal to the 'gravitational mass density'[5] of the sources at time t. We do not include in the definition of the potential of ρ_G the Newtonian gravitational constant G_N. We will introduce G_N later after proving the proportionality of inertial mass and so-called gravitational mass.

Classical results on elliptic equations, in particular on the Laplace operator, imply that the Newtonian gravitational potential U is given at each instant of time and each point $x \in E^3$ by the space integral[6] over E^3 at this time:[7]

$$U(x) = \int_{R^3} \frac{\rho_G(y)}{|x - y|} d^3 y, \ d^3 y \equiv dy^1\, dy^2\, dy^3. \tag{III.3.4}$$

The equivalence principle, the identity of the gravitational mass and the inertial mass up to a choice of units, can now be proved through the Newtonian principle of equality of action and reaction, as follows. Let the source of gravitation be a pointlike particle P located at the origin O with gravitational mass m_G; that is, a density $\rho_G := m_G \delta$, where δ is the Dirac density at O. This particle generates a potential with value $U = m_G/r$ at a point x at distance r from O. The attractive Newtonian force generated by P on a particle P' of inertial mass m'_i located at x has intensity $m_G m_I / r^2$. We can infer from the principle of action and reaction that $m_G m'_I = m'_G m_I$, and hence the quotient m_G/m_I is independent of the considered particle. Following the usual notation, we denote the constant by G_N, and call G_N the Newtonian gravitational constant, and we have

$$m_G = G_N m_I. \tag{III.3.5}$$

The value of this universal constant depends on the choice of units for time, length and mass. It is approximately, in CGS units (see Section III.5), the very small number[8]

$$G_N \doteqdot 6,67259 \times 10^{-8}\, \mathrm{cm^3\, g^{-1}\, s^{-2}}.$$

The **Newton equivalence principle** (the equality $m_G = m_I$ up to a choice of units of inertial and gravitational masses) has been verified by experiments on the motion of pendulums by Galileo and Newton (1610–1680), followed by several subsequent accurate experiments by Eötvös (1860–1908) and in modern times by Dicke, Braginski and especially Aidelberger and his group, comparing with a torsion balance the gravitational force and the inertial, centrifugal, force due to rotation. There have been also accurate tests of this equality from celestial mechanics, starting from work of Newton and Laplace, up to modern tests based on laser ranging to the Moon.

III.4   General Relativity

We have in Chapter II expressed the metric of Minkowski spacetime, the arena of the dynamics of Special Relativity, in arbitrary coordinates. We have seen that in the equations of motion of test particles, the inertial forces appear through the Christoffel symbols of the metric.[9] This fact, the Galileo–Newton equivalence principle and his genius inspired Einstein to replace the Minkowski spacetime of Special Relativity by a general Lorentzian manifold (hence the name General Relativity).

[9]Inertial forces (for example Coriolis forces) appear also in Newtonian mechanics through non-inertial (i.e. non-Galilean) frames.

The Newtonian gravitational potential had no influence on Newtonian absolute spacetime structure. The new revolutionary idea due to Einstein is that space and time are not a priori given structures but are united in a four-dimensional curved Lorentzian manifold whose metric g is linked with the energy content of spacetime. This Lorentzian metric governs the **spacetime causality structure** (see Chapter I). The length of a timelike curve measures the intrinsic, called **proper, time** along that curve: in Relativity, Special or General, *the basic observable quantity is proper time*. It is assumed that quantum phenomena, namely vibrations of atoms, permit its measure, independently of everything else. The official clock is at present the caesium atom.

In General Relativity, as was already the case in Special Relativity, there is no intrinsic notion of simultaneity, or, therefore, of spatial distances. Velocities are defined mathematically in the tangent space at the location of an observer at a point of the spacetime (see Chapter II). Assuming that *the speed of light is an absolute constant* is physically sensible because it is consistent with the experimental observation that at any given point, the velocity of light is independent of direction, and when measured on Earth with standard of length a solid body, it appears to be constant with extremely great accuracy. Taking the velocity c of light equal to 1 for any observer is no restriction—it is a definition of the unit of length in terms of the unit of time.[10]

[10]Evaluating astronomical distances in a time unit, namely the light-year, and saying that this is the time it takes light to travel from one object to another is misleading since the proper time is zero along a light ray. The distance is not the length of a spacelike geodesic joining two points of spacetime, but is defined in a more subtle way and depends on the trajectories of the objects in the Lorentzian manifold that represents the spacetime (see Section III.5).

III.4.1   Einstein equivalence principles

Unification of gravitation and inertia is at the root of Einstein General Relativity.

In General Relativity, massive pointlike objects in free fall are assumed to follow timelike geodesics of the metric, and their equations of motion are therefore independent of their mass, reading, in arbitrary coordinates,

$$\frac{d^2 x^\alpha}{ds^2} + \Gamma^\alpha_{\beta\gamma} \frac{dx^\beta}{ds} \frac{dx^\gamma}{ds} = 0,$$

with s the proper time (see Chapter I). The connection Γ of the metric represents in the coordinates x^α both gravitation and inertia. This is often called the **weak Einstein equivalence principle.** This priciple is used in the formulation of equations of motion in a given gravitational field of massive bodies small enough at the considered scale to be taken as pointlike, with mass negligible compared with the masses that produce the gravitational field: this is the case for the motion of the planet Mercury in the gravitational field of the Sun. The advance of the perihelium of its elliptic orbit was the first spectacular confirmation of Einstein General Relativity (see Chapter V).

Light rays are null geodesics, trajectories of particles with zero rest mass. Their deviation by gravitation was first verified during the 1919 solar eclipse.

The weak Einstein equivalence principle can be shown to be a particular case, modulo the Einstein equations (see Chapter IV), of the **Einstein equivalence principle**, which says that in the dynamical equations previously expressed in Special Relativity, the spacetime Minkowski metric η must be replaced, if the gravitatonal field is not negligible, by the spacetime Lorentzian metric, usually denoted by g. One sometimes speaks of the strong Einstein equivalence principle when self-gravitational phenomena are important.

We will treat in Chapter IV the coupling of gravitation and matter.

III.4.2   Conclusion

The Einstein equivalence principle has its root in Einstein's basic idea of identifying gravitation with inertia. In particle physics in Minkowski spacetime, there appear phenomenological constants that are dimensionless, and hence cannot be given a geometric value by choice of units: their variation in spacetime would violate the Einstein equivalence principle. No such violation has been found up to now. The fine-structure constant $\alpha := e^2/hc$ has been estimated to have a fractional variation of no more than order 10^{-7} from analysis of a natural fission reactor phenomenon that took place in Oklo, Gabon two billion years ago. Another dimensionless constant, the ratio $\mu = m_e/m_p$ of the rest masses of electron and proton, has been estimated with a precision of 10^{-6} through measurements of absorption lines in astronomical spectra. The value of these constants is puzzling. Some physicists have remarked that if the constants of nature were not what they are, life as we know it would not exist and that this is enough of an explanation for our observations.[11]

[11] This is known under the name of the 'Anthropic principle'.

III.5   Constants and units of measurement

This section recalls definitions, common to Newtonian and Einsteinian mechanics, that permit us to express quantitative facts, as required for the confrontation of theory and reality. The three familiar items of everyday life are distance, mass and time. The fundamental requirement here is the definition of units that are both common to a group of people and reliable. The system of units used in nearly all countries in science and technology, and in most of them in everyday life, is the **CGS system**.[12] The Convention, an assembly of the French Republic, defined in 1795 the metre (100 centimetres) as the $1/(4 \times 10^6)$ part of the meridian of the Earth and the gram as the mass of $1\,\mathrm{cm}^3$ of water. Standards of mass and length made of platinum were deposited at an Institute in Sèvres—a kilogram (1000 grams) for the mass, and were taken as the definitions of the metre and kilogram, replacing the physical quantities previously used to define them. The unit of time, the second, was defined first as a fraction of the solar day, then as a fraction of a specific year and then of a sideral day.[13]

Modern science has defined fundamental universal dimensional[14] constants. Two such constants can, in principle, be used to define two units in terms of a third by giving an arbitrary fixed value to these constants. If this arbitrary value is 1, it is called the geometric value and the units deduced from it are called geometric units. However, one of the units must be defined by an experimental device with verifiable reproducibility.

An international committee fixes the definition of the officially chosen units, called SI for 'Système International'. Of course, multiples of these units can equivalently be used when they lead to values that are easier to grasp at the relevant scale, as is the case with the mks (metre, kilograms, second) and cgs systems.

The velocity of light, c, observed experimentally to be constant to very great accuracy, is, in metres per second,

$$c_{\mathrm{m\,s^{-1}}} = 299\,792\,458 \doteq 3 \times 10^8. \qquad (\text{III.5.1})$$

This equality is now taken as fixing a relation between the units of length and time. It is the unit of time that is actually chosen[15] as basic for formulation of experimental results; it is defined at present by the period T_{caesium} of a particular atomic transition of caesium, predicted to be constant by quantum mechanics and observed experimentally with a very high accuracy to be the number

$$T_{\mathrm{caesium}} = \frac{1}{9\,192\,631\,770}\ \text{second.} \qquad (\text{III.5.2})$$

The second is now defined by this equality. In the future, it will probably be defined with still higher precision by an optical clock.

The velocity of light will be equal to 1 if we choose as unit of length the distance[16] covered by light in 1 second. We denote[17] this distance

---

[12] centimetre, gram, second.

[13] Defined by observation of the stars.

[14] A constant is called dimensional if its value depends on the units chosen. The dimensional equation of a function depending on positive or negative powers p_L, p_M and p_T of length, mass and time is

$$L^{p_L} M^{p_M} T^{p_T}.$$

For example, an acceleration has dimensional equation LT^{-2}, a force MLT^{-2}. Under a change by a proportionality factor k of a unit with power p in the dimensional equation, the function changes by a proportionality factor k^{-p}.

[15] For a while, it was the unit of length that was chosen, namely the wavelength of a specified ray of the radiation emitted by a specific transition of the krypton atom, which replaced the definition of the metre by the standard in Sèvres. But it was less precise and less convenient to use. Remember that astronomers have long been using time to measure distance, namely light-years.

[16] Note that, in Relativity, the definition of distance is ambiguous since there is no absolute definition of simultaneity.

[17] One could call it a light-second.

by L_s, and it holds that

$$c_{L_\mathrm{s},\mathrm{s}^{-1}} = 1.$$

The metre is defined, using (5.1), by the formula

$$1\,\mathrm{metre} = (299\,792\,458)^{-1} L_\mathrm{s}.$$

A unit of mass could be defined through the previously chosen units by taking the gravitational constant equal to 1; in cgs units, it is observed to be,[18]

$$G_{N,\mathrm{cgs}} = 6.67259 \times 10^{-8}.$$

However, the lack of sufficient accuracy[19] in the knowledge of G_N has so far prevented the use of this formula to define a unit of mass in terms of the length or time unit. It has recently been decided (July 2011) by the BIPM (Bureau International des Poids et Mesures) that Planck's constant h, which is known with a very high accuracy, should be used to officially define a unit of mass-energy. It is now h that is a geometric quantity.[20]

III.6   Classical fields in General Relativity

We have written in Chapter II the expressions for the stress–energy tensors of the usual and simplest forms of matter and fields in Special Relativity. The Einstein equivalence principle leads immediately to their expressions in General Relativity. We give them in a general Lorentzian metric g on a manifold V of arbitrary dimension $n + 1$, although the physical case on a macroscopic scale is $n = 3$.

III.6.1   Perfect fluid

The tensor representing the energy, stress and momentum of a perfect fluid is, in arbitrary coordinates, the following symmetric 2-tensor on V:

$$T_{\alpha\beta} = \mu u_\alpha u_\beta + p(g_{\alpha\beta} + u_\alpha u_\beta) \tag{III.6.1}$$

where u is the unit timelike flow vector, i.e. such that

$$g_{\alpha\beta} u^\alpha u^\beta = -1,$$

while μ and p are scalar functions, respectively the pointwise energy and pressure density of the fluid, both of which are non-negative in classical situations.

Exercise III.6.1 *Show that in a proper frame of the fluid, i.e. an orthonormal frame for the metric g with timelike vector u, the components of the tensor T are respectively the fluid pointwise densities*

$$T_{00} = \mu(\text{energy}),\ \ T_{0i} = 0(\text{momentum}), T_{ij} = p\delta_{ij}(\text{stresses}).$$

In the case of dust, $p \equiv 0$.

[18]The dimensions of G_N can be deduced from the fact that $G_N m/r^2$ is an acceleration, and hence has dimensions LT^{-2}. The dimensions of G_N are therefore $L^3 M^{-1} T^{-2}$.

[19]This lack of accuracy is due to the extreme weakness of the gravitational force between bodies of laboratory size. Another difficulty appears with masses of planetary size: this is that only the product of G_N with their mass appears in their equations of motion, and these masses must be determined from the planetary orbits—which requires knowledge of the value of G_N.

[20]This official definition does not prevent mathematicians from setting the gravitational constant equal to 1 in the equations on which they work. It must only be remembered what this means when one wants to make numerical estimates using various specified units.

The Einstein equivalence principle implies that a relativistic perfect fluid satisfies the conservation equations

$$\nabla_\alpha T^{\alpha\beta} = 0,$$

which have been found in Special Relativity as consequence of the classical conservation laws of energy and momentum.

III.6.2   Electromagnetic field

The electromagnetic field is an exterior 2-form on the manifold V. The equivalence principle says that on the Lorentzian manifold (V, g), the electromagnetic field satisfies, in the absence of electric charges, the vacuum **Maxwell equations**

$$dF = 0, \quad \nabla \cdot F = 0. \tag{III.6.2}$$

Its stress energy tensor is the **Maxwell tensor**

$$\tau_{\alpha\beta} := F_\alpha{}^\lambda F_{\beta\lambda} - \frac{1}{4}g_{\alpha\beta}F^{\lambda\mu}F_{\lambda\mu}. \tag{III.6.3}$$

As foreseen from the equivalence principle, if an electromagnetic field satisfies the sourceless Maxwell equations, then its Maxwell tensor is divergence-free:

$$\nabla_\alpha \tau^{\alpha\beta} = 0. \tag{III.6.4}$$

The covariant derivatives that appear in these equations of motion are a symptom of the influence of gravitation on electromagnetism.

III.6.3   Charged fluid

In an electrically charged fluid, there is an electromagnetic field generated by the fluid motion, which in turn is influenced by the electromagnetic field $F_{\alpha\beta}$. In the simplest classical model, the fluid is a perfect fluid with unit velocity u, and energy and presure densities μ and p, and the electric current is the sum of convection and conduction currents:

$$J^\alpha = qu^\alpha + \sigma^\alpha$$

The stress–energy tensor is the sum of the fluid stress–energy tensor and the Maxwell tensor:

$$T_{\alpha\beta} = (\mu + p)u_\alpha u_\beta + pg_{\alpha\beta} + \tau_{\alpha\beta}.$$

Exercise III.6.2   *Given the metric g, write the coupled system of equations satisfied by u, μ, p, q and F.*

Hint: Use the equations

$$\nabla_\alpha T^{\alpha\beta} = 0 \quad \text{and} \quad \nabla_\alpha \tau^{\alpha\beta} = u_\alpha J^\alpha.$$

III.7   Gravitation and curvature

There is no intrinsic splitting between gravity and inertial-type forces. In a general spacetime (V, g), it is always possible at a given point (see Chapter I) to choose local coordinates such that the Christoffel symbols vanish at that point; gravity and relative acceleration are then, at that point, exactly balanced. It is even possible to choose local coordinates such that the Christoffel symbols vanish along a given geodesic—astronauts spacecrafts have made popular knowledge the fact that in free fall one feels neither acceleration nor gravity; in a small enough neighbourhood of a geodesic, the relative accelerations of objects in free fall are approximately zero.

Nevertheless, gravity is a physical reality that cannot be assimilated with the old notion of inertia. In a general Lorentzian manifold, there is no coordinate system (unless the metric is locally flat) in which all the Christoffel symbols vanish in the domain of a chart; a family of geodesics cannot in general be represented by straight lines. In the presence of a non-everywhere-vanishing mass or energy density, gravity manifests itself by the non-vanishing of the Riemann curvature tensor $Riem(g)$. If $Riem(g)$ does not vanish in an open set U of a spacetime (V, g), then the gravitational field in U cannot be identified with an inertial field. The existence of a non-zero curvature of the metric is revealed, beyond the first approximation, by the relative acceleration of test particles, as is predicted by the equation of geodesic deviation (Chapter I, Section 7).

III.8   Observations and experiments

The following observations have been made and experiments performed to confirm the predictions of General Relativity.[21]

[21] For details, see Will (2014).

III.8.1   The Einstein equivalence principle

The fact that the acceleration at a point in a gravitational field of bodies small at the considered scale is independent of their nature is common to both Newton's and Einstein's theories of gravitation. It has been verified with an accuracy of the order of 10^{-13} in several experiments—in particular in the laboratory with atoms of titanium and beryllium and a very sophisticated apparatus based on a torsion balance, and also at an astronomical scale by refined analyses of results of lunar laser ranging (**LLR**) to determine precisely the orbit of the Moon together with sophisticated approximation methods to show the compatibility of this orbit with the equivalence principle applied to the system Sun–Earth–Moon.

III.8.2   Deviation of light rays

A consequence of Einstein's postulate of the equivalence of mass and energy is that particles of zero mass should also follow geodesics of the spacetime metric. Light rays in particular are null geodesics, and light is deflected by a gravitational field.[22] This was confirmed by observation as early as 1919 by Eddington and Dyson during a total solar eclipse (though not with very high precision, because of large experimental errors). A solar eclipse makes it possible for a photograph of the sky to be taken in which stars are visible, so that their positions can be compared with those in a photograph taken in the absence of the Sun. Greater precision is now obtained for the deflection of radio waves, which does not need an eclipse to be observed.

A spectacular verification of the deflection of light rays by a gravitational field is provided by the observed phenomena called **gravitational lensing**, in which the light rays from a star or galaxy, i.e. the null geodesics issuing from some point in curved spacetime, may intersect again such that an observer can see several images simultaneously. Several gravitational lenses have now been observed,[23] in particular the 'Einstein cross', obtained by the Hubble telescope, consists of four images of the same quasar; in the centre appears a fainter image of the galaxy whose gravitational field acts as the lens.[24]

[22]Newtonian gravitational theory also predicts light deflection, with light treated as small but may be massive particles. But the Newtonian light deflection is exactly half of that predicted by Einstein and verified by Eddington.

[23]See, for instance, Schneider et al. (1992).

[24]See for instance Figure 4.16 in Chapter 4 of Ohanian and Ruffini (2013).

III.8.3   Proper time, gravitational time delay

As in Special Relativity, the physically measurable quantity is **proper time,** which is the length in the spacetime metric of timelike curves. It is again postulated that the period of specific spectral radiation from atoms provides universal clocks for the measure of proper time.

A consequence of the reality of the dependence of this physically measurable time on the presence of a gravitational field is the observation of a **redshift** in the spectrum of a given atom in a gravitational field, as seen by a distant observer for whom this field is weaker. Such shifts add to the Doppler effect for bodies in motion. The reality of the **Einstein effect** was verified in the years 1926–1928 with convincing accuracy by careful analysis[25] of the lines of various elements in the solar spectrum by St John, an astronomer working in the Mount Wilson observatory. In 1960, the **Mossbauer resonant effect** (emission of γ rays with an extremely narrow profile, reabsorbed in crystals with a very sharp resonance) enabled Pound and Rebka to measure the shift of spectral lines due to the variation of the gravitational field with height above the Earth's surface in their laboratory. We give below a brief account of the theoretical prediction in this case.

[25]See details on pp. 126–130 of Bruhat (1931).

Inspired by the Newtonian approximation (see Section IV.5 in Chapter IV), we assume the spacetime metric to be, in the laboratory,

$$g = -N^2(x^1)dt^2 + g_{11}(x^1)(dx^1)^2 + g_{ab}dx^a dx^b, \quad a, b = 2, 3, \quad \text{(III.8.1)}$$

where x^1 varies with the height of a point while x^a, $a = 2, 3$, label the position of this point in a horizontal plane. Consider two identical clocks, one at rest on the ground, say at $x^1 = 0$, the other at rest on the same vertical, with height labelled by $x^1 = h$. The hypotheses are therefore that, in the coordinates x^a, the representatives of the world lines of the clocks lie in the plane (x^1, t); they are the lines where only t varies, given respectively by $x^1 = 0$ and $x^1 = h$. Let $T(0)$ and $T(h)$ be the periods, in proper time, of these clocks. The signals of the beginning and end of a period of the clock on the ground are transmitted by light rays to an observer handling the other clock. Such light rays obey the differential equation

$$\frac{dt}{dx^1} = \frac{\sqrt{g_{11}}}{N}.$$ (III.8.2)

A signal emitted at parameter time $t(0)$ reaches the clock at height h at parameter time

$$t(h) = \int_0^h \frac{\sqrt{g_{11}}}{N}(x^1)\, dx^1 + t(0).$$ (III.8.3)

The difference in the parameter times between two emitted and two received signals is therefore the same. However, this statement is not true for the proper times, which are the physically measured quantities. We denote by $T(0)$ (respectively $T(h)$) the period in proper time of the clocks. Using the relation between the parameter time and the proper time for each of the clocks, namely $dT = N dt$, we find that

$$t_2(0) - t_1(0) = N^{-1}(0)T(0) = t_2(h) - t_1(h) = N^{-1}(h)T(h).$$ (III.8.4)

The corresponding lapse of proper time marked by the clock of this observer is

$$T(h) = \frac{N(h)}{N(0)}T(0).$$ (III.8.5)

A longer period is observed (redshift) if $N(h) > N(0)$, which is the case in our laboratory example where the source of gravitation is the Earth with mass m. Then (see the Newtonian approximation)

$$N \cong 1 - \frac{m}{x^1}.$$ (III.8.6)

Theory and experimental results have proven to be in excellent agreement.

A related effect (which involves the geometry of null geodesics connecting the timelike world lines of various massive bodies) has been measured directly for celestial objects, and called **time delay** by Shapiro, who measured it first, in 1966, by sending radar signals to planets (i.e. in the gravitational field of the Sun) and measuring the time elapsed on Earth between their emission and their return after reflection. We will give the calculations leading to the theoretical prediction in Chapter V, devoted to the Schwarzschild spacetime.

The reality of the proper time defined on spacetime by a Lorentzian metric that is not locally flat in the presence of a gravitational field has been checked directly by carrying caesium clocks on satellites around the Earth and observing that they gain some hundred nanoseconds[26] over identical clocks that remain on the ground. The experiment has now been done with clocks on satellites orbiting Mars and on its surface.

Finally, it has been found that the accuracy of the GPS positioning system depends upon General Relativistic correction of the physical (proper) time, both because the satellites that receive the signals from Earth are far above the Earth's surface and because, for high-precision results, it is necessary to take into account that the Earth is not exactly spherical, and has plains and mountains that induce variations in the gravitational field.[27]

III.8.4 Conclusion

All experimental results to date[28] are in agreement with the predictions of Einstein's General Relativity.

[26] After correction for the kinematic effect of Special Relativity. An approximately drag-free satellite describes a geodesic of the Lorentzian (Schwarzschild) metric, but the observer on Earth, not in free fall because held by the ground, does not.

[27] We compute in Problem IV.11.5 in Chapter IV the first correction (quadrupole moment) to the Newtonian potential due to non-sphericity of the source. In Newton's theory of gravity, the potential has no influence on time. We will study in Chapter IV the Newtonian approximation of the Einstein theory of gravity, in which time is dependent of the gravitational field.

[28] For a description of basic experiments, their results and discussions, see, for instance, Chapter 4 of Ohanian and Ruffini (2013). For up to date results, see papers by Clifford Will, for instance his recent review (Will, 2014).

III.9 Problems

III.9.1 Newtonian gravitation theory in absolute space and time $E^n \times R$

1. By analogy with the case $n = 3$, the Newtonian potential on the Euclidean space E^n of a gravitational mass density ρ_G is a scalar function U given by

$$U(x) := \int_{R^n} \frac{\rho_G(y)}{|x - y|^{n-2}} \, dy, \quad \text{with} \quad |x - y| = \left[\sum_{i=1,\dots n} (x^i - y^i)^2 \right]^{\frac{1}{2}}.$$

Show that U satisfies the elliptic partial differential equation

$$\Delta U := \sum_{i=1,\dots,n} \frac{\partial^2}{(\partial x^i)^2} U = -(n-2)(\text{area } S^{n-1}) \rho_G$$

2. Show that the Newtonian potential of a pointlike particle with gravitational mass density $m_G \delta$, where δ is the Dirac measure at the origin O, is

$$U = \frac{m_G}{r^{n-2}}.$$

Hint: The Dirac measure at O is by definition such that if f is a continuous function at O, then

$$\int_{R^n} f(y)\delta \, d^n y = f(0)(\text{area } S^{n-1}).$$

3. Assuming the laws of Newtonian dynamics and gravitation,

$$F = m_I \gamma, \quad \text{with } m_I \text{ the inertial mass and} \quad \gamma = \text{grad } U,$$

show by using Newton's principle of action and reaction that m_G is proportional to m_I by a fixed constant.

III.9.2   Mass in length units (case $n = 3$)

The value of the Newton gravitational constant G_N is, in cgs units with a 10^{-13} accuracy,

$$G_{N,\text{cgs}} = 6.67259 \times 10^{-8} \stackrel{\sim}{=} 7 \times 10^{-8}.$$

Compute the unit of mass in the centimetre–gram system (in which the second is expressed in terms of the centimetre by taking $c = 1$) under the condition that the Newton gravitational constant is equal to 1.

Compute in this unit the mass of the Earth, whose mass is about 6×10^{24} cgs grams in the cgs system.

Solution

This uses only the dimensions $L^3 M^{-1} T^{-2}$ of G_N. Recall that

$$1\,\text{s} \stackrel{\sim}{=} 3 \times 10^{10} \text{ cm-s.} \tag{III.9.1}$$

Set $1\,g = \mu\,\text{cm-g}$. The trivial fact that the quotient of the values of a quantity in different units is the inverse of the quotient of the units gives

$$G_{N,\text{geom,cm}} = G_{N,\text{cgs}}\mu^{-1}\frac{1}{9} \times 10^{-20} = 1,$$

with

$$G_{N,\text{cgs}} \stackrel{\sim}{=} 7 \times 10^{-8}.$$

Hence, $G_{N,\text{geom}} = 1$ if $\mu = \frac{7}{9} \times 10^{-28}$. Then $m_{\text{Earth}} = 6 \times 10^{27}$ g implies[29]

$$m_{\text{Earth,geom}} \stackrel{\sim}{=} \frac{42}{90} \text{ cm.}$$

III.9.3 Planck units

The dimensional constant that has been measured with the greatest accuracy is the Planck constant h. Planck remarked that it is possible to define from G_N, h and c units for time, length and mass, called Planck units or geometric units, such these constants all take the value 1.

1. Show that the dimensions of h are ML^2/T^{-1}.

2. Assume that the three constants c, G_N and h are known in cgs units. Denote by $L_P = x\,\mathrm{cm}$, $M_p = y\,\mathrm{grams}$ and $T_P = z\,\mathrm{seconds}$ new units such that these constants are equal to 1. Such units are called Planck units. Write the three equations satisfied by x, y and z.

3. Compute the unique solution giving the Planck units.

Solution

1. $h = E/\nu$. The dimensions of energy E are ML^2/T^{-2} and those of frequency ν are T^{-1}. The result follows immediately.

2. The relation between cgs and Planck units and the dimensions of the given constants imply

$$c_P = c_{\mathrm{cgs}}\frac{z}{x}, \quad G_P = G_{\mathrm{cgs}}\frac{yz^2}{x^3}, \quad h_P = h_{\mathrm{cgs}}\frac{z}{yx^2},$$

and setting these constants equal to 1 gives

$$\frac{x}{z} = c_{\mathrm{cgs}}, \quad \frac{x^3}{yz^2} = G_{\mathrm{cgs}}, \quad \frac{yx^2}{z} = h_{\mathrm{cgs}},$$

which imply trivially

$$\frac{x}{z} = c_{\mathrm{cgs}}, \quad \frac{x}{y} = c_{\mathrm{cgs}}^{-2}G_{\mathrm{cgs}}, \quad \frac{z}{y} = c_{\mathrm{cgs}}^{-3}G_{\mathrm{cgs}}, \quad c_{\mathrm{cgs}}^3 G_{\mathrm{cgs}}^{-1}x^2 = h_{\mathrm{cgs}}.$$

Therefore, the unique values of the unknowns x, y and z are

$$x = \sqrt{\frac{h_{\mathrm{cgs}}G_{\mathrm{cgs}}}{c_{\mathrm{cgs}}^3}}, \quad y = xc_{\mathrm{cgs}}^2 G_{\mathrm{cgs}}^{-1}, \quad z = xc_{\mathrm{cgs}}^{-1},$$

which give

$$x = \sqrt{\frac{h_{\mathrm{cgs}}}{c_{\mathrm{cgs}}^3 G_{\mathrm{cgs}}}}, \quad y = \sqrt{\frac{h_{\mathrm{cgs}}c_{\mathrm{cgs}}}{G_{\mathrm{cgs}}}}, \quad z = \sqrt{\frac{h_{\mathrm{cgs}}G_{\mathrm{cgs}}}{c_{\mathrm{cgs}}^5}}.$$

Of course, the use of Planck units in formulas does not eliminate the need for at least one of them to be defined through a physical phenomenon, to link equations with reality.

The Einstein equations

IV.1   Introduction

The previous chapter replaced the Minkowski spacetime (R^4, η) of Special Relativity by a general Lorentzian manifold, (V^4, g)—hence the name 'General Relativity'. The connection coefficients of the metric g replaced both the gravitational and the inertial forces. There are spacetime frames where at a given point, or even along a given spacetime curve, they cancel each other. This statement is in agreement with the weak Einstein equivalence principle, which concerns the motion of one test particle. However, in contrast to inertial forces in Newtonian theory, which disappear in inertial frames (without canceling gravitation), there are no reference frames defined on an open set of a spacetime of General Relativity where all the connection coefficients vanish, except if the metric is flat, i.e. free of gravitation. In Einstein gravity, the existence of a non-zero gravitational field manifests itself by the curvature of the spacetime metric.

The non-Minkowskian metric g replaces in a sense Newton's gravitational potential, which satisfies the Poisson equation, a linear elliptic second-order partial differential equation with mass density as a source. The metric g should satisfy equations that give in first approximation the same results as Newton's law for the motions of bodies in a weak gravitational field and with velocities small with respect to the speed of light. Indeed, in these circumstances, Newton's law had been verified with excellent agreement.[1] Einstein's geometric foresight looked for geometric relations between the metric g of General Relativity and possible sources of gravitation. The equivalence of mass and energy that he had discovered before, the symmetric 2-tensor character of the metric g whose curvature he wanted to link with the gravitation sources, and his genius all led Einstein, after various unsuccessful attempts,[2] to the famous Einstein equations that are at present more fundamental than ever for theoretical physics at all scales.

We will write the equations in spacetimes of arbitrary dimension $n+1$, since they are now used by physicists in a wider context than the original dimension $n = 3$. We will specify that $n = 3$ only when this leads to special properties.

---

[1] Except for the advance of the perihelion of the orbit of Mercury, which behaves approximately like a slowly rotating ellipse whose perihelion advances $42''$ per century more than that computed using Newton's law when taking into account the gravitational fields due to other planets.

[2] Einstein was helped in obtaining the final result by his friend, the mathematician Marcel Grossmann.

IV.2 The Einstein equations

IV.2.1 The Einstein equations in vacuum

The vanishing in a domain of spacetime of the Riemann curvature tensor of the metric g implies that this metric is locally flat in this domain (see Chapter I), and hence without gravitational effects. It is therefore too strong a condition to impose in domains empty of matter or field energy sources in a spacetime that is not globally empty. A natural candidate for an equation to impose in vacuo[3] on the Lorentzian metric g, a symmetric 2-tensor, is the vanishing of the symmetric 2-tensor $Ricci(g)$, which is related to the Riemann curvature tensor by contraction. The tensor $Ricci(g)$ is a second-order partial differential operator on g, like the Laplace equation for the Newtonian potential, though it has quite different properties, as we will see in the following sections: first, it is nonlinear;[4] second, it presents both elliptic properties like the Newtonian potential and hyperbolic ones like the wave equation, corresponding to propagation of the gravitational field with the speed of light. We formulate the basic proposition[5] of Einsteinian gravitation.

Proposition IV.2.1 *In an open set U devoid of energies other than gravitation, the Lorentzian metric g of an Einsteinian spacetime (V, g) satisfies the following tensorial* **Einstein equations in vacuum:**

$$Ricci(g) = 0. \tag{IV.2.1}$$

The equations (2.1) are represented in each chart (see Chapter I) with domain included in U by the following system of second-order hyperquasilinear[6] partial differential equations for the components $g_{\alpha\beta}$ of the metric g:

$$R_{\alpha\beta} \equiv \frac{\partial}{\partial x^\lambda}\Gamma^\lambda_{\alpha\beta} - \frac{\partial}{\partial x^\alpha}\Gamma^\lambda_{\beta\lambda} + \Gamma^\lambda_{\alpha\beta}\Gamma^\mu_{\lambda\mu} - \Gamma^\lambda_{\alpha\mu}\Gamma^\mu_{\beta\lambda} = 0, \tag{IV.2.2}$$

where the $\Gamma^\lambda_{\alpha\beta}$ are the Christoffel symbols of the metric g, given in local coordinates (see I.7.13) by

$$\Gamma^\lambda_{\alpha\beta} := g^{\lambda\mu}[\alpha\beta, \mu], \quad \text{with} \quad [\alpha\beta, \mu] := \frac{1}{2}\left(\frac{\partial g_{\alpha\mu}}{\partial x^\beta} + \frac{\partial g_{\beta\mu}}{\partial x^\alpha} - \frac{\partial g_{\alpha\beta}}{\partial x^\mu}\right).$$

These equations are not independent; they satisfy the identities deduced in Chapter I from the Bianchi identities:

$$\nabla_\alpha S^{\alpha\beta} \equiv 0, \quad S_{\alpha\beta} := R_{\alpha\beta} - \frac{1}{2}g_{\alpha\beta}R, \quad R := g^{\lambda\mu}R_{\lambda\mu}. \tag{IV.2.3}$$

The tensor with components[7] $S_{\alpha\beta}$ is called the **Einstein tensor**, it is the symmetric 2-tensor with geometric expression

$$Einstein(g) := Ricci(g) - \frac{1}{2}gR(g).$$

[3]The notion of vacuum seems clear to common sense: it is a region of spacetime where there is neither matter nor electromagnetic field. Modern physics has shown that what we called the vacuum is indeed full of strange things—pairs of particles and antiparticles constantly annihilating each other and liberating virtual photons or neutrinos, which in turn give birth to pairs of particles. But these stange phenomena are in the realm of quantum field theory and do not concern us in this book.

[4]Gravitation is also its own source.

[5]It is known now that the Einstein equations are the only tensorial ones (up to addition of a 'cosmological term', see Section IV.3) that are of second order and quasilinear for the metric g in four dimensions.

[6]A partial differential equation of order m is called quasilinear if it is linear in the derivatives of order m. It is called hyperquasilinear if, in addition, the coefficients of these derivatives do not contain derivatives of order $m - 1$.

[7]Denoted by $G_{\alpha\beta}$ by many authors.

IV.2.2   Equations with sources

It took a considerable time for Einstein to settle on the choice of equations connecting the metric g with sources. From a physical point of view, he was inspired on the one hand by the relativity to observers of the splitting beween energy and momentum (see Chapter II) and by the conservation laws for the various stress–energy–momentum tensors T found in Special Relativity and on the other hand by the equivalence principle. From a mathematical point of view Einstein's choice of equations was motivated by the contracted Bianchi identities.

The geometric equations found by Einstein in 1915, with the help of the mathematician Marcel Grossmann, are the very simple ones

$$Einstein(g) := Ricci(g) - \frac{1}{2}gR(g) = G_E T;$$

i.e. in local coordinates and arbitrary dimension $n + 1 \geq 3$,

$$S_{\alpha\beta} := R_{\alpha\beta} - \frac{1}{2}g_{\alpha\beta}R = G_E T_{\alpha\beta}. \tag{IV.2.4}$$

Exercise IV.2.1 *Show that in a spacetime of dimension $n + 1 > 2$, the Einstein equations with sources are equivalent to the following:*[8]

$$R_{\alpha\beta} = G_E \rho_{\alpha\beta}, \quad \text{with} \quad \rho_{\alpha\beta} \equiv T_{\alpha\beta} - \frac{T^\lambda_\lambda}{n-1}g_{\alpha\beta}. \tag{IV.2.5}$$

The source T is the stress–energy symmetric 2-tensor, supposed to represent the pointwise value of all the energies, momenta and stresses present in the spacetime. It is a phenomenological tensor whose choice is not always easy,[9] even in the classical dimension $n + 1 = 4$ and even without taking quantum mechanical considerations into account.

The factor[10] G_E is a phenomenological dimensional constant. We will explain later, by using the Newtonian approximation, why in spacetime dimension $n + 1 = 4$ physicists take

$$G_E = 8\pi G_N,$$

with G_N the Newtonian gravitational constant.

Note that the gravitational constant does not appear in Special Relativity—hence, the Einstein equivalence principle does not imply that it is a constant. In fact, several physicists (Jordan, Thiry, Dicke, Dirac, and others) have conjectured that it is a spacetime scalar function. However, no experiment has confirmed this conjecture.

The geometric units of time and length have been defined in Chapter III. In these units, the speed of light is equal to 1.

In this book, as in most mathematical studies, except when making numerical estimates, we assume units of mass–energy chosen such that G_E takes its geometric value: $G_E = 1$.

---

[8]It is only in the classical dimension $n + 1 = 4$ that $d - 2 = 2$—a fact that should not be forgotten when working in other dimensions.

[9]In particular when there is a density of linear momentum, which is naturally represented by a non-symmetric 2-tensor. This is the case for electromagnetic fields with induction.

[10]Denoted by κ by some authors.

The **contracted Bianchi identities** (2.3) show that the Einstein equations are compatible only if the tensor T satisfies the equations, called **conservation laws**,

$$\nabla_\alpha T^{\alpha\beta} = 0. \qquad (IV.2.6)$$

We have written in Chapter III the equations of motion of the simplest, most common energy sources on a general Lorentzian metric of arbitrary dimension by using their expression in Special Relativity and the Einstein equivalence principle.

IV.2.3   Matter sources

In the case of **dust (incoherent matter,** also called **pure matter)**, the stress–energy tensor is

$$T_{\alpha\beta} = r u_\alpha u_\beta, \qquad (IV.2.7)$$

with u the unit flow velocity and r the rest-mass density.

In the case of a **perfect fluid**, the stress–energy tensor is

$$T_{\alpha\beta} = \mu u_\alpha u_\beta + p(g_{\alpha\beta} + u_\alpha u_\beta), \qquad (IV.2.8)$$

with u again the unit flow velocity and with μ and p respectively the energy and the pressure densities, both of which are non-negative in classical situations. The equations of motion of the phenomenological quantities μ, p and u (see Chapter III) imply that $T^{\alpha\beta}$ satisfies the conservation law

$$\nabla_\alpha T^{\alpha\beta} = 0. \qquad (IV.2.9)$$

Note that we could have made the converse argument and deduced the equations of motion of the sources from the conservation laws implied by the Einstein equations satisfied by the spacetime metric. In Einstein's theory, in contrast to Newton's, the equations of motion are a consequence of the equations satisfied by the gravitational potential.

IV.2.4   Field sources

For an **electromagnetic field**, the stress–energy tensor is the **Maxwell tensor**

$$\tau_{\alpha\beta} := F_\alpha{}^\lambda F_{\beta\lambda} - \frac{1}{4}\eta_{\alpha\beta} F^{\lambda\mu} F_{\lambda\mu}. \qquad (IV.2.10)$$

The Maxwell tensor satisfies the **conservation laws**

$$\nabla_\alpha \tau^{\alpha\beta} = 0 \qquad (IV.2.11)$$

if the 2-form F that represents the electromagnetic field satisfies the Maxwell equations of General Relativity

$$dF = 0 \quad \text{and} \quad \nabla \cdot F = 0. \qquad (IV.2.12)$$

IV.3   The cosmological constant

It is known that the only tensorial operator for a metric g on a four-dimensional[11] spacetime that is second-order and quasilinear is the Einstein tensor with the possible addition of a linear term Λg, with Λ an arbitrary constant, called the cosmological constant, which Einstein did not include, because it has no a priori physically fixed value. The Einstein equations with cosmological constant Λ read

$$Einstein(g) + \Lambda g \equiv Ricci(g) - \frac{1}{2}gR(g) + \Lambda g = T. \qquad (IV.3.1)$$

In local coordinates, these take the form

$$S_{\alpha\beta} + \Lambda g_{\alpha\beta} \equiv R_{\alpha\beta} - \frac{1}{2}g_{\alpha\beta}R + \Lambda g_{\alpha\beta} = T_{\alpha\beta}, \qquad (IV.3.2)$$

where T is the stress–energy tensor considered before, supposed to represent the density of all the energies, momenta and stresses of the matter and classical field sources. Einstein introduced the cosmological constant when looking for a stationary model for the cosmos (see Problem IV.11.1) and removed it[12] after the interpretation of astronomical observations as showing the universe to be expanding. It is again included by cosmologists, although its value is controversial and may have been different at different epochs. It is generally considered to be very small now but to have been large in the early universe. In fact, the tendency now is to put Λ among the sources, interpreting $-\Lambda g_{\alpha\beta}$ as an energy–momentum tensor of the vacuum generated by quantum-particle processes; the constant Λ is then replaced by a function of a new field ϕ satisfying a wave equation with a non-linear potential. A 'cosmological' term also appears in supergravity and string theories.

The equations (3.2) on an $n + 1 = d$-dimensional spacetime are equivalent to the following:

$$R_{\alpha\beta} = \rho_{\alpha\beta}, \quad \text{with} \quad \rho_{\alpha\beta} \equiv T_{\alpha\beta} + \left(\frac{d}{d-2}\Lambda - \frac{T^\lambda_\lambda}{d-2}\right)g_{\alpha\beta}. \qquad (IV.3.3)$$

IV.4   General Einsteinian spacetimes

A spacetime of General Relativity is a pair (V, g), with V a differentiable manifold and g a Lorentzian metric on V, both a priori arbitrary. It is called an **Einsteinian spacetime**[13] if the metric g satisfies on V the Einstein equations with source a phenomenological 2-tensor T assumed to represent all the non-gravitational energies, momenta and stresses:

$$Einstein(g) = T, \quad \text{i.e.} \quad S_{\alpha\beta} = T_{\alpha\beta}, \qquad (IV.4.1)$$

$$\nabla \cdot T = 0, \quad \text{i.e} \quad \nabla_\alpha T^{\alpha\beta} = 0. \qquad (IV.4.2)$$

The definition of T is clear in the cases mentioned in Section IV.2.3 and IV.2.4, but is not so clear and is still controversial for more complicated sources, even in the classical cases of dissipative fluids with or without heat current and of electromagnetic fields with inductions. Coupling with quantum fields raises the deep problem of quantum gravity. New and mysterious fields are now being discovered in cosmological studies (see Chapter VII).

Two isometric[14] *spacetimes (V, g) and (V', g') are considered as identical.*

The spacetime is called a vacuum spacetime if $T \equiv 0$.

[14]That is, such that there exists a diffeomorphism f of V onto V' such that $f * g' = g$.

IV.4.1   Regularity

In an Einsteinian spacetime, the manifold V can be assumed to be C^∞ without restricting its generality.[15] We have already said that we assume in this book that the metric is as smooth as necessary for any given statement to be true. It is clear that if the source T is discontinuous, then the metric g cannot be C^2. Using the definition of generalized derivatives, it may be possible to define the Einstein tensor[16] for metrics that do not admit second derivatives in the usual sense.

The largest simple Sobolev Hilbert space for the components of the metric g for which the components of the Riemann tensor in four dimensions are almost everywhere defined functions on a coordinate patch diffeomorphic to a ball Ω of R^4 is $H_3(\Omega)$, the space of functions that, together with their generalized partial derivatives of order up to 3, are square-integrable on Ω.

For physical interpretation, one is led to split, at least locally, space and time and consider coordinate patches $\Omega = \omega \times I$, with ω a ball in R^3 and I an interval of R. The restriction to ω of $g_{\alpha\beta}$ is then assumed to be in $H_2(\omega)$, and hence uniformly continuous, while the restrictions of the first derivatives $\partial g_{\alpha\beta}/\partial x^\lambda$ and second derivatives $\partial^2 g_{\alpha\beta}/\partial x^\lambda \partial x^\mu$ belong respectively to $H_1(\omega)$ and $L^2(\omega)$. The restriction of the Riemann tensor is then defined as a function in $L^2(\omega)$. For more comments, see Chapter VIII on the Cauchy problem.

[15]It is known that any C^1 manifold can be given a C^∞ structure that induces this C^1 structure.

[16]This possibility is limited, however, by the nonlinearity of the Riemann curvature tensor.

IV.4.2   Boundary conditions

The Einstein equations are partial differential equations, and so their solutions are linked with the data of boundary conditions. We will see in Chapter VIII on the Cauchy problem that, even in a vacuum, the Einstein equations present both elliptic aspects like the Poisson equation of Newtonian gravity and hyperbolic aspects like the wave equation of light propagation. We just mention here that in the primary cases the manifold V of an Einsteinian spacetime (V, g) is a product $M \times R$, with M spacelike and R timelike.

For cosmological studies, the case of a compact manifold M without boundary has been used, although this seems to be in conflict with recent observations hinting at a flat universe—but cosmology is a very debated subject.

[17]This model is only an approximation, like all mathematical models of physical situations. The system considered as isolated is in fact embedded in a curved universe.

[18]Higher-dimensional asymptotically Euclidean manifold can also be defined.

[19]Asymptotically Euclidean manifolds with several ends can also be considered, with just some additional complications in writing down the appropriate equations.

In the astrononphysical context, particularly relevant situations that can be confronted with observation are isolated systems of a few bodies far from any other source of gravitation, for example the Solar System and binary star systems. In these cases, it is legitimate to assume that $(M, \bar{g})$, with $\bar{g}$ the Riemannian metric induced on M by the spacetime metric, tends to be flat far away from the studied system. It is then assumed to be an asymptotically Euclidean manifold, which we define below.[17]

A three-dimensional[18] **asymptotically Euclidean Riemannian manifold** $(M, \bar{g})$ is defined as a smooth manifold M union of a compact set and an[19] end M^{end} diffeomorphic to the exterior of a ball of R^3. This end is given local coordinates x^1, x^2, x^3 by this diffeomorphism. We denote by r, θ, ϕ the corresponding polar (pseudo) coordinates, $r := \left[\sum_i (x^i)^2 \right]^{\frac{1}{2}}$. The Riemannian manifold (M, g) is called asymptotically Euclidean if in M^{end} it holds that

$$g_{ij} = \delta_{ij} + O(r^{-1}), \quad \partial^k g_{ij} = O(r^{-|k|+1}),$$

with the notation

$$\partial^k := \frac{\partial^{k_1}}{(\partial x^1)^{k_1}} \frac{\partial^{k_2}}{(\partial x^2)^{k_2}} \frac{\partial^{k_3}}{(\partial x^3)^{k_3}}, \quad |k| = k_1 + k_2 + k_3,$$

and where a function f is said to be $O(r^{-k})$ if $r^k f$ is uniformly bounded on M^{end}. The useful maximum order of the derivatives appearing in the definition depends on the circumstances. Variants of hypotheses on the behaviour of metrics at infinity appear in the solution of problems on asymptotically Euclidean manifolds.[20]

[20]For metrics approaching the Euclidean metric in weighted Hölder spaces, see Choquet-Bruhat (1974) and Chaljub-Simon and Choquet-Bruhat (1979), and for those in weighted Sobolev spaces, see Cantor (1979), Cantor and Brill (1981), and Choquet-Bruhat and Christodoulou (1981).

IV.4.3   Physical comment

Even in the vacuum case, there exist many Einsteinian spacetimes. They have underlying manifolds with different topologies, as well as different metrics on each of these manifolds. A great number of exact solutions of Einstein equations possessing isometry groups have been constructed, in vacuum or with various sources. Some of these spacetimes have (at least at present) a purely mathematical interest, but a few of them are models of known physical situations, at different time or space scales. There is no universal Einsteinian spacetime as a model for reality— this is in disagreement with Newton's concepts, and also with Special Relativity.

Before giving the main lines of construction and properties of general Einsteinian spacetimes, we justify their validity by their possible approximation by the Newton spacetime for sources with low velocity, and we give for weak gravitation the approximation of Einsteinian spacetimes by Minkowski spacetime. The Minkowskian approximation reveals a property of gravitational fields that has no analogue in Newton's theory, namely the existence of gravitational waves and their propagation with the speed of light—this property eliminates the need for 'action at a distance' that was so puzzling for Newton.

IV.5 Newtonian approximation

We have said that Einstein's theory of gravitation must nearly coincide with Newton's in the physical situations where the latter proved to be accurate. These situations are slow (compared to the speed of light) velocity of the gravitating bodies with respect to Newton spacetime and weak gravitational fields. In Einstein's theory, the spacetime coincides with Minkowski spacetime in the absence of gravitation, that is with the manifold $R^3 \times R$ endowed with the flat metric, which, in inertial coordinates and geometrical units, reads as follows:

$$-(dx^0)^2 + \sum_{i=1}^{3}(dx^i)^2.$$

An Einsteinian spacetime with a weak gravitational field will be $R^3 \times R$ with a metric such that

$$g_{00} = -1 + h_{00}, \quad g_{0i} = h_{0i}, \quad g_{ij} = \delta_{ij} + h_{ij},$$

with $h_{\alpha\beta}$ of order $\epsilon \ll 1$.

In a spacetime with masses moving slowly (compared with 1, the speed of light), the components $h_{i\alpha}$ are small with respect to h_{00} and the time derivatives $\partial_0 h_{\alpha\beta}$ are small with respect to the space derivatives $\partial_i h_{\alpha\beta}$; these are considered to be of the same order as $h_{\alpha\beta}$.

Exercise IV.5.1 *Justify these statements by considering units where $c \doteq 3 \times 10^5 km/s$.*

Hint: Change the timescale by seting $dx^0 = c\,dt$.

IV.5.1 Determination of G_E

The component R_{00} in the considered approximation is equivalent to

$$R_{00} \doteq \partial_i \Gamma^i_{00} \doteq -\frac{1}{2}\Delta h_{00},$$

with Δ the Laplace operator of Euclidean space. In Newtonian theory, the gravitational field is the gradient of the Newtonian potential, a function U that satisfies the Poisson equation:

$$\Delta U = -\rho_G \equiv -4\pi G_N \rho \qquad\qquad (\text{IV.5.1})$$

where ρ is a positive scalar function on R^3 equal to the mass density of the sources and G_N is the Newtonian gravitational constant (see Chapter III).

In Einstein's theory, the energy sources are represented by the stress–energy tensor. Under usual circumstances, the most important energy source is pure matter, i.e. the stress–energy tensor is of the following form, with μ some positive scalar function and u the unit flow vector:

$$T_{\alpha\beta} \hat{=} \mu u_\alpha u_\beta.$$

In the case where the matter has a small velocity (with respect to the speed of light, taken to be $c = 1$) for observers following the time lines of the coordinate system, it holds that u_0 is equivalent to -1 and u_i to zero; therefore, $T_{i\alpha}$ is negligible and

$$\rho_{00} \equiv G_E(T_{00} - \frac{1}{2}g_{00}T) \hat{=} \frac{1}{2}G_E\mu. \tag{IV.5.2}$$

We see that in the approximation that we are making, called the Newtonian approximation, the equation $R_{00} = G_E\rho_{00}$ gives

$$\Delta h_{00} \hat{=} -G_E\mu,$$

that is, the Poisson equation of Newton's theory for the gravitational potential if h_{00} is identified with $2U$ and one sets $G_E := 8\pi G_N$. This is the reason why many physicists write the Einstein equations in dimension $3 + 1$ as

$$S_{\alpha\beta} = 8\pi G_N T_{\alpha\beta}. \tag{IV.5.3}$$

The value of the Newtonian gravitational constant depends on the choice of units for time, length, and mass. As we said before, G_N is approximately, in CGS units,

$$G_{N,\text{cm}^3\text{g}^{-1}\text{s}^{-2}} = 6,67259 \times 10^{-8}.$$

Of course, making $G_E = 1$ instead of $G_N = 1$ implies a change by a factor 8π in the unit of mass–energy.

In this book, we write the Einstein equations in geometric units, i.e. $S_{\alpha\beta} = T_{\alpha\beta}$, introducing the relevant scale factor to the usual physical units only when this is useful for interpretation of observations.

IV.5.2 Equations of motion

In Newton's theory, the equation of motion of a test particle is

$$\frac{d^2x^i}{dt^2} = \frac{\partial U}{\partial x^i}.$$

In General Relativity, a test particle follows a geodesic of the spacetime metric, i.e.

$$\frac{d^2x^\alpha}{ds^2} + \Gamma^\alpha_{\mu\lambda}\frac{dx^\lambda}{ds}\frac{dx^\mu}{ds} = 0.$$

For small velocities, we have already seen that $dx^0/ds \sim 1$ and $dx^i/ds \sim 0$; therefore,

$$\frac{d^2x^i}{(dx^0)^2} \sim -\Gamma^i_{00} \sim \frac{1}{2}\partial_i h_{00}. \tag{IV.5.4}$$

We see that the *Einstein law coincides with the Newton law in the Newtonian approximation where* $h_{00} = 2U$.

IV.5.3   Post-Newtonian approximation

One can improve the Newtonian approximation by retaining further terms in the expansions. This leads to the so-called **post-Newtonian (PN) approximation**. To make these approximations, it is more convenient not to replace the gravitational constant and speed of light by their geometric value 1.

Let us briefly explain the structure of the first post-Newtonian (1PN) approximation, where one keeps all the corrections to the Newtonian approximation containing one power of $1/c^2$. There are three sources of such $1/c^2$ corrections: (i) in algebraic $(v/c)^2$ terms, where v is a velocity variable; (ii) in U/c^2 terms, where U is a Newtonian-potential-like variable; or (iii) in time-derivative terms $\partial_0^2 = c^{-2}\partial_t^2$. At the 1PN approximation level, one must go beyond the linearized approximation to the Einstein equations, and keep some of the quadratically nonlinear terms in the time–time Einstein equation

$$R^{00} = 8\pi\, G_N c^{-4}\left(T^{00} - \frac{1}{n-1}\,T g^{00}\right).$$

A convenient way of doing so is to parametrize the metric in the following exponential form:[21]

$$g_{00} = -\exp\left(-\frac{2}{c^2}\,V\right),$$

$$g_{0i} = -\frac{4}{c^3}\,V_i, \qquad\qquad\text{(IV.5.5)}$$

$$g_{ij} = \exp\left(+\frac{2}{c^2}\,V\right)\gamma_{ij}.$$

[21]Blanchet and Damour (1989).

Inserting this form in Einstein's equations, and using for simplicity wave coordinates, one finds (in dimension $n = 3$) that the auxiliary metric γ_{ij} must be flat modulo corrections of the next PN order $O(1/c^4)$, i.e.

$$\gamma_{ij} = \delta_{ij} + O\left(\frac{1}{c^4}\right). \qquad\qquad\text{(IV.5.6)}$$

In addition, one finds that the gravitational 'scalar potential' V and the gravitational 'vector potential' V_i satisfy the following simple, *linear* equations:

$$\Delta V - \frac{1}{c^2}\,\partial_t^2 V = -4\pi\, G_N \sigma,$$

$$\Delta V_i = -4\pi\, G_N \sigma^i, \qquad\qquad\text{(IV.5.7)}$$

where the source terms σ and σ^i are simply given in terms of the contravariant components of the stress–energy tensor by

$$\sigma := \frac{T^{00} + T^{ss}}{c^2}, \qquad (\text{IV.5.8})$$

$$\sigma^i := \frac{T^{0i}}{c}.$$

The linear form of the equations (5.7) for V and V_i allows one to solve them explicitly in terms of σ and σ_i. At the 1PN approximation, one can write

$$V(t, x) = G_N \int d^3 x' \left[\frac{\sigma(t, x')}{|x - x'|} + \frac{1}{2c^2} \partial_t^2 \sigma(t, x') |x - x'| + O\left(\frac{1}{c^4}\right) \right],$$

$$V_i(t, x) = G_N \int d^3 x' \frac{\sigma^i(t, x')}{|x - x'|} + O\left(\frac{1}{c^2}\right).$$

$$(\text{IV.5.9})$$

The approximate form of the metric obtained by inserting Eqs. (IV.5.9) in to the expressions (5.5), the first of which explicitly reads (upon expanding the exponential to show the level of nonlinearity actually contained in the 1PN approximation)

$$g_{00} = -\left[1 - \frac{2}{c^2} V + \frac{2}{c^4} V^2 + O\left(\frac{1}{c^6}\right)\right], \qquad (\text{IV.5.10})$$

is sufficiently accurate for describing all current experiments and observations in the Solar System, including perihelion advances of planetary orbits, the bending and delay of electromagnetic signals exchanged between the Earth and planets or satellites, and the very accurate laser ranging data to the Moon. Let us, however, note several different extensions of the above 1PN formalism.

First, to compare not only General Relativity but also different (non-Einsteinian) gravitational theories with experiments and observations, an extension of the above general relativistic 1PN metric, containing a whole collection of adjustable parameters, has been introduced, and is called the parametrized post-Newtonian (PPN) formalism.[22]

Second, when developing a general relativistic theory of the motion of an N-body system (such as the Solar System), it has been found useful to generalize the PN formalism into a **multichart** approach to general relativistic celestial mechanics.[23] In contrast to the traditional one-chart approach to the general relativistic N-body problem (in which a single coordinate system is used to describe both the gravitational field and the motion of N bodies), the multichart approach uses $N + 1$ coordinate systems: a global coordinate system x^μ and N local coordinate systems X_A^α (with $A = 1, 2, \ldots, N$), each of which is attached to one of the N bodies. This multichart approach has been found useful not only in the Solar System, but also in binary systems containing strongly self-gravitating objects such as neutron stars or black holes (see Section VI.6 in Chapter VI).

[22] See, for instance, Will (2014).

[23] See Brumberg and Kopejkin (1989) and Damour, Soffel, and Xu (1991, 1992).

IV.6   Minkowskian approximation

To compute the Minkowskian approximation of solutions of the Einstein equations, one considers Lorentzian metrics on R^{n+1} with components in some coordinate frames:

$$g_{\alpha\beta} = \eta_{\alpha\beta} + h_{\alpha\beta},$$

with $\eta_{\alpha\beta} = \mathrm{diag}(-1, 1, \ldots, 1)$. One assumes $h_{\alpha\beta}$ and its derivatives to be small with respect to 1; their products are then even smaller.

IV.6.1   Linearized equations at η

A first approximation is obtained by neglecting in the Einstein equations all terms containing products of the perturbation h, that is, by considering a solution of the Einstein equations linearized at η. The following linearizations are particular cases of those computed in Chapter I and result easily from the expressions for the Christoffel symbols and the Ricci tensor when products of h's and of their derivatives are neglected:

$$(\delta\Gamma^{\lambda}_{\alpha\beta})_{\eta}(h) = \frac{1}{2}\eta^{\lambda\mu}[\mu, \alpha\beta] = \frac{1}{2}\eta^{\lambda\mu}\left(\partial_{\alpha}h_{\mu\beta} - \partial_{\mu}h_{\alpha\beta}\right)$$

$$(\delta R_{\alpha\beta})_{\eta}(h) = -\frac{1}{2}\Box h_{\alpha\beta} + \partial_{\alpha}f_{\beta} + \partial_{\beta}f_{\alpha}, \qquad \text{(IV.6.1)}$$

where $\Box := \eta^{\lambda\mu}\partial^2_{\lambda\mu}$, with $\partial_{\alpha} := \partial/\partial x^{\alpha}$, is the Minkowskian d'Alembertian operator and f_{α} is the set of functions (indices are raised with the Minkowski metric η)

$$f_{\alpha} := \frac{1}{2}\left(\partial_{\lambda}h^{\lambda}_{\alpha} - \frac{1}{2}\partial_{\alpha}h^{\lambda}_{\lambda}\right). \qquad \text{(IV.6.2)}$$

The decomposition (6.1) is non-tensorial; the conditions $f^{\alpha} = 0$ can be satisfied by choice of coordinates.

Exercise IV.6.1   *Show that the equations $f^{\alpha} = 0$ are equivalent to the linearized wave equations for the coordinates x^{α}, i.e.*

$$f^{\lambda} \equiv (\delta F^{\lambda})_{\eta}(h), \quad with \quad F^{\lambda} := \Box_{g}x^{\lambda} \equiv g^{\alpha\beta}\Gamma^{\lambda}_{\alpha\beta}.$$

Hint:

$$\delta(g^{\alpha\beta}\Gamma^{\lambda}_{\alpha\beta})|_{\eta} = \eta^{\alpha\beta}(\delta\Gamma^{\lambda}_{\alpha\beta})|_{\eta}.$$

IV.6.2   Plane gravitational waves

The formula (6.1) shows that if $f^{\alpha} = 0$ the first approximation is solution of the linearized Einstein equations at η in vacuum, $(\delta R_{\alpha\beta})_{\eta}(h^{(1)}) = 0$ which reduce to

$$\Box h^{(1)}_{\alpha\beta} := \eta^{\lambda\mu}\partial^2_{\lambda\mu}h^{(1)}_{\alpha\beta} \equiv \left[-\frac{\partial^2}{(\partial t)^2} + \sum_{i}\frac{\partial^2}{(\partial x^i)^2}\right]h^{(1)}_{\alpha\beta} = 0. \qquad \text{(IV.6.3)}$$

These equations are a system of ordinary wave equations for the perturbation $h^{(1)}$; the solutions propagate on Minkowski spacetime with velocity 1, i.e. the speed of light.

In the case of a Minkowski background, the equations $\Box h^{(1)}_{\alpha\beta} = 0$ imply trivially $\Box f_\alpha = 0$. However, $\Box f_\alpha = 0$ implies $f_\alpha = 0$ only if f_α satisfies appropriate initial data. We will study such problems in Section IV.7 for the full Einstein equations. Here, for weak fields, we will proceed directly, considering the general **plane waves**

$$h^{(1)}_{\alpha\beta} = c_{\alpha\beta}\varphi(a_\lambda x^\lambda), \quad \text{with } c_{\alpha\beta} \quad \text{and} \quad a_\lambda \text{ constants.}$$

Such a perturbation of the Minkowski metric is a solution of the linearized Einstein equations at η if

$$\Box h^{(1)}_{\alpha\beta} \equiv c_{\alpha\beta}\Box\varphi \equiv c_{\alpha\beta}\left[(-a_0)^2 + \sum_i (a_i)^2\right]\varphi''(a_\lambda x^\lambda) = 0 \quad \text{(IV.6.3a)}$$

and (linearized gauge condition)

$$2f_\alpha \equiv a^\lambda \chi_{\alpha\lambda}\varphi'(a_\lambda x^\lambda) = 0, \quad \text{with } \chi_{\alpha\lambda} := c_{\alpha\lambda} - \frac{1}{2}\eta_{\alpha\lambda}c. \quad \text{(IV.6.3b)}$$

Equation (6.3a) implies that a_α is a null vector,

$$(a_0)^2 - \sum_i (a_i)^2 = 0, \quad \text{(IV.6.4)}$$

while (6.3b) implies that the constant tensor $c_{\alpha\beta}$ satisfies the **polarization conditions**

$$a^\lambda \chi_{\alpha\lambda} = 0,$$

which are $n+1$ linear and homogeneous equations with $(n+1)(n+2)/2$ unknowns $\chi_{\alpha\lambda}$. The wave is said to be **transverse-traceless,** abbreviated to TT, because by a first-order change of frame one can reduce the polarization conditions to the equations

$$c_{00} = c_{0i} = 0, \quad c^i_i = 0, \quad \text{and} \quad a^i c_{ij} = 0.$$

Exercise IV.6.2   *Prove this statement.*

The polarized $c_{\alpha\beta}$ span a vector space of dimension $(n+1)(n-2)/2$: in the classical dimension $n = 3$, weak gravitational waves have two degrees of polarization. For example, choose a frame such that $a_2 = a_3 = 0$. The equations above then imply $c_{1i} = 0$ and $c_{22} + c_{33} = 0$. Examples of oscillating weak gravitational waves are

$$h^{(1)}_{\alpha\beta} = 0, \quad \text{except for } h^{(1)}_{23}, \ h^{(1)}_{22}, \ \text{and} \ h^{(1)}_{33},$$

with

$$h^{(1)}_{23} = c_{23}\sin(\omega t - \omega x^1), \quad h^{(1)}_{22} = -h^{(1)}_{33} \equiv c_{22}\sin(\omega t - \omega x^1).$$

IV.6.3   Further results on gravitational waves

We have seen that in situations where one can apply the linearized approximation to Einstein's equations, a system emits gravitational waves at infinity, which are given, to the lowest approximation, in terms of the quadrupole moment[24] of the system. More precisely, in a TT coordinate system, the metric perturbation far from the system reads, to lowest order,

$$
h_{ij}^{\rm TT}(t,x) \simeq \frac{2\,G_N}{c^4}\frac{1}{r}\left(P_{ik}\,P_{jl} - \frac{1}{2}\,P_{ij}\,P_{kl}\right)\frac{d^2}{dt^2}\,I_{kl}\left(t - \frac{r}{c}\right) + O\left(\frac{1}{r^2}\right),
$$
$$(IV.6.5)$$

where $P_{ij} \equiv \delta_{ij} - n_i\,n_j$ is a spatial projector orthogonal to the unit vector $n^i \equiv x^i/r$ and where the quadrupole moment of the system, $I_{ij}(t)$, is given (to lowest approximation) by

$$
I_{ij}(t) = \int d^3x'\,\frac{T^{00}(t,x')}{c^2}\left(x'^i\,x'^j - \frac{1}{3}x'^2\,\delta^{ij}\right).
$$
$$(IV.6.6)$$

The recent development of kilometre-size interferometric detectors of gravitational waves (LIGO, VIRGO, etc.) has provided strong motivation for improving the theoretical treatment of the generation of gravitational waves by astrophysically realistic systems. As many of these systems (such as binary neutron stars or binary black holes) contain strongly self-gravitating systems, one needs to use the multi-chart formalism sketched in Section IV.5.3. In addition, it has been found necessary to (i) include higher multipoles than the quadrupole I_{ij}, (ii) include higher post-Newtonian corrections to the Newtonian-level quadrupole formula, and (iii) push the post-Newtonian expansion to the highest possible approximation. These developments, however, are outside the scope of this book.[25]

IV.6.4   Tidal force

The influence of a gravitational wave on the trajectory of a single isolated particle, as noted before, cannot be observed. The **tidal force** due to a gravitational wave acting on a pair of particles, i.e. the geodesic deviation equation (see Chapter I), depends on the Riemann tensor. Considering two nearby particles with initially parallel unit velocity $u^\alpha = \delta_0^\alpha$ and spatial separation vector X^i, the equation for the geodesic deviation force reads

$$
u^\alpha u^\beta \nabla_\alpha \nabla_\beta X^i = X^\alpha u^\lambda u^\mu R^{(1)\beta}_{\lambda\alpha\ \mu} \equiv X^j R_0^{(1)i}{}_{0j},
$$
$$(IV.6.7)$$

with, at first order,

$$
u^\alpha u^\beta \nabla_\alpha \nabla_\beta X^i \simeq \partial_{00}^2 X^i
$$

and

$$
R_0^{(1)i}{}_{0j} \equiv -R_{0i}^{(1)0}{}_j \simeq \frac{\partial}{\partial x^0}\delta\Gamma_{ij}^0 - \frac{\partial}{\partial x^i}\delta\Gamma_{0i}^0.
$$
$$(IV.6.8)$$

[24]See Problem IV.11.5.

[25]For the generalization of (6.5) to the infinite sequence of higher multipoles see Damour and Iyer (1991). For a review of the current theory of gravitational radiation from post-Newtonian sources, see Blanchet (2014). See also Damour and Nagar (2011).

In the TT gauge, where $h_{\alpha 0}^{(1)} = 0$,

$$(\delta\Gamma_{\alpha\beta}^0)_\eta(h^{(1)}) \equiv \frac{1}{2}\partial_0 h_{\alpha\beta}^{(1)}, \tag{IV.6.9}$$

and hence

$$\delta R_{0iJ}^0 = -\frac{1}{2}\partial_{00}^2 h_{ij}$$

We see that the equation of geodesic deviation implies at first order of the Minkowskian approximation that

$$\partial_{00}^2(X^i + X^j h_j^{(1)i}) = 0.$$

IV.6.5   Gravitational radiation

The word 'radiation' usually refers to energy transfer without material support.

In the preceding subsections, the approximations used were mathematically well defined and physically well understood. By analogy with other fields, in particular electromagnetism, one expects gravitational waves to carry energy. However, as we have already mentioned, there is no pointwise intrinsically defined gravitational energy. At best, it is possible to define some non-local quantities depending also in general on another a priori given metric—quantities that possess some properties analogous to those of the energies of other fields. Much important and very complex work, analytical as well as numerical, has been (and is continuing to be) devoted to the problem of **gravitational radiation energy,** but some fundamental questions are still open.

IV.7   Strong high-frequency waves

IV.7.1   Introduction

It is of course possible to define weak gravitational waves as perturbations of a given non-flat Lorentzian metric and to obtain for them linear equations. It is more interesting to study nonlinear effects by extending to nonlinear equations the WKB[26] anzatz, first used to study approximate rapidly oscillating solutions of the Schrödinger equation in the form

$$u(x) = a(x)e^{i\omega\phi(x)}, \tag{IV.7.1}$$

[26] Wentzel–Kramers–Brillouin.

where x is a spacetime point and a and ϕ are functions on spacetime. The function ϕ, called the phase of the wave, and ω, a number called the frequency, are real. The number ω is assumed to be large compared with the values of the functions a and ϕ and their derivatives with respect to x. Such an approximation is called in mechanics a **high-frequency wave,** or a **progressive wave** with wave fronts $\phi(x) = $ constant; the function u varies more rapidly in directions transverse to the wave fronts than on the wave fronts. The study of progressive waves, also called the **two-timing method**, has many applications in classical as well as in quantum physics.

The WKB ansatz, or its generalization by Lax to asymptotic series, is not well adapted for the study of high-frequency waves associated with nonlinear equations, because the product of two functions like (7.1) is not of the same type.[27] J. Leray[28] and his collaborators introduced for linear equations a more general anzatz, replacing $e^{i\omega\phi(x)}$ by a general function of x and $\omega\phi(x)$. For quasilinear equations, it is then possible[29] to construct high-frequency waves that are approximate solutions in a well-defined sense. *The effects of the nonlinearity on the waves are possible distortion of signals and the appearance of singularities similar to shocks.*

In the case of General Relativity, where the unknown is a Lorentzian metric on a manifold V, a high-frequency gravitational wave in vacuum is a Lorentzian metric that is the sum of a non-oscillating part $\underline{g}$ and a rapidly varying one depending on a large parameter ω called the frequency:

$$g(x, \omega\phi(x)) = \underline{g}(x) + h, \quad h := \{\omega^{-1}v(x,\xi) + \omega^{-2}w(x,\xi)\}_{\xi=\omega\phi(x)}.$$
(IV.7.2)

We say that (7.2) is an asymptotic solution of order p of the vacuum Einstein equations if it is such that $\omega^p Ricci(g(x, \omega\phi(x)))$ remains uniformly bounded[30] as ω tends to infinity.

To write the asymptotic expansion of the Ricci tensor of the metric (7.2), we set[31]

$$h := \delta g := g - \underline{g}, \quad \text{hence} \quad \partial h = \partial(g - \underline{g}), \quad \partial^2 h = \partial^2(g - \underline{g}). \quad \text{(IV.7.3)}$$

The Taylor formula gives the expansion

$$Ricci(g) - Ricci(\underline{g}) = \underline{\delta Ricci} + \frac{1}{2}\underline{\delta^2 Ricci} + \dots, \quad \text{(IV.7.4)}$$

with

$$\underline{\delta Ricci} \equiv \underline{Ricci'_g} \cdot h + \underline{Ricci'_{\partial g}} \cdot \partial h + \underline{Ricci'_{\partial^2 g}} \cdot \partial^2 h.$$

The coefficients $\underline{Ricci'_g}$, $\underline{Ricci'_g}$, $\underline{Ricci'_{\partial g}}$, and $\underline{Ricci'_{\partial^2 g}}$ depend only on the background metric $\underline{g}$. The second derivative $\underline{\delta^2 Ricci}$, a quadratic form in $(h, \partial h, \partial^2 h)$, is computed analogously. We remark that $\underline{\delta^2 Ricci}$ does not contain the square of $\partial^2 h$, because $Ricci(g)$ is linear in $\partial^2 g$.

To compute ∂h and $\partial^2 h$, we use the definition of h. We denote by a prime a derivative with respect to ξ and underline partial derivatives with respect to x. We set $\varphi_\lambda := \partial_\lambda\varphi \equiv \underline{\partial}_\lambda\varphi$. Elementary calculus implies

$$\partial_\lambda h_{\alpha\beta}(x, \omega\phi(x)) \equiv \{\phi_\lambda v'_{\alpha\beta}(x,\xi) + \omega^{-1}[\underline{\partial}_\lambda v_{\alpha\beta}(x,\xi) + \phi_\lambda w'_{\alpha\beta}(x,\xi)]$$
$$+ \omega^{-2}\underline{\partial}_\lambda w_{\alpha\beta}(x,\xi)\}_{\xi=\omega\phi(x)}$$

[27] Isaacson applied the original WKB method to the linearized Einstein equations and then looked for a solution of equations with source a stress–energy tensor obtained by averaging the obtained perturbation. Progressive waves of the type (7.2) for the nonlinear Einstein equations are constructed in Choquet-Bruhat (1969a).

[28] Gårding, Kotake, and Leray (1966).

[29] Choquet-Bruhat (1969b). For coupling of high frequency gravitational and fluid waves see Choquet-Bruhat and Greco (1983).

[30] The added term in ω^{-2} is introduced to ensure rigorously the asymptotic character of the constructed solution.

[31] It is also possible to write a coordinate-independent formula by introducing a given metric e. Note that $Ricci''_{g\partial^2 g}$ and $Ricci_{\partial g\partial g}$ are independent of the choice of the given metric e.

and a corresponding formula for the second derivative,

$$\partial_\lambda \partial_\mu h_{\alpha\beta} = \omega \varphi_\lambda \varphi_\mu v''_{\alpha\beta} + \varphi_\lambda \underline{\partial}_\mu v'_{\alpha\beta} + \varphi_\mu \underline{\partial}_\lambda v'_{\alpha\beta} + v'_{\alpha\beta} \underline{\partial}_\lambda \varphi^\lambda$$
$$+ \varphi_\lambda \varphi_\mu w''_{\alpha\beta} + \omega^{-1} \mathcal{R}_{\lambda\mu,\alpha\beta}$$

with

$$\mathcal{R}_{\lambda\mu,\alpha\beta} := \{ [\partial^2_{\lambda\mu} v_{\alpha\beta} + \phi_{\lambda\mu} w'_{\alpha\beta} + (\varphi_\mu \underline{\partial}_\lambda + \varphi_\lambda \underline{\partial}_\mu) w'_{\alpha\beta}$$
$$+ \omega^{-1} \underline{\partial}_\lambda \underline{\partial}_\mu w_{\alpha\beta}](x,\xi) \}_{\xi=\omega\phi(x)}$$

IV.7.2   Phase and polarization

The asymptotic expansion in ω of the Ricci tensor is obtained by replacing the metric and its first and second derivatives by the above expressions. We see that this expansion starts with a term in ω^1, whose coefficient must vanish for g to be an asymptotic solution of order zero. We thus obtain the equations

$$Ricci'_{\partial^2_{\lambda\mu} g} \varphi_\lambda \varphi_\mu v'' = 0. \tag{IV.7.5}$$

This is a linear homogeneous system for the second derivative with respect to the parameter ξ of the tensor v, which reads, in coordinates,

$$-\frac{1}{2} \varphi^\lambda \varphi_\lambda v''_{\alpha\beta} + \frac{1}{2} [\varphi_\alpha P_\beta(v') + \varphi_\beta P_\alpha(v')] = 0, \tag{IV.7.6}$$

with

$$P_\alpha(v') \equiv \varphi_\lambda v'^\lambda{}_\alpha - \frac{1}{2} \varphi_\alpha v'^\lambda{}_\lambda.$$

By analogy with what we saw in the Minkowskian approximation, we call P_α the **polarization operator.** We state as a theorem the result we have obtained.

Theorem IV.7.1 *If the phase is isotropic,[32] then the necessary and sufficient condition for the progressive wave to satisfy the Einstein equations at order zero in ω is that the tensor v satisfies the four polarization conditions*

$$P_\alpha(v) = 0.$$

Exercise IV.7.1 *Prove that the polarization conditions express the vanishing at order zero of the perturbation of the harmonicity functions $g^{\lambda\mu} \Gamma^\alpha_{\lambda\mu}$.*

Hint: For an arbitrary metric, the following identity can be proved by straightforward computation (see Chapter VIII on the Cauchy problem):

$$R_{\alpha\beta} \equiv R^{(h)}_{\alpha\beta} + L_{\alpha\beta}, \tag{IV.7.7}$$

with

$$L_{\alpha\beta} \equiv \frac{1}{2} \left(g_{\alpha\lambda} \partial_\beta F^\lambda + g_{\beta\lambda} \partial_\alpha F^\lambda \right), \tag{IV.7.8}$$

[32] That is, if its gradient is a null vector. This condition is necessary for the wave to be significant.

$$F^\alpha \equiv g^{\lambda\mu}\Gamma^\alpha_{\lambda\mu\cdot} \equiv g^{\lambda\mu}\nabla_\lambda\partial_\mu x^\alpha \equiv \Box_g x^\alpha,$$

while the $R^{(h)}_{\alpha\beta}$ are a system of quasilinear, quasidiagonal (i.e. linear and diagonal in the principal, second-order, terms) wave operators,

$$R^{(h)}_{\alpha\beta} \equiv -\frac{1}{2}g^{\lambda\mu}\partial^2_{\lambda\mu}g_{\alpha\beta} + P_{\alpha\beta}(g)(\partial g, \partial g), \qquad \text{(IV.7.9)}$$

where P is a quadratic form in the components of ∂g, with coefficients polynomial in the components of g and its contravariant associate.

IV.7.3   Propagation and backreaction

Using previous results, or by direct computation, we find that the coefficients at zeroth order in ω in the expansion of $Ricci(g)$ are

$$R^{(0)}_{\alpha\beta} = \underline{R}_{\alpha\beta} + (\mathcal{P}v')_{\alpha\beta} + L(v', w'')_{\alpha\beta} + N_{\alpha\beta}(v, v', v'),$$

with $\mathcal{P}$ a linear propagation operator along the rays of $\partial\varphi$, namely

$$(\mathcal{P}v')_{\alpha\beta} \equiv -\left(\varphi^\lambda\underline{\partial}_\lambda v'_{\alpha\beta} + \frac{1}{2}v'_{\alpha\beta}\underline{\partial}_\lambda\varphi^\lambda\right),$$

while L reads

$$L_{\alpha\beta} \equiv \frac{1}{2}\left[\varphi_\alpha Q_\beta(v') + \varphi_\beta Q_\alpha(v')\right] + \frac{1}{2}\left[\varphi_\alpha P_\beta(w'') + \varphi_\beta P_\alpha(w'')\right],$$

with

$$Q_\alpha(v') := \underline{\partial}_\lambda v'^\lambda_\alpha - \frac{1}{2}\underline{\partial}_\alpha v'^\lambda_\lambda.$$

The nonlinear term $N(v, v', v'')$ comes from $\underline{\delta}^2 R_{\alpha\beta}$. We find that for polarized v it reduces to

$$N_{\alpha\beta}(v, v', v'') \equiv \frac{1}{2}\varphi_\alpha\varphi_\beta\left[v^{\lambda\mu}v''_{\lambda\mu} - \frac{1}{2}v^\lambda_\lambda v''^\mu_\mu + \frac{1}{2}\left(v'^{\lambda\mu}v'_{\lambda\mu} - \frac{1}{2}v'^\lambda_\lambda v'^\mu_\mu\right)\right].$$

We can now prove the following theorem.

Theorem IV.7.2   *The progressive wave*

$$\underline{g}_{\alpha\beta}(x) + \{\omega^{-1}v_{\alpha\beta}(x, \xi) + \omega^{-2}w_{\alpha\beta}(x, \xi)\}_{\xi=\omega\varphi(x)}$$

is an asymptotic solution of order one of the vacuum Einstein equations on a manifold V under the following hypotheses:

(1) The phase φ is isotropic for the background $\underline{g}$.

(2) v satisfies the linear, non-homogeneous, propagation system

$$\mathcal{P}(v') = 0$$

along the rays[33] of the phase φ and also satisfies the polarization conditions on a hypersurface Σ transverse to these rays, assumed to span V. The field v is periodic in ξ on Σ.

[33]Null curves tangent to $\partial\varphi$.

(3) w is a solution of the linear system

$$P_\alpha(w'') = Q_\alpha(v') + \frac{1}{4}\varphi_\alpha(v^{\lambda\mu}v'_{\lambda\mu} - v^\lambda_\lambda v'^\mu_\mu)' + \frac{1}{2}\varphi_\alpha(E - \underline{E}).$$

(4) The background metric $\underline{g}$ satisfies the Einstein equations with source a null fluid:

$$\underline{R}_{\alpha\beta} = \underline{E}\phi_\alpha\phi_\beta,$$

where, with T the period of v,

$$\underline{E} \equiv \frac{1}{T}\int_0^T E(.,\xi)d\xi, \quad with \quad E \equiv \frac{1}{4}\left(v'^{\lambda\mu}v'_{\lambda\mu} - \frac{1}{2}v'^\lambda_\lambda v'^\mu_\mu\right).$$

Proof.

(1) If v' satisfies the propagation equations $\mathcal{P}(v') = 0$ on $V \times R$ and the polarization conditions on Σ transverse to rays that span V, then it satisfies the polarization conditions on $V \times R$ because the equation $\mathcal{P}(v') = 0$ implies the propagation both of $\varphi^\alpha v_{\alpha\beta} = 0$ and $v^\alpha_\alpha = 0$. Indeed,

$$\underline{g}^{\alpha\beta}(\mathcal{P}v')_{\alpha\beta} \equiv -\varphi^\lambda\underline{\partial}_\lambda v'^\lambda_\lambda + \frac{1}{2}v'^\lambda_\lambda\underline{\partial}_\lambda\varphi^\lambda$$

and also

$$\varphi^\alpha(\mathcal{P}v')_{\alpha\beta} \equiv -\varphi^\lambda\underline{\partial}_\lambda(\varphi^\alpha v'_{\alpha\beta}) - \frac{1}{2}\varphi^\alpha v'_{\alpha\beta}\underline{\partial}_\lambda\varphi^\lambda,$$

because if φ_α is a gradient and isotropic, then

$$\varphi^\lambda\underline{\partial}_\lambda\varphi_\alpha = \varphi^\lambda\underline{\partial}_\alpha\varphi_\lambda = 0.$$

The coefficient of ω in the asymptotic expansion of $Ricci(g)$ is therefore zero.

(2) The function $x \to v'(x,\xi)$, a solution of the linear differential equation $\mathcal{P}(v') = 0$ with coefficients independent of ξ, is determined on $V \times R$ if it is known on a submanifold transverse to rays that span V. It has period T in ξ if its data on the submanifold have period T.

(3) When v is known, the equations for w'' are non-homogeneous linear equations, namely

$$L_{\alpha\beta} + N_{\alpha\beta} + \underline{R}_{\alpha\beta} = 0.$$

(4) If v and w' have period T in ξ, then the following relation holds, because linear terms in v' or w'' integrate to zero:

$$T\underline{R}_{\alpha\beta} = -\int_0^T N_{\alpha\beta}(.,\xi)\,d\xi.$$

An elementary computation gives

$$N_{\alpha\beta} = -\varphi_\alpha\varphi_\beta\left[E - \frac{1}{2}\left(v'^{\lambda\mu}v_{\lambda\mu} - \frac{1}{2}v^\lambda_\lambda v'^\mu_\mu\right)'\right]$$

from which there follows the expression given for $\underline{R}$ and hence the linear system for w'' given in the theorem. These linear equations have solutions on $V \times R$, periodic with period T in ξ, because the linear transposed homogeneous system has an empty kernel and the coefficients have period T. The tensor w can be chosen also of period T in ξ and the right-hand side has a zero integral on ξ on the interval $0 \leq \xi \leq T$.

∎

Note a most remarkable fact, which is not shared by other nonlinear fields: *the vector v obeys linear propagation equations; that is, there is no distortion under propagation of gravitational signals, in spite of the nonlinearity of the Einstein equations.*

IV.7.4   Observable displacements

As in the case of weak gravitational waves, the observable displacements due to a strong gravitational wave are governed by the tidal force determined by the highest-order terms in ω of the Riemann tensor.

Exercise IV.7.2 *Consider two nearby particles with initially parallel unit velocity $u^\alpha = \delta_0^\alpha$ and spatial separation vector X^i. Write the first approximation of the geodesic deviation due to a strong high-frequency wave.*

IV.8   Stationary spacetimes

IV.8.1   Definition

We call **stationary** an $(n+1)$-dimensional spacetime (V, g) that admits a one-parameter group $G_1 \equiv R$ of isometries with timelike Killing vector ξ whose orbits are diffeomorphic[34] to R and span the manifold V. More precisely,[35] we assume that V is a product $M \times R$, a point of V is a pair (x, t), $x \in M$, $t \in R$, the subspaces $M \times \{t\}$ of constant t are spacelike submanifolds, and the Killing vector ξ is represented by $\partial/\partial t$. In frames adapted to the product structure, the spacetime metric g reads

$$g \equiv -\psi^2 (dt + a)^2 + \bar{g}, \qquad (IV.8.1)$$

with $t \in R$ a time coordinate on the orbits of the vector $\xi = \partial/\partial t$; ψ, a, and $\bar{g}$ are respectively a t-dependent scalar, a 1-form, and the Riemannian metric on M. In local coordinates x^i in the domain of a chart of M, one has

$$a \equiv a_i dx^i, \qquad \bar{g} \equiv g_{ij} dx^i dx^j, \qquad (IV.8.2)$$

with all coefficients in g being independent of t.

[34]We exclude the case of orbits diffeomorphic to S^1, i.e. the case of closed timelike curves, because this is considered non-physical.

[35]The assumptions that V is a product $M \times R$ could be relaxed in the definition of stationarity, but the uniqueness theorems given below would not necessarily hold.

A stationary spacetime is called **static** if the orbits are orthogonal to the n-dimensional space manifolds. Its metric then reads

$$g \equiv -N^2 dt^2 + \bar{g}. \tag{IV.8.3}$$

Static spacetimes are considered to represent equilibrium situations, while stationary spacetimes model permanent motions. Both play an important role in relativistic dynamics.

Lemma IV.8.1   *A stationary spacetime is static if the 1-form a is an exact differential:*

$$a = d\phi, \quad i.e. \quad a_i = \partial_i \phi,$$

with ϕ a t-dependent scalar function on M.

Proof. For a stationary spacetime, assume that $a = d\phi$, and consider the change of its time and space factorization under the change of time parameter

$$t' = t + \phi. \tag{IV.8.4}$$

This puts its metric into the static form

$$g \equiv -N^2 dt'^2 + g_{ij} dx^i dx^j, \tag{IV.8.5}$$

with g and $N \equiv \psi$ independent of t'.   ∎

Static spacetimes are invariant under time reversal, $t \to -t$. This property can also be taken as a definition to distinguish static spacetimes among stationary ones.

One calls **locally static** those stationary spacetimes for which the 1-form a is closed, $da = 0$, but is not globally an exact form, i.e. there does not exist on the whole of M a function such that $a = d\phi$. This can happen only[36] if M is not diffeomorphic to R^n.

[36]See, for instance, CB-DMI IV.

IV.8.2   Equations

We denote by $\bar{\nabla}$ and $\overline{Ricci}$ the covariant derivative and Ricci tensor in the Riemannian metric $\bar{g}$. One can show by straightforward computation that the components of the Ricci tensor of the spacetime metric

$$g := -\psi^2 (dt + a_i dx^i)^2 + \bar{g}, \quad \text{with} \quad \bar{g} := g_{ij} dx^i dx^j,$$

reduce, with $f = da$, to

$$R_{ij} \equiv \bar{R}_{ij} + \frac{\psi^2}{2} f_i{}^h f_{jh} - \psi^{-1} \bar{\nabla}_i \partial_j \psi, \quad \text{with} \quad f_i^j := g^{jk}(\bar{\nabla}_k a_i - \bar{\nabla}_i a_k), \tag{IV.8.6}$$

$$R_{i0} \equiv -\frac{1}{2\psi} \bar{\nabla}_j (f_i{}^j \psi^3), \tag{IV.8.7}$$

$$R_{00} \equiv \frac{1}{4} \psi^4 f_i{}^j f_j{}^i + \psi \Delta_{\bar{g}} \psi. \tag{IV.8.8}$$

Exercise IV.8.1   *Prove these formulas using the general formulas for connection and curvature given in Chapter I. More general formulas are derived in Chapter IX.*

In the case of a *locally static spacetime*, the components of the Ricci tensor reduce to

$$R_{ij} \equiv \bar{R}_{ij} - \psi^{-1}\bar{\nabla}_i\partial_j\psi, \tag{IV.8.9}$$

$$R_{i0} \equiv 0, \tag{IV.8.10}$$

$$R_{00} \equiv \psi\Delta_{\bar{g}}\psi. \tag{IV.8.11}$$

The second identity[37] shows that the source must have zero momentum for an Einsteinian spacetime to be locally static, justifying its name. A reciprocal theorem is easy to prove under a physically meaningful hypothesis, namely the following:

[37]For a consequence of the last identity when M is compact, see Problem IV.11.9.

Theorem IV.8.1   *A stationary Einsteinian spacetime with sources of zero momentum, i.e. $\rho_{0i} \equiv 0$, is locally static if either*

(1) M is compact.

(2) The spacetime is asymptotically Euclidean.

Proof. If $\rho_{0i} = 0$, a previous identity gives

$$R_{i0} \equiv -\frac{1}{2\psi}\bar{\nabla}_j(f_i{}^j\psi^3) = 0,$$

and hence

$$\bar{\nabla}_j(a^i f_i{}^j\psi^{3\cdot}) = \frac{1}{2}\psi^3 f_i{}^j f_j{}^i. \tag{IV.8.12}$$

If M is compact (and hence without boundary according to our definitions) integration of the above equality on M gives, by Stokes's formula,

$$\int_M \psi^3 f_i{}^j f_j{}^i \mu_M = 0,$$

and $f_j{}^i \equiv 0$ since $f_i{}^j f_j{}^i \geq 0$,

If M is asymptotically Euclidean, then we have, with B_r a ball of radius r, and using again Stokes's formula,

$$\int_M \psi^3 f_i{}^j f_j{}^i \mu_M = \frac{1}{2}\lim_{r\to\infty}\int_{B_r} \psi^3 f_i{}^j f_j{}^i \mu_{B_r} = \lim_{r\to\infty}\int_{\partial B_r} a^i f_i{}^j \psi^3 n_j \mu_{\partial B_r}. \tag{IV.8.13}$$

When the radius r of the ball tends to infinity, the right-hand side tends to zero if $n > 2$, because if the spacetime is asymptotically Euclidean, then the product af falls off like $r^{-(2p+1)}$, $2p+1 > n-1$, while $\mu_{\partial B_r} \doteq r^{n-1}\mu_{S^2}$, where μ_{S^2} is the volume element of the sphere S^{n-1}. We conclude again that $f_i^j \equiv 0$ on M. ∎

IV.8.3   Non-existence of gravitational solitons

One calls a gravitational **soliton** a complete non-trivial (i.e. non-flat) stationary solution of the vacuum Einstein equations.

We extend to arbitrary dimensions the proof of a theorem given in 3+1 dimensions by Einstein and Pauli in the static case and by Lichnerowicz[38] in the stationary case.

[38]Lichnerowicz (1939).

Theorem IV.8.2   *In $(n+1)$-dimensional spacetimes, the gravitational solitons $(M \times R, g)$ with M compact or asymptotically Euclidean are locally static with Ricci-flat space metric and ψ a constant.*

Proof.   Since gravitational solitons are solutions of the vacuum Einstein equations, Theorem IV.8.1 shows that they are locally static: $f_i^j \equiv 0$. The remaining vacuum Einstein equations read

$$\Delta_{\bar{g}}\psi = 0, \qquad (IV.8.14)$$

$$\bar{R}_{ij} = \psi^{-1}\nabla_i\partial_j\psi. \qquad (IV.8.15)$$

By well-known theorems resulting from the maximum principle applied to solutions of elliptic partial differential equations, the equality (8.14) implies that $\psi = $ constant on a manifold M that is compact or asymptotically Euclidean (if ψ is smooth and uniformly bounded—assumptions we implicitly include in the definition of gravitational solitons). The last equation concludes the proof.   ∎

Corollary IV.8.1   *Under the hypotheses of the theorem, the only (3+1)-dimensional gravitational solitons are locally flat.*

Proof.   In 3 dimensions, a Ricci-flat manifold is flat.   ∎

[39]By a theorem of Schoen and Yau, a locally flat asymptotically Euclidean (complete—a property included in our definition) Riemannian 3-manifold is isometric to R^3.

In the classical case considered by Einstein where M is an asymptotically Euclidean[39] manifold diffeomorphic to R^3, the only gravitational soliton is Minkowski spacetime M^4.

In the compact case, there are, besides the flat torus T^3, five non-isometric (i.e. non-diffeomorphic) Riemannian orientable compact flat manifolds, quotients of the Euclidean space E^3 by discrete isometry groups.[40]

[40]See Chapter 3 of Wolf (2011).

Remark IV.8.1   *Corollary IV.8.1 is not true in higher dimensions.*

IV.8.4   Gauss's law

A property of stationary spacetimes deduced easily from the previous identities is the following, which we have called[41] the **relativistic Gauss law**.

[41]Fourès (Choquet)-Bruhat (1948).

Theorem IV.8.3   *The time–time component R_t^t of the Ricci tensor of a stationary spacetime $(M \times R, \hat{g})$ in a natural frame with time axis tangent to the timelines and space axis tangent to the manifold M_t satisfies the divergence identity*

$$\psi R_t^t \equiv \psi R_t^t \equiv \tilde{\nabla}^i \left[\partial_i \psi + \frac{1}{2} (a^j f_{ji} \psi^3) \right]. \qquad \text{(IV.8.16)}$$

Proof. The component R_t^t is immediately deduced from the components previously computed in the frame $\theta^0 := dt + a_i dx^i$. The change-of-frame formula gives

$$R_t^t \equiv R_0^0 + a_i R_0^i, \qquad \text{(IV.8.17)}$$

with

$$R_0^0 \equiv -\psi^{-2} R_{00}, \quad R_0^i \equiv \tilde{g}^{ij} R_{0j}. \qquad \text{(IV.8.18)}$$

Therefore,

$$R_t^t \equiv - \left[\frac{1}{4} \psi^2 f_i{}^j f_j{}^i + \psi^{-1} \Delta_{\tilde{g}} \psi + \frac{1}{2} \psi^{-1} a^i \nabla_j (f_i{}^j \psi^3) \right], \qquad \text{(IV.8.19)}$$

which gives, using the definition of f_{ij},

$$R_t^t \equiv -\psi^{-1} \left[\Delta_{\tilde{g}} \psi + \frac{1}{2} \tilde{\nabla}_j (a^i f_i{}^j \psi^3) \right], \qquad \text{(IV.8.20)}$$

that is, the given identity. ■

Corollary IV.8.2 *If the space manifold of a stationary Einsteinian spacetime with source ρ is asymptotically Euclidean, then it holds that*

$$\text{flux}_{\text{infinity}} \operatorname{grad} \psi := \lim_{\rho \to \infty} \int_{\partial B_\rho} n^i \partial_i \psi \mu_{\partial B_\rho} = \int_M \psi \rho_t^t \mu_{\tilde{g}}.$$

Exercise IV.8.2 *Prove this corollary, which has an analogue in Newtonian mechanics, called Gauss's law.*

IV.9   Lagrangians

Lagrangians, arising from energies, play a fundamental role in physics. The Lagragian formulation of the Einstein equations, found independently by Einstein and Hilbert, stands apart, being unrelated to a pointwise intrinsically defined gravitational energy. However, it plays an important role in many modern developments, in particular for equations with field sources.

IV.9.1   Einstein–Hilbert Lagrangian in vacuo

In this section we use the results of Chapter I, and its physicists' notation.

Theorem IV.9.1   *The Einstein equations in vacuo are the Euler equations*[42] *of the Einstein–Hilbert Lagrangian, defined by*

[42] One equivalently says that a solution of the Einstein equations in vacuo is a critical point of the Einstein–Hilbert Lagrangian.

$$\mathcal{L}_{\mathrm{grav}}(g) := \int R(g)\mu_g, \qquad (\text{IV.9.1})$$

with $R(g)$ the scalar curvature and μ_g the volume element of the Lorentzian metric g.

Proof. We have

$$\delta\mathcal{L}_{\mathrm{grav}}(g) = \int \left[\delta R(g)\mu_g + R(g)\delta\mu_g\right], \qquad (\text{IV.9.2})$$

with

$$\delta R \equiv g^{\alpha\beta}\delta R_{\alpha\beta} + R_{\alpha\beta}\delta g^{\alpha\beta}$$

and

$$\delta\mu_g \equiv -\frac{1}{2}g_{\alpha\beta}\delta g^{\alpha\beta}\mu_g.$$

We have found in Chapter I that $g^{\alpha\beta}\delta R_{\alpha\beta}$ is the divergence of a vector,

$$g^{\alpha\beta}\delta R_{\alpha\beta} \equiv -\nabla_\lambda v^\lambda, \quad v^\lambda := \nabla^\lambda h_\alpha^\alpha - \nabla_\alpha h^{\lambda\alpha}, \quad h_{\alpha\beta} := \delta g_{\alpha\beta}, \quad (\text{IV.9.3})$$

and hence the integral of this term vanishes if v^λ vanishes on the boundary of the integration domain, as is assumed in computing the Euler equations. We therefore have

$$\delta L_{\mathrm{grav}}(g) = \int \left[\delta R(g)\mu_g + R(g)\delta\mu_g\right] = \int \left(R_{\alpha\beta} - \frac{1}{2}g_{\alpha\beta}R\right)\delta g^{\alpha\beta}\mu(g). \qquad (\text{IV.9.4})$$

∎

Exercise IV.9.1   *Prove the formula (IV.9.4).*

Hint: The derivative of a determinant is $\partial_\lambda \det g \equiv g^{\alpha\beta}\partial_\lambda g_{\alpha\beta} \det g$.

IV.9.2   Lagrangians for Einstein equations with sources

A Lagrangian for the Einstein equations with sources (with the latter being generically denoted by ϕ) is the sum of the vacuum Lagrangian and a Lagrangian for sources. By the equivalence principle, the Lagrangian in General Relativity of a source, $\mathcal{L}(g,\phi) = \int L(g,\phi)\mu_g$, should be deduced from the Lagrangian in Special Relativity, invariant under diffeomorphism, by replacing the Minkowski metric η by g.

General theorem

The vanishing of the variation with respect to g and ϕ of the sum of the Einstein–Hilbert Lagrangian and the Lagrangian of the sources gives the Einstein equations with source a tensor T, together with the equations of motion for the sources.

The first variation of the total Lagrangian

$$\mathcal{L}_{\text{tot}}(g, \phi) := \int [R(g) + L(g, \phi)]\mu_g \qquad (\text{IV.9.5})$$

reads (the dot below denotes some algebraic linear form)

$$\delta\mathcal{L}_{\text{tot}}(g, \phi) \equiv \int \{[S^{\alpha\beta}(g) - M^{\alpha\beta}(g, \phi)]\delta g_{\alpha\beta} - \Phi(g, \phi) \cdot \delta\phi\}\mu_g, \quad (\text{IV.9.6})$$

where $\Phi = 0$ are the equations of motion of the fields ϕ on (V, g) and $M^{\alpha\beta}$ coincides with their stress–energy tensor.

Remark IV.9.1 *The conservation laws $\nabla_\alpha M^{\alpha\beta} = 0$, necessary for the solvability of the Einstein equations with source ϕ, are a consequence of the invariance under diffeomorphisms of the source Lagrangian, which implies that if $\Phi(g, \phi) = 0$, then it holds that for any vector X,*

$$\int M^{\alpha\beta}(\nabla_\alpha X_\beta + \nabla_\beta X_\alpha)\mu_g = 0.$$

Matter and field sources

In the case of **matter sources**, it is somewhat difficult to find a Lagrangian in Special Relativity, and even in Newtonian mechanics. A simple-looking Lagrangian for barotropic perfect fluids was proposed by Taub.[43] It reads[44]

$$\mathcal{L}_{\text{fluid}} := \int p(\mu)\mu_g.$$

The Taub Lagrangian leads to the Euler equations through a somewhat involved process, introducing first a Lagrangian with the constraint $g(u, u) = -1$ and a Lagrange multiplier λ, and using Lagrangian-type coordinates where the timelines are the flow lines.

Another proposal introducing velocity potentials is due to Schutz.[45]

None of these Lagrangians has been effectively used in the study of properties of relativistic fluids.

In the case of the usual **field sources**, the Lagrangians in Special Relativity are well known and naturally transferred to General Relativity.

The Maxwell equations in vacuo are the Euler equations with respect to a closed 2-form F of the Lagrangian

$$\mathcal{L}_{\text{e.m.}}(F, g) := -\frac{1}{4}\int F^{\alpha\beta}F_{\alpha\beta}\mu_g. \qquad (\text{IV.9.7})$$

The Lagrangian of a scalar field f with potential $V(f)$ is

$$\mathcal{L}_{\text{scal}}(f, g) := -\int \left[\frac{1}{2}g^{\alpha\beta}\partial_\alpha f\partial_\beta f + V(f)\right]\mu_g. \qquad (\text{IV.9.8})$$

[43] Taub (1954, ?).

[44] Here μ denotes the energy density and μ_g the volume element.

[45] Schutz (1971).

[46]See YCB-OUP 2009, Chapter III, Section 6.5.

The Lagrangians for Yang–Mills fields and wave maps are obtained analogously by replacing ordinary products by scalar products in the Lie algebra in the case of Yang–Mills fields and in the metric of the target in the case of wave maps.[46]

In particle physics at a classical (i.e. non-quantum) level the dynamics is ruled by the **standard model Lagrangian** with a structure of the type

$$\mathcal{L}_{\text{standard}}(\phi) := -\int \left[\frac{1}{4} F^{\alpha\beta} . F_{\alpha\beta} + \bar{\psi}\gamma^\mu D_\mu \psi + \frac{1}{2} g^{\mu\nu} D_\mu H \cdot D_\nu H \right.$$
$$\left. + V(H) + \lambda \bar{\psi} H \psi \right] \mu_g,$$

with $\phi := (F, \psi, H, g)$, F a Yang–Mills field with values in the Lie algebra of $U(1) \times SU(2) \times SU(3)$, ψ various spinor fields, and γ^μ Dirac gamma matrices, while H is a complex-valued scalar doublet, the Higgs field; V is a potential for H and λ is some constant coefficient.

Exercise IV.9.2   *Check the general theorem in the Maxwell and scalar field cases.*

Hint:

$$\delta \int g^{\alpha\beta} \partial_\alpha \phi \partial_\beta \phi \mu_g = \int \delta g^{\alpha\beta} \partial_\alpha \phi \partial_\beta \phi \mu_g + 2 \int g^{\alpha\beta} \partial_\alpha \delta\phi \partial_\beta \phi \mu_g$$
$$+ \int g^{\alpha\beta} \partial_\alpha \phi \partial_\beta \phi \delta\mu_g$$

$$= \int \delta g^{\alpha\beta} \left(\partial_\alpha \phi \partial_\beta \phi - \frac{1}{2} g_{\alpha\beta} g^{\lambda\mu} \partial_\lambda \phi \partial_\mu \phi \right) - 2 \int \delta\phi g^{\alpha\beta} \nabla_\alpha \partial_\beta \phi \mu_g.$$

The theorem is taken to define the equations for sources and the stress–energy tensor when only the Lagrangian is known.

Remark IV.9.2   *A constant factor depending on the units in which the sources are measured must be put in front of the Lagrangian for the sources when it is added to the vacuum Einstein Lagrangian in the case of interpretation of observations in non-geometric units.*

IV.10   Observations and experiments

In Chapter V, we will study the spherically symmetric vacuum Eisteinian spacetime, the first exact solution constructed by Schwarzschild in 1916. We will explain most of its observable consequences, and briefly quote the results of observations, which are all in very good agreement with the predictions.

We will discuss the reality of the revolutionary prediction of Einstein's equations—black holes—in Chapter VI.

The recently operational gravitational wave detectors VIRGO and LIGO have not so far been able to detect any gravitational wave signal.

This is not so surprising, given the predicted weakness of such signals. However, the reality of **energy loss by gravitational radiation** was confirmed as early as 1979 by the observation by Taylor and collaborators of the slowing down of the period of a binary pulsar,[47] which was interpreted as a shrinkage of its orbit due to gravitational radiation. Many other binary pulsars have now been discovered. More than 30 years of observations and recording of data from the original Hulse–Taylor pulsar have led (not without hard work from theoreticians and physicists) to information being obtained on several physical properties, including the masses, orbits, rotation, advance of the periastron, and other parameters, of the pulsar and its invisible companion (both are now thought to be neutron stars). All the results obtained agree with those predicted by Damour and Deruelle[48] from the Einstein equations, rather than those predicted by alternative gravitation theories.[49]

General Relativity and the Einstein equations are universally adopted as the best model for gravitation at planetary and astronomical scales. Their use at the atomic, quantum, scale requires new tools, still the subject of much research and discussion. The sub-Planckian and cosmological scales pose serious problems that probably require new ideas.

If gravity is to become a quantum field theory like others, there must exist an elementary particle called the **graviton**. This particle should be massless like the photon, since gravitation propagates with the speed of light. The putative graviton[50] is thought to be spin[51] 2 for the same reasons why the photon is called spin 1 and the electron and the neutrino spin $\frac{1}{2}$.

[47] A binary pulsar is a system of two stars gravitating near enough to each other to be considered as an isolated system, one of them being a pulsar. A pulsar is a star from which we receive strong electromagnetic signals at regular intervals. It is now thought that the signals are emitted in the direction of the axis of a rapidly rotating neutron star.

[48] Damour and Deruelle (1981, 1986) and Damour (1982).

[49] For observational and experimental results up to the end of 2013, see the review by Damour for the Particle Data Group (Damour, 2013b).

[50] The observation of a graviton is very unlikely in the near future.

[51] Spin is a quantum notion whose precise definition is outside of the scope of this book. It takes only integer or half-integer values. A supersymmetric partner of the graviton, the gravitino, would have spin $\frac{3}{2}$.

IV.11 Problems

IV.11.1 The Einstein cylinder

The **Einstein cylinder**[52] is the manifold $S^3 \times R$ endowed with the static metric

$$- dt^2 + a_0^2 \gamma_+, \qquad (IV.11.1)$$

with a_0 a constant and γ_+ the metric of the unit 3-sphere S^3, which is, in the usual angular coordinates,

$$\gamma_+ := d\theta^2 + \sin^2 \theta \, (d\phi^2 + \sin^2 \phi \, d\psi^2).$$

1. Show that setting $\sin \theta = r$ gives the form familiar to geometers:

$$\gamma_+ = \frac{dr^2}{1 - r^2} + r^2(d\phi^2 + \sin^2 \phi d\psi^2). \qquad (IV.11.2)$$

[52] The Einstein cylinder, also called the Einstein static universe, was the first Einsteinian cosmological model to be constructed.

Show that the components of the Ricci tensor are (whatever coordinates are used)

$$R_{0\alpha} \equiv 0, \quad R_{ij} \equiv 2\gamma_{ij}^+. \tag{IV.11.3}$$

2. Show that the Einstein cylinder is a solution of the Einstein equations with perfect fluid source of constant positive energy and negative pressure:

$$\mu_0 = 3a_0^{-2}, \quad p_0 = -a_0^{-2} < 0. \tag{IV.11.4}$$

3. A negative pressure is unacceptable on classical physical grounds. Show that the Einstein cylinder is a solution of the Einstein equations with a positive cosmological constant and source a perfect fluid with positive energy and pressure.

4. Show that the Einstein cylinder is locally conformally flat. Determine the image of Minkowski spacetime under a diffeomorphism into the Einstein cylinder.

Solution

1. This is a straightforward computation.

2. The Einstein equations with cosmological constant Λ are

$$S_{\alpha\beta} = T_{\alpha\beta} - \Lambda g_{\alpha\beta}.$$

We deduce from (11.3) and (11.4) that

$$R \equiv 6a_0^{-2}, \quad \text{and} \quad S_{00} \equiv 3a_0^{-2}, \quad S_{ij} \equiv -\gamma_{ij}^+.$$

The stress–energy tensor $T_{\alpha\beta} \equiv (\mu + p)u_\alpha u_\beta + p g_{\alpha\beta}$ of a perfect fluid in the Einstein static universe with static flow vector $u^0 = 1$, $u^i = 0$ is

$$T_{00} \equiv \mu_0, \ T_{0i} \equiv 0, \quad T_{ij} \equiv p_0 a_0^2 \gamma_{ij}^+,$$

and hence (recall that on the Einstein cylinder, $g_{00} \equiv -1$, $g_{ij} \equiv a_0^2 \gamma_{ij}^+$)

$$S_{00} = T_{00} + \Lambda, \quad S_{ij} = T_{ij} - a_0^2 \gamma_{ij}^+ \Lambda$$

if

$$\mu_0 = 3a_0^{-2} - \Lambda, \quad p_0 = \Lambda - a_0^{-2}.$$

IV.11.2   de Sitter spacetime

The de Sitter spacetime has for supporting manifold $S^3 \times R$, like the Einstein cylinder. Its metric, analogous to the Einstein cylinder metric, is

$$-dt^2 + a^2 \gamma_+, \tag{IV.11.5}$$

but the coefficient a is now time-dependent.

1. Show that the de Sitter metric is a solution of the vacuum Einstein equations with positive cosmological constant Λ if and only if the coefficient a is such that

$$\ddot{a} - k^2 a = 0, \quad \text{with} \quad \dot{a} := \frac{\partial a}{\partial t}, \quad \ddot{a} := \frac{\partial^2 a}{\partial t^2}, \quad k^2 := \frac{\Lambda}{3}, \quad \text{(IV.11.6)}$$

and

$$\dot{a}^2 - k^2 a^2 = -1. \quad \text{(IV.11.7)}$$

2. Show that the general solution for (11.5) is, with A and B a pair of constants,

$$a = A e^{kt} + B e^{-kt}. \quad \text{(IV.11.8)}$$

Show that such a function a satisfies (11.6) if $4AB = k^{-2}$. Show that the metric, called the **de Sitter** metric,[53]

$$- dt^2 + k^{-2}(\cosh^2 kt)[d\alpha^2 + \sin^2 \alpha(d\theta^2 + \sin^2 \theta \, d\phi^2)] \quad \text{(IV.11.9)}$$

is a solution of the vacuum Einstein equations with positive cosmological constant.

[53] The de Sitter spacetime is time-symmetric. The radius of its spherical space sections expands to infinity in both time directions.

IV.11.3   Anti-de Sitter spacetime

1. Show that the spacetime metric defined on $R^3 \times (-\pi/2, \pi/2)$ by

$$- dt^2 + (\cos^2 t) \, \gamma_-, \quad \text{(IV.11.10)}$$

where γ_- is the metric of the hyperbolic 3-space of constant negative curvature,

$$\gamma_- := d\chi^2 + (\sinh^2 \chi)(\sin^2 \theta \, d\phi^2 + d\theta^2),$$

satisfies the vacuum Einstein equations with negative cosmological constant $\Lambda = -3$.

2. The spacetime classically called **anti-de Sitter** spacetime is the manifold $R^3 \times R$ with the static metric

$$g_{\text{AdS}} := -dt^2(\cosh^2 \chi) + d\chi^2 + \sinh^2 \chi(\sin^2 \theta \, d\phi^2 + d\theta^2), \quad 0 \le \chi < \infty. \quad \text{(IV.11.11)}$$

Show that it is conformal[54] to the half $0 < \theta < \pi/2$ of the Einstein cylinder.

3. Show that no light ray emitted at time t_0 will go beyond a space slice at finite time $t = t_0 + T$.

[54] This property has been used by Choquet-Bruhat to prove the global existence of Yang–Mills fields in anti-de Sitter spacetime. It plays an important role in recent work by H. Friedrich.

Remark IV.11.1   *Five-dimensional anti-de Sitter spacetime plays an important role in supersymmetric theories conformal field, in particular with regard to the so-called AdS–CFT correspondence.*

Solution

1. This is straightforward computation.
2. Set

$$\cosh \chi = \frac{1}{\cos \alpha}, \quad \text{hence} \ \sinh \chi \, d\chi = -\frac{\sin \alpha}{\cos^2 \alpha} \, d\alpha.$$

Computation using the identities satisfied by sines and cosines gives the announced identity:

$$g_{\text{AdS}} \equiv \frac{1}{\cos^2 \alpha} [-dt^2 + d\alpha^2 + \sin^2 \alpha (\sin^2 \theta \, d\phi^2 + d\theta^2)].$$

3. We have on a radial light ray

$$dt = \frac{d\chi}{\cosh \chi}, \quad \text{hence} \quad t = t^0 + \int_{\chi_0}^{\chi} \frac{d\chi}{\cosh \chi}, \qquad \text{(IV.11.12)}$$

and hence, whatever χ_0 is,

$$t - t_0 \leq \int_0^{\chi} \frac{d\chi}{\cosh \chi} < 2 \int_0^{\infty} \frac{d\chi}{e^{\chi}} := T < \infty. \qquad \text{(IV.11.13)}$$

Therefore, no light ray emitted at time t_0 will go beyond the space slice $t = t_0 + T$. The anti-de Sitter spacetime has no global causality properties; it is not globally hyperbolic (see Chapter VIII).

IV.11.4   Taub–NUT spacetime

The underlying manifold of the Taub spacetime is $S^3 \times I$ with S^3 a three-dimensional topological sphere and $I := (t_-, t_+)$ an interval of R. This manifold is endowed with the Lorentzian metric

$$-U^{-1} dt^2 + (2\ell)^2 U (d\psi + \cos \theta \, d\phi)^2 + (t^2 + \ell^2)(d\theta^2 + \sin^2 \theta d\phi^2), \quad \text{(IV.11.14)}$$

where

$$t_{\pm} = m \pm (m^2 + \ell^2)^{\frac{1}{2}} \qquad \text{(IV.11.15)}$$

and

$$U \equiv -1 + \frac{2(mt + \ell^2)}{t^2 + \ell^2}. \qquad \text{(IV.11.16)}$$

The numbers ℓ and m are positive constants. The coordinates ψ, θ, ϕ are the Euler coordinates on the sphere S^3.

1. Show that the Taub spacetime is a solution of the vacuum Einstein equations.

2. Show that U tends to zero as t tends to t_+ or t_-. Show that the metric is still Lorentzian when t is in the complement in R of the closure of I, but ψ is now a time variable and t a space variable.

3. Show that by setting

$$\psi' = \psi + \frac{1}{2\ell} \int_0^t \frac{d\tau}{U(\tau)}, \qquad \text{(IV.11.17)}$$

the Taub metric extends to a C^∞ metric on $S^3 \times R$, called Taub–NUT:[55]

$$4\ell^2 \, d\psi'^2 - 4\ell \, d\psi' \, dt + 4\ell^2 U \cos\theta \, d\phi \, d\psi' - 2\ell \cos\theta \, dt \, d\phi$$
$$+ (t^2 + \ell^2)(d\theta^2 + \sin^2\theta \, d\phi^2). \qquad \text{(IV.11.18)}$$

[55]For its discoverers, Newman, Unti, and Tamburino.

4. Show that the two hypersurfaces, diffeomorphic to S^3, $t = m \pm (m^2 + \ell^2)^{\frac{1}{2}}$, which bound the Taub spacetime in **Taub–NUT** are null manifolds, generated by closed null geodesics where ψ only varies. They are the Cauchy horizons of the maximal development of any Cauchy surface (see Chapter VIII) in the Taub spacetime.

5. Show that one family of null geodesics issuing from a point of the Taub region crosses both horizons, and that another family spirals near these surfaces and is incomplete. Show that in the NUT region there are closed timelike curves (the surfaces $t = $ const are timelike there).

6. Construct another extension of the Taub spacetime with analogous properties, though not isometric to it, by the change of coordinates

$$\psi'' = \psi - \frac{1}{2\ell} \int_0^t \frac{d\tau}{U(\tau)}. \qquad \text{(IV.11.19)}$$

Remark IV.11.2 *The Taub–NUT spaces give counterexamples to several conjectures (see Chapter VIII). The consolation is that they are not generic, because of their symmetries.*

IV.11.5 The quadrupole formula

Many approximation methods in General Relativity use approximations of a Newton-type potential U generated by a mass of density ρ given in geometrical units ($\kappa = 1$) by the integral

$$U(x) = \int_{R^3} \frac{\rho(y)}{|x - y|} \, d^3y, \quad d^3y \equiv dy^1 \, dy^2 \, dy^3.$$

1. Show that $1/|x - y|$ admits the following Taylor expansion about $y = 0$:

$$\frac{1}{|x - y|} = \frac{1}{r} + \sum_{i=1,2,3} y^i \frac{x^i}{r^3} + \frac{1}{2} \sum_{i,j=1,2,3} (3y^i y^j - r'^2) \delta_i^j \frac{x^i x^j}{r^5} + \dots,$$

with $r^2 := \sum_{i=1,2,3} (x^i)^2$, and $r'^2 := \sum_{i=1,2,3} (y^i)^2$.

2. Write, for a point x outside the support of the function ρ,

$$U(x) = \frac{M}{r} + \sum_i D^i \frac{x^i}{r^3} + \frac{1}{2} \sum_{i,j} Q^{ij} \frac{x^i x^j}{r^5} + \cdots .$$

Give the expressions for M, D^i, and Q^{ij}, which are called respectively the mass, the dipole moment, and the quadrupole moment of the matter density ρ. Show that U reduces to $-M/r$ if ρ is spherically symmetric. Show that D^i can be made zero by choosing for origin the Newtonian centre of gravity of the matter.

3. Show that for a non-spherical mass, the gradient of U will in general differ from $-M/r^2$ by a term in $1/r^4$.

IV.11.6   Gravitational waves

Give details and draw images of the displacement of the plane-wave perturbation of a Minkowski metric.

IV.11.7   Landau–Lifshitz pseudotensor

By analogy with other fields, it was thought long ago that a locally defined energy for the gravitational field should be defined in a post-Minkowskian approximation by a quadratic form in the first derivatives $\partial_\lambda h_{\alpha\beta}$, $h_{\alpha\beta} := \delta g_{\alpha\beta} \hat{=} g_{\alpha\beta} - \eta_{\alpha\beta}$. A simple-minded way to obtain such a quadratic form is inspired on the one hand by the Einstein equations with right-hand side the stress–energy tensor of matter or other fields and on the other hand by the classical Picard iteration method for solving quasilinear equations.

1. Show that in vacuum the first approximation $h := \delta g$ at η of a metric-g solution of the Einstein equations satisfies the linear equations

$$\delta R_{\alpha\beta} \equiv L_{\alpha\beta}(h) \equiv -\frac{1}{2}\partial^\lambda \partial_\lambda h_{\alpha\beta} + \frac{1}{2}\left[\partial_\lambda \partial_\alpha h_\beta^\lambda + \partial_\lambda \partial_\beta h_\alpha^\lambda - \partial_\alpha \partial_\beta h_\lambda^\lambda\right] = 0.$$
$$\text{(IV.11.20)}$$

2. Show that the Picard method gives for the approximation $h^{(2)}$ of order two the following linear equations with source quadratic in the first approximation h:

$$L_{\alpha\beta}(h^{(2)}) = h^{\lambda\mu}\left[\partial_\lambda(\partial_\alpha h_{\beta\mu} + \partial_\beta h_{\alpha\mu} - \partial_\mu h_{\alpha\beta}) - \partial_\alpha \partial_\beta h_{\lambda\mu}\right] + t_{\alpha\beta},$$

with $t_{\alpha\beta}$ quadratic in first derivatives, namely

$$t_{\alpha\beta} := -\partial_\lambda h^{\lambda\mu}(\partial_\alpha h_{\beta\mu} + \partial_\beta h_{\alpha\mu} - \partial_\mu h_{\alpha\beta}) + \frac{1}{2}\partial_\beta h^{\lambda\mu}\partial_\alpha h_{\lambda\mu} \quad \text{(IV.11.21)}$$

$$+ \frac{1}{2}\partial^\lambda h_\rho^\rho(\partial_\alpha h_{\beta\lambda} + \partial_\beta h_{\alpha\lambda} - \partial_\lambda h_{\alpha\beta}) + \partial_\lambda h_\alpha^\mu \partial^\lambda h_{\beta\mu} - \partial_\lambda h_\alpha^\mu \partial_\mu h_\beta^\lambda.$$

The $t_{\alpha\beta}$ are the components of the Landau–Lifshitz pseudotensor. This was considered by Landau and Lifshitz as representing the stress–energy of the first-order perturbation h.

IV.11.8 High-frequency waves from a spherically symmetric star

Take as background the spherically symmetric **Vaidya metric**, which in the coordinates $x^0 = u$, $x^1 = r$, $(x^A) = (\theta, \phi)$ reads

$$\underline{g} := -\left[1 - \frac{2m(u)}{r}\right](du)^2 - 2\,du\,dr + r^2(d\theta^2 + \sin^2\theta\,d\phi^2).$$

1. Check that the Vaidya metric coincides with the Schwarzschild metric (see Chapter V) when m is constant. Write down its contravariant components.

2. Write the conditions that must be satisfied by the high-frequency wave

$$g(x, \omega u) := \underline{g} + \omega^{-1}v(x, \omega u), +\omega^{-2}w(x, \omega u), \quad \text{with} \quad x := (u, r, \theta, \phi)$$

   for it to be an asymptotic solution of the vacuum Einstein equations with phase u

   (a) of order zero;

   (b) of order 1, in a gauge such that $v_{01} = 0$.

3. Show that the first term v in the wave falls off at infinity like r^{-1}, as the background metric. Show that the second term w falls off like r^{-2}.

Solution

1. Elementary computation gives

$$\mathrm{Det}\,g = -\left[1 - \frac{2m(u)}{r}\right]r^4\sin^2\theta,$$

$$\underline{g}^{00} = \underline{g}^{0A} = \underline{g}^{1A} = 0, \quad \underline{g}^{01} = \frac{r^4\sin^2\theta}{\mathrm{Det}\,g}, \quad \underline{g}^{11} = 1.$$

2. The coordinate function u is isotropic for $\underline{g}$ because $\underline{g}^{uu} = 0$. The gradient of the phase $\varphi \equiv u \equiv x^0$ is $\varphi_\alpha = \delta^0_\alpha$.

   (a) For a solution of order zero, the polarization conditions

$$P_i(v) \equiv v\,^0_{\,i} \equiv v_{i1} = 0$$

   and

$$P_0(v) \equiv -\frac{1}{2}v^i_i = 0$$

must be satisfied; that is, using previous relations,

$$g^{i\alpha}v_{i\alpha} = g^{10}v_{10} + g^{AB}v_{AB},$$

$$g^{AB}v_{AB} \equiv \frac{1}{r^2}\left(v_{22} + \frac{1}{\sin^2\theta}v_{33}\right) = -v_{10}. \qquad \text{(IV.11.22)}$$

(b) To have an asymptotic wave of order one, we must have $R^{(0)}_{\alpha\beta} = 0$. If we choose the gauge such that $v_{10} = 0$, then the propagation equations, $R^{(0)}_{ij} = 0$ are found to reduce to

$$\partial_1 v_{AB} - \frac{1}{r}v_{AB} = 0.$$

These equations integrate to

$$v_{AB}(x,\xi) = r\gamma_{AB}(u,\theta,\phi,\xi).$$

We set

$$\gamma_{22} =: \alpha, \quad \gamma_{23} =: \beta.$$

The polarization conditions are then equivalent to

$$\gamma_{33} = -\alpha\sin^2\theta.$$

To have a progressive-wave solution of order one of the vacuum Einstein equations, it remains to satisfy the equations $R^{(0)}_{0\alpha} = 0$. The equations $R^{(0)}_{0i} = 0$ are found to be

$$R^{(0)}_{01} \equiv -\frac{1}{2}w''_{11} = 0,$$

$$R^{(0)}_{02} \equiv -\frac{1}{2}w''_{12} + \frac{1}{2r}\left(\partial_2\alpha' + \frac{1}{\sin^2\theta}\partial_3\beta' + 2\alpha'\frac{\cos\theta}{\sin\theta}\right) = 0,$$

$$R^{(0)}_{03} \equiv -\frac{1}{2}w''_{13} + \frac{1}{2r}\left(\partial_2\beta' - \partial_3\alpha' + \frac{\cos\theta}{\sin\theta}\beta'\right) = 0.$$

We see that, given α and β uniformly bounded functions of ξ as well as their primitives with respect to ξ, these linear equations in w''_{ij}, also linear in α' and β', are satisfied by functions w''_{1i} such that there exist functions w_{1i} that are uniformly bounded in ξ.

It remains to satisfy the equation $R^{(0)}_{00} = 0$, which reads

$$\frac{1}{2}\left[g^{ij}w''_{ij} + \frac{1}{r^2}\left(\alpha^2 + \frac{\beta^2}{\sin^2\theta}\right)\right]'' = -\frac{1}{2r^2}\left(\alpha'^2 + \frac{\beta'^2}{\sin^2\theta}\right) - \frac{2}{r^2}\frac{dm}{du}. \qquad \text{(IV.11.23)}$$

This equation can be satisfied by a w uniformly bounded in ξ only if the right-hand side is also the second derivative of a function uniformly bounded in ξ. This will be the case if

$$\frac{1}{4}\left(\alpha'^2 + \frac{\beta'^2}{\sin^2\theta}\right) + \frac{dm}{du} = 0, \qquad \text{(IV.11.24)}$$

for example, by assuming $dm/du < 0$ and setting $\mu(u) :=$
$2(-dm/du)^{\frac{1}{2}}$,

$$\alpha = \mu(u) \sin \xi \sin \theta, \quad \beta = \mu(u) \cos \xi \sin \theta \cos \theta. \qquad \text{(IV.11.25)}$$

One finds bounded w_{ij} satisfying all the required equations:

$$w_{12} = \frac{\mu(u)}{r} \cos \xi \, (\cos \theta + 2), \qquad \text{(IV.11.26)}$$

$$w_{13} = \frac{\mu(u)}{r} \sin \xi \sin^2 \theta, \qquad \text{(IV.11.27)}$$

and

$$w_{AB} = 0. \qquad \text{(IV.11.28)}$$

Remark IV.11.3   *The asymptotic metric obtained is axially sym-metric (the coefficients do not depend on φ), but not spherically symmetric. It is conjectured and partially proved that, in agree-ment with physical intuition, no progressive gravitational wave with spherically symmetric background is spherically symmetric.*

3. This is left to the reader.

IV.11.9   Static solutions with compact spacelike sections

Show that in dimension 3+1, the only locally static solution of the Ein-stein equations with compact spacelike sections and source such that $\rho_{00} \geq 0$ is vacuum and locally flat.

Solution

The Einstein equations for a locally static spacetime metric are

$$\bar{R}_{ij} - \psi^{-1} \bar{\nabla}_i \partial_j \psi = \rho_{0i}, \ \rho_{i0} \equiv 0, \ \psi \Delta_{\bar{g}} \psi = \rho_{00}. \qquad \text{(IV.11.29)}$$

If M is a compact manifold (without boundary) and $\Delta_{\bar{g}} \psi \equiv \bar{\nabla}_i \partial^i \psi$, then it holds that

$$0 = \int_M \Delta_{\bar{g}} \psi \mu_{\bar{g}} = \int_M \psi^{-1} \rho_{00} \mu_{\bar{g}}, \text{ hence } \rho_{00} \equiv 0 \text{ if } \psi > 0 \text{ and } \rho_{00} \geq 0.$$

Therefore, $\Delta_{\bar{g}} \psi = 0$. On a compact manifold, this implies $\psi = \text{constant}$, and the equations imply $Ricci(\bar{g}) = 0$. We know that Ricci-flat three-dimensional manifolds are locally flat.

IV.11.10   Mass of an asymptotically Euclidean spacetime

The ADM[56] mass of an asymptotically Euclidean manifold (M, g) is defined by a flux integral through the sphere at infinity of its end:

$$m_{\text{ADM}} := \frac{1}{16\pi} \lim_{r \to \infty} \int_{\partial B_r} (\partial_i g_{ij} - \partial_j g_{ii}) n^i \, d\mu.$$

Consider the asymptotically Euclidean metric on $R^3 \cap r > 0$, $r^2 = x^2 + y^2 + z^2$,

$$\left(1 + \frac{m}{2r}\right)^4 (dx^2 + dy^2 + dz^2),$$

isometric with the space part of the Schwarzschild metric in standard coordinates (see Chapter V).

Show that its **ADM mass** is equal to m.

Solution

For the given metric $g_{ij} = (1 + m/2r)^4 \delta_{ij}$, $\delta_{ij} = 0$ for $i \neq j$,

$$\partial_i g_{ij} = 4\left(1 + \frac{m}{2r}\right)^3 \left(-\frac{m}{2r^2}\frac{x^j}{r}\right), \quad \partial_j g_{ii} = 12\left(1 + \frac{m}{2r}\right)^3 \left(-\frac{m}{2r^2}\frac{x^j}{r}\right),$$

$$(\partial_i g_{ij} - \partial_j g_{ii}) = 8\left(1 + \frac{m}{2r}\right)^3 \left(\frac{m}{2r^2}\frac{x^j}{r}\right) \simeq 4\frac{mx^i}{r^3} = 4m\frac{n_i}{r^2},$$

and hence, with $n^i n_i = 1$ and 4π the surface of the unit sphere S^2,

$$m_{\text{ADM}} = \frac{1}{16\pi} \int_{S_1} (\partial_i g_{ij} - \partial_j g_{ii}) n^i \, d\mu = m.$$

IV.11.11   Taub Lagrangian

Consider the Lagrangian with constraint $g(u, u) = -1$ and Lagrange multiplier λ:

$$\mathcal{L} = \int \{R(g) + [\mu + p + \lambda g(u, u)]\} \mu_g.$$

Prove by choosing coordinates adapted to the fluid flow that the Euler equations for such a Lagrangian lead, with the choice $\lambda = \mu$, to the usual Euler equations for perfect fluids and the Einstein equations with source their stress–energy tensor.

The Schwarzschild spacetime

V.1   Introduction

In 1916, soon after the publication of Einstein's equations, an exact solution was constructed by Schwarzschild that could model the gravitational field outside a spherically symmetric isolated body such as the Sun. It was shown that, in first approximation, the relativistic planetary trajectories, i.e. timelike geodesics of the Schwarzschild metric, coincide with the Kepler orbits. However, in the case of Mercury, the planet nearest to the Sun, an additional advance of its perihelion of about 42″ per century beyond the Newtonian calculuation was found,[1] in agreement with astronomers' observations. Other effects predicted by the Einsteinian gravitation theory, namely deflection of light rays, redshift, and time delay, have also been found to be in remarkable agreement with observations and experiments (see Sections V.8 and V.9). In Chapter VI, we will treat a revolutionary property of Einsteinian spacetimes, the existence of black holes, which first appeared through investigations of the Schwarzschild metric.

V.2   Spherically symmetric spacetimes

We start with the following natural and elementary definitions.

Definition V.2.1   *A three-dimensional Riemannian manifold $(M, \bar{g})$ is said to be spherically symmetric if*

1. *The manifold M is represented by one chart (U, Φ) with $\Phi(U) = R^3$ or the exterior of a ball B of R^3 centred at some point O. We denote by ρ, θ, ϕ spherical (pseudo) coordinates in $\Phi(U)$, linked to the canonical coordinates x, y, z of R^3 by the usual relations*

$$x = \rho \sin\theta \sin\phi, \quad y = \rho \sin\theta \cos\phi, \quad z = \rho \cos\theta. \qquad (V.2.1)$$

2. *In $\Phi(U)$, given by $\rho \geq \rho_0 \geq 0$, $0 \leq \theta < \pi$, $0 \leq \varphi < 2\pi$, $\bar{g}$ is represented by a metric of the form*

$$e^{h(\rho)}d\rho^2 + f^2(\rho)(d\theta^2 + \sin^2\theta\, d\varphi^2), \qquad (V.2.2)$$

[1] Actually, this was first obtained by Einstein, in November 1915, by a perturbative calculation where he solved the Einstein equations by successive approximations, going to second order in G_E.

with f a monotonically increasing function of ρ. The interpretation is that $\Phi(U)$ is foliated by metric 2-spheres $\rho = $ constant, centred at O; their areas in the metric (2.2) are $4\pi f^2$. The metric (2.2) is the general form of a metric invariant under rotations in R^3, centred at O. It is defined on the whole of R^3 if the ball B is empty.

Remark V.2.1 *In the preceding definition, $\rho = 0$ represents, by definition, a single point O. The vanishing of $f^2(0)$ does not imply a singularity in the metric, but reflects the fact that spherical 'coordinates' are not admissible coordinates at $\rho = 0$.*

The choice of the coordinate r given by $r = f(\rho)$ is called the **standard choice**.

Definition V.2.2 *Consider a spacetime (V, g) with V contained in the product $R^3 \times R$, a point of V being labelled (x, t). Suppose the subsets M_t of constant t are spacelike submanifolds; then we denote by g_t the Riemannian metric induced by g on M_t. The trajectories of the vectors $\partial/\partial t$ are supposed timelike. The spacetime is said to be spherically symmetric if the following hold:*

1. *Each manifold M_t has a representation as the exterior $R^3 - B_t$ of a ball B_t of R^3 centred at the origin O. Each manifold (M_t, g_t) is spherically symmetric. In $R^3 - B_t$ the metric g_t reads, in standard coordinates,*

$$g_t = e^{\lambda(r,t)} \, dr^2 + r^2(d\theta^2 + \sin^2\theta \, d\varphi^2). \tag{V.2.3}$$

2. *For each t, the g length and the representative of the projection on M_t of the vector $\partial/\partial t$ tangent to the timeline[2] are both invariant under the rotation group defined above.*

[2] Equivalently, the lapse and the shift of the slicing (see Chapter VIII).

Lemma V.2.1 *A spherically symmetric spacetime (V, g) admits a metric of the form*

$$g = -e^{\nu} \, dt^2 + e^{\lambda} \, dr^2 + r^2(d\theta^2 + \sin^2\theta \, d\varphi^2), \tag{V.2.4}$$

where λ and ν are functions of t and r only.

Proof. A scalar on $R^3 - B$ invariant under the rotation group is necessarily a function of r and t alone. A vector field invariant under this group is tangent to the radial lines (lines where only the r coordinate varies) and its magnitude depends only on r for each t. Therefore, the given definition implies that

$$g = -a^2(r,t) \, dt^2 + 2b(r,t) \, dt \, dr + e^{\lambda(r,t)} \, dr^2 + r^2(d\theta^2 + \sin^2\theta \, d\varphi^2). \tag{V.2.5}$$

We can eliminate the diagonal term in $dt \, dr$ by computing an integrating factor for the 1-form in two variables $\omega := a \, dt - b \, dr$, i.e. a function of t and r, which we denote by $e^{-2\nu}$, such that its product by the 1-form is the differential of a function τ:

$$e^{-2\nu}\omega \equiv e^{-2\nu}(t,r)(a \, dt - b \, dr) \equiv d\tau.$$

We take τ as a new time coordinate, keeping r, θ, φ as space pseudo-coordinates. Writing the metric (2.4) with this new time coordinate, renamed as t, gives the formula (2.3). The coordinates t, r are called **standard coordinates**. ∎

Exercise V.2.1   *Given a and b, determine ν.*

Remark V.2.2   *If the point with coordinate $r = 0$ belongs to the manifolds M_t, it describes a timelike line called the central world line.*

V.3   Schwarzschild metric

We will prove the following theorem:

Theorem V.3.1   *A smooth spherically symmetric metric is a solution of the vacuum Einstein equations if and only if it is the Schwarzschild metric, which reads, in standard coordinates, with m a constant*

$$g_{\text{Schw}} = -\left(1 - \frac{2m}{r}\right)dt^2 + \left(1 - \frac{2m}{r}\right)^{-1}dr^2 + r^2(d\theta^2 + \sin^2\theta\, d\varphi^2).$$

$$\text{(V.3.1)}$$

Proof. We set $t = x^0$, $r = x^1$, $\theta = x^2$, $\varphi = x^3$ and denote by a prime the derivative with respect to r. We find that the only non-zero Christoffel symbols of the spherically symmetric metric (2.4) are

$$\Gamma^0_{00} = \frac{1}{2}\partial_t\nu, \ \Gamma^1_{00} = \frac{1}{2}e^{\nu-\lambda}\nu', \ \Gamma^0_{01} = \frac{\nu'}{2}, \ \Gamma^0_{11} = \frac{\partial_t\lambda}{2}e^{\lambda-\nu}, \ \Gamma^1_{01} = \frac{\partial_t\lambda}{2},$$

$$\text{(V.3.2)}$$

$$\Gamma^1_{11} = \frac{\lambda'}{2}, \ \Gamma^1_{22} = -re^{-\lambda}, \ \Gamma^2_{12} = \Gamma^3_{13} = r^{-1}, \ \Gamma^1_{33} = -r\sin^2\theta\, e^{-\lambda},$$

$$\text{(V.3.3)}$$

$$\Gamma^2_{33} = -\sin\theta\cos\theta, \ \Gamma^3_{23} = \cot\theta. \qquad \text{(V.3.4)}$$

Computing the components of the Ricci tensor, we find

$$R_{10} \equiv r^{-1}\partial_t\lambda, \quad R_{22} \equiv -e^{-\lambda}\left[1 + \frac{r}{2}(\nu' - \lambda')\right] + 1, \quad R_{33} \equiv \sin^2\theta\, R_{22}.$$

$$\text{(V.3.5)}$$

We deduce from these identities that for a solution of the vacuum Einstein equations, $R_{\alpha\beta} = 0$, λ is independent of t. Therefore, ν' is also independent of t, $\partial_t\nu' = 0$, and ν is of the form

$$\nu(t,r) = \nu(r) + f(t). \qquad \text{(V.3.6)}$$

We set $\tau := \int e^{\frac{1}{2}f(t)^2}\, dt$ and we rename τ as t; we then have $\partial_t\nu = 0$. The other non-identically zero components of the Ricci tensor then reduce to

$$R_{00} \equiv e^{\nu-\lambda}\left(\frac{\nu''}{2} + \frac{\nu'^2}{4} - \frac{\nu'\lambda'}{4} + \frac{\nu'}{r}\right), \qquad \text{(V.3.7)}$$

$$R_{11} \equiv -\frac{\nu''}{2} - \frac{\nu'^2}{4} + \frac{\lambda'\nu'}{4} + \frac{\lambda'}{r}. \qquad (V.3.8)$$

The vacuum Einstein equations therefore imply

$$r(e^{\lambda-\nu}R_{00} + R_{11}) \equiv \nu' + \lambda' = 0. \qquad (V.3.9)$$

Modulo this relation, the equations $R_{22} = 0$ and $R_{33} = 0$ reduce to

$$-e^{-\lambda}(1 - r\lambda') + 1 = 0;$$

that is,

$$(e^{-\lambda})' + \frac{e^{-\lambda}}{r} = \frac{1}{r}. \qquad (V.3.10)$$

The general solution of this linear equation for $e^{-\lambda}$ is

$$e^{-\lambda} = 1 + \frac{A}{r}, \quad \text{hence } e^{\nu} = B\left(1 + \frac{A}{r}\right), \qquad (V.3.11)$$

with A and B arbitrary constants. The constant B can be made equal to 1 by a rescaling of t. The constant A is denoted by $-2m$; we will see in Section V.6, Equation (6.8), that the Newtonian approximation corresponding to the Schwarzschild metric then coincides with Newton's gravity, with the coefficient m being the gravitational mass expressed in units of length. We have thus obtained the metric (3.1).

To show that this metric satisfies the full vacuum Einstein equations, we must check that it satisfies also the equations $R_{00} = 0$ and $R_{11} = 0$; this can be done by using (3.9) and (3.7), (3.8), (3.11) to show that $e^{\lambda-\nu}R_{00} - R_{11} = 0$. ∎

Exercise V.3.1 *Prove this last statement by using the Bianchi identities.*

Remark V.3.1 *We have supposed in this theorem that the areas of the orbits of the symmetry group are monotonically increasing along their orthogonal trajectories.*

In the course of the proof, we have obtained **Birkhoff's theorem:**

Theorem V.3.2 *(Birkhoff) A smooth spherically symmetric metric solution of the vacuum Einstein equations is necessarily static.*

V.4   Other coordinates

In some problems, it is useful to use alternative, non-standard, coordinates for the Schwarzschild metric.

V.4.1   Isotropic coordinates

One defines new coordinates X, Y, Z, called isotropic, on $R^3 \times R$ that are related to the standard r, θ, ϕ (i.e. x, y, z) by setting

$$r = R \left(1 + \frac{m}{2R}\right)^2, \tag{V.4.1}$$

$$X = Rr^{-1}x, \quad Y = Rr^{-1}y, \quad Z = Rr^{-1}z. \tag{V.4.2}$$

In terms of these coordinates, the metric reads, with $R^2 = X^2 + Y^2 + Z^2$,

$$g_{\text{Schw}} = -\left(\frac{2R - m}{2R + m}\right)^2 dt^2 + \left(1 + \frac{m}{2R}\right)^4 (dX^2 + dY^2 + dZ^2). \tag{V.4.3}$$

Exercise V.4.1 *Prove this formula (see the solution of Problem V.12.4).*

It is clear from the above expression that the spaces $t = \text{constant}$ of the Schwarzschild metric are conformal to Euclidean space.

Remark V.4.1 *r is a monotonically increasing function of R, from $2m$ to infinity, when R increases from $m/2$ to infinity; r and R are equivalent for large r.*

V.4.2   Wave (also called harmonic) coordinates

Theorem V.4.1 *The $3 + 1$ metric,[3] defined for $\bar{r} > m$,*

$$-\frac{\bar{r} - m}{\bar{r} + m} dt^2 + \frac{\bar{r} + m}{\bar{r} - m} d\bar{r}^2 + (\bar{r} + m)^2 (d\theta^2 + \sin^2 \theta \, d\phi^2) \tag{V.4.4}$$

is isometric to the Schwarzschild metric by the mapping $r = \bar{r} + m$. The corresponding Cartesian coordinates t and x^i defined by

$$x^1 = \bar{r} \sin \theta \sin \phi, \quad x^2 = \bar{r} \sin \theta \cos \phi, \quad x^3 = \bar{r} \cos \theta$$

are wave coordinates.

Proof. We write an arbitrary spherically symmetric, static, metric on $R^3 \times R$ in the form

$$- A^2 dt^2 + B^2 d\bar{r}^2 + r^2 (d\theta^2 + \sin^2 \theta \, d\phi^2), \tag{V.4.5}$$

where θ, ϕ are spherical coordinates on S^2. The coefficients A, B, and r are functions of $\bar{r}$ only. The coordinate t is obviously a wave coordinate. We look for a function $r = f(\bar{r})$ such that the coordinates x^i defined by the above relations are wave coordinates, i.e. such that the functions x^i satisfy the wave equation. For an arbitrary function ψ, the wave equation in the metric (4.4) reads

$$-\frac{1}{A^2} \frac{\partial^2 \psi}{\partial t^2} + \frac{1}{r^2} \left[\frac{1}{AB} \frac{\partial}{\partial \bar{r}} \left(AB^{-1} r^2 \frac{\partial \psi}{\partial \bar{r}} \right) + \Delta^* \psi \right], \tag{V.4.6}$$

[3]Note that while the standard form of the Schwarzschild metric generalizes to higher values of n, its expression in wave coordinates does not (see Problem V.12.5).

where Δ^* is the Laplacian on the sphere S^2. If x^i is one of the functions given above, then

$$\frac{\partial x^i}{\partial \bar{r}} = \frac{x^i}{\bar{r}}, \quad \frac{\partial^2 x^i}{\partial \bar{r}^2} = 0 \quad \text{and} \quad \Delta^* x^i = -2x^i. \tag{V.4.7}$$

Hence the condition that the x^i be wave coordinates reduces to

$$\frac{1}{AB}\frac{d}{d\bar{r}}(AB^{-1}r^2) - 2\bar{r} = 0. \tag{V.4.8}$$

In the case of the Schwarzschild metric, we have for $r > 2m$

$$A \equiv \sqrt{1 - \frac{2m}{r}}, \quad B \equiv A^{-1}\frac{dr}{d\bar{r}}, \tag{V.4.9}$$

and hence the harmonicity condition is

$$\frac{d}{dr}\left(\frac{d\bar{r}}{dr}A^2 r^2\right) - 2\bar{r} \equiv \frac{d}{dr}\left[\frac{d\bar{r}}{dr}(r^2 - 2mr)\right] - 2\bar{r} = 0. \tag{V.4.10}$$

We set

$$r = m(1 + z), \tag{V.4.11}$$

and (4.10) becomes

$$\frac{d}{dz}\left[(z^2 - 1)\frac{d\bar{r}}{dz}\right] - 2z = 0, \quad \text{for} \ z > 1. \tag{V.4.12}$$

This linear second-order differential equation is a Legendre equation, whose general solution is, with C_1 and C_2 arbitrary constants

$$\bar{r} = C_1 z + C_2\left(\frac{z}{2}\ln\frac{z+1}{z-1} - 1\right). \tag{V.4.13}$$

To avoid introducing an additional logarithmic singularity on the horizon $r = 2m$, we take

$$\bar{r} = mz, \quad \text{i.e.} \quad r = \bar{r} + m. \tag{V.4.14}$$

Using this value of r, we can check that the Schwarzschild metric in the wave coordinates x^i defined above is the given metric. ∎

V.4.3   Painlevé–Gullstrand-like coordinates

The following stationary but non-static form of the Schwarzschild metric has been used in numerical computations:[4]

$$-\left(1 - \frac{2m}{r}\right)dt^2 + dx^2 + dy^2 + dz^2 + \frac{2}{r}\sqrt{\frac{2m}{r}}(x\,dx + y\,dy + z\,dz)\,dt. \tag{V.4.15}$$

It is called a **boosted Schwarzschild metric**.

[4]Pretorius (2005a).

V.4.4   Regge–Wheeler coordinates

We define, in the region $r > 2m$, a tortoise radial coordinate

$$\rho = r + 2m \log(r - 2m). \qquad (V.4.16)$$

The metric then takes a form in which the timelike sections $\theta = $ constant, $\phi = $ constant are conformal to two-dimensional Minkowski space:

$$g_{\text{Schw}} = \left(1 - \frac{2m}{r}\right)(-dt^2 + d\rho^2) + r^2(d\theta^2 + \sin^2\theta\, d\phi^2), \qquad (V.4.17)$$

where r is the function of ρ defined by (4.16).

V.5   Schwarzschild spacetime and event horizon

If $m = 0$, the Schwarzschild metric reduces to the Minkowski metric; if $m \neq 0$, it is singular for $r = 0$ and has a coordinate singularity for $r = 2m$ (see below).

The sign of the constant $m \neq 0$ is very important in determining the properties of the Schwarzschild metric.

If $m < 0$, the Schwarzschild metric defines a spherically symmetric spacetime on the whole of $R^3 \times R$, except at $\{0\} \times R$, where the metric is singular. At present, the Schwarzschild metric with $m < 0$ has no physical interpretation. It can be shown that there exists no asymptotically Euclidean Einsteinian spacetime $(R^3 \times R, g)$ with sources of positive energy that coincides for $r > a \geq 0$ with a Schwarzschild metric with $m < 0$; this is a particular case of the **positive-mass theorem**.[5]

If $m > 0$, the Schwarzschild metric is a regular Lorentzian metric with t timelike and r spacelike and $r > 2m$; $2m$ is called the **Schwarzschild radius**. For r large with respect to m, we will see in the following sections that the resulting physical properties coincide in first approximation to those of the Newtonian theory for a spherical body of gravitational[6] mass m centred at $r = 0$.

It is possible to construct so-called interior Schwarzschild solutions that are spherically symmetric spacetimes, not necessarily static, smooth on a manifold $R \times R^3 \cap \{r \leq a\}$, satisfy the Einstein equations with sources of positive energy, and lead to a complete admissible Einsteinian spacetime on the whole manifold $R \times R^3$, with, for $r > a$, a Schwarzschild spacetime of mass $2m < a$ (see Section V.10).

Since $1 - 2m/r$ vanishes for $r = 2m$, the Schwarzschild metric in standard coordinates appears to be singular there: g_{00} vanishes and g_{rr} becomes infinite. For $r < 2m$, $r \neq 0$, the Schwarzschild metric is again a regular Lorentzian metric, but the timelike and spacelike character of the coordinates t and r are interchanged.

We will construct later, whatever the value of $a > 0$, vacuum Einsteinian spacetimes that are regular for $r \geq a > 0$ and isometric for

[5] See, for instance, YCB-OUP2009 and references therein.

[6] Recall that the gravitational and inertial masses differ by a factor equal to the gravitational constant. In this book, unless otherwise stated, we work in units such that this constant as well as the speed of light are equal to 1.

$r > 2m$ to a Schwarzschild spacetime in canonical coordinates with radius $2m$. They are interpreted as spacetimes modelling the exteriors of bodies with Schwarzschild radius $2m$. However, the apparent 'Schwarzschild singularity' $r = 2m > a$ has a deep physical meaning that we will discuss in Chapter VI: no light ray or other classical signal (i.e. one not due to a quantum effect) can escape from the regions $r < 2m$. For this reason, the hypersurface $R \times \{r = 2m\}$ is called an **event horizon** and a spacetime with source of radius $a < 2m$ is called a black hole (see Chapter VI).

If a is identified with the radius of the body in a classical CGS length unit and the gravitational mass m determining the Schwarzschild radius is expressed in the same unit of length, but with the units of time and mass chosen so that the speed of light and the gravitational constant are both equal to 1, as we do in writing the equations (see Problem III.9.3 in Chapter III), then the Schwarzschild radii of the solar planets and normal stars are very much inside these bodies:[7] $2m << a$.

In the rest of this chapter, we will study properties of Schwarzschild spacetimes in the region $r > 2m$. We will compare their predictions with the Newtonian ones and unravel new effects of gravitation predicted by Einstein's General Relativity. We will also quote results of observations and experiments, all in agreement with Einstein's theory.

[7]For the Sun, $2m_{\text{Sun}} \doteq 2.96\,\text{km}$ and for the Earth, $2m_{\text{Earth}} \doteq 8.87\,\text{mm}$.

V.6   The motion of the planets and perihelion precession

V.6.1   Equations

The trajectories of bodies of small size and mass in a spherically symmetric gravitational field, for instance the trajectories of the solar planets, are timelike geodesics of this field, i.e. of the Schwarzschild spacetime.

We denote by ds the element of proper time on a timelike curve; i.e. with our signature convention,

$$ds^2 \equiv -g_{\alpha\beta}dx^\alpha dx^\beta. \tag{V.6.1}$$

As in the Newtonian case, we find that the orbits remain in a 'plane' of R^3 by considering first the equation (cf. the expressions for the Christoffel symbols)

$$\frac{d^2\theta}{ds^2} + \frac{2}{r}\frac{dr}{ds}\frac{d\theta}{ds} - \sin\theta\cos\theta\left(\frac{d\varphi}{ds}\right)^2 = 0. \tag{V.6.2}$$

We choose the coordinate θ such that at some initial instant s_0 we have for the considered motion of the planet $\theta(s_0) = \pi/2$ and $(d\theta/ds)(s_0) = 0$. The equation satisfied by θ then implies that the orbit remains in the 'plane' $\theta = \pi/2$. With this choice of the coordinate θ, the equation for φ reduces to

$$\frac{d^2\varphi}{ds^2} + \frac{2}{r}\frac{dr}{ds}\frac{d\varphi}{ds} = 0, \tag{V.6.3}$$

which integrates to an analogue of the Newtonian area law, with ℓ some constant such that

$$r^2 \frac{d\varphi}{ds} = \ell. \tag{V.6.4}$$

Remark V.6.1 *Equation (6.4) is a consequence of the invariance of the metric under rotations. The constant ℓ can be interpreted as the angular momentum per unit mass, as seen at large distance.*

The equation

$$\frac{d^2 t}{ds^2} + \frac{d\nu}{dr} \frac{dr}{ds} \frac{dt}{ds} = 0, \quad \text{with} \quad \nu := \log\left(1 - \frac{2m}{r}\right) \tag{V.6.5}$$

integrates to ('energy' integral due to t-translation invariance)

$$\left(1 - \frac{2m}{r}\right) \frac{dt}{ds} = E, \quad E \text{ a constant.} \tag{V.6.6}$$

The remaining geodesic equation is

$$\frac{d^2 r}{ds^2} + \frac{1}{2} \frac{d\lambda}{dr} \left(\frac{dr}{ds}\right)^2 - e^{-\lambda} \left(\frac{d\varphi}{ds}\right)^2 + \frac{e^{\nu - \lambda}}{2} \frac{d\nu}{ds} \left(\frac{dt}{ds}\right)^2. \tag{V.6.7}$$

Using the expressions for λ and ν together with the previous integrals (6.4) and (6.6), we see that, when $\ell \neq 0$, this equation reduces to

$$\frac{d^2 u}{d\varphi^2} + u = \frac{m}{\ell^2} + 3mu^2, \quad \text{with} \quad u \equiv \frac{1}{r}. \tag{V.6.8}$$

This equation is formally the same as the linear equation in u of Newtonian mechanics, except for the addition of the nonlinear term $3mu^2$. Therefore, it holds that

$$u(\varphi) = u_{\text{Newton}}(\varphi) + v(\varphi), \tag{V.6.9}$$

where v satisfies the equation

$$\frac{d^2 v}{d\varphi^2} + v = 3mu^2. \tag{V.6.10}$$

As is well known, the solutions of the Newtonian equation (for $\ell \neq 0$) are the conics

$$\frac{1}{r} = m\ell^{-2}(1 + e \, \cos \varphi) =: u_{\text{Newton}}, \tag{V.6.11}$$

where e is the eccentricity and where the longitude of the perihelion, a constant in this Newtonian case, has been taken to be equal to zero. The constant e depends on the initial position and velocity of the planet. It is not far from 0 for the solar planets—their orbits are almost circular.

For the solar planets, r is large with respect to ℓ, and hence $3mu^2$ can be considered as a small correction to $m\ell^{-2}$; the correction v will therefore be small with respect to u_{Newton}. An approximate solution of

(6.11), a small correction to the Newtonian expression, is obtained by replacing (6.11) by the linear equation

$$\frac{d^2 v}{d\varphi^2} + v = 3m u^2_{\text{Newton}} \equiv 3m^3 \ell^{-4}(1 + 2e \cos\varphi + e^2 \cos^2\varphi). \quad \text{(V.6.12)}$$

This is the differential equation for a forced oscillation. The general solution is the sum of the general solution of the associated homogeneous equation, i.e. an arbitrary periodic function of period 2π in φ, and a particular solution, for instance

$$3m^3 \ell^{-4}\left[1 + e\varphi \sin\varphi + e^2\left(\frac{1}{2} - \frac{1}{6}\cos 2\varphi\right)\right]. \quad \text{(V.6.13)}$$

The term that will give the observationally most significant contribution to the correction is the 'secular' (non-periodic) term in φ; that is, we consider the Einsteinian approximation:

$$u_{\text{Einstein}} \sim u_{\text{Newton}} + 3m^2 \ell^{-4} e\varphi \sin\varphi$$
$$= m\ell^{-2}(1 + e\cos\varphi + e3m^2\ell^{-2}\varphi \sin\varphi). \quad \text{(V.6.14)}$$

In geometric units, $m^2\ell^{-2}$ is small, and therefore $3m^2\ell^{-2}\varphi$ is equivalent to $\sin(3m^2\ell^{-2}\varphi)$ and u_{Einstein} is equivalent to

$$u_{\text{Einstein}} \sim m\ell^{-2}\left\{1 + e\cos[(1 - 3m^2\ell^{-2})\varphi]\right\}. \quad \text{(V.6.15)}$$

The orbit is no longer a closed curve (unless it is circular, i.e. if $e = 0$), because u_{Einstein} does not have period 2π in φ. It is only approximately an ellipse; the closest point to the centre (perihelion), attained for $\varphi = 0$, is attained successively after each increase of φ by $2\pi(1 - 3m^2\ell^{-2})^{-1} \sim 2\pi + 6\pi m^2\ell^{-2}$. The additional $6\pi m^2\ell^{-2}$ is the famous Einsteinian perihelion precession.

Remark V.6.2 *Elementary calculus gives a perihelion precession per orbit*

$$\frac{6\pi m}{p}, \qquad p := a(1 - e^2).$$

p is called the parameter of the Newtonian ellipse, a is its semi-major axis, and e its eccentricity. For equivalent a's, the precession of the perihelion is greatest for eccentricities approaching 1.

V.6.2   Results of observations

It was observed long ago by astronomers that the orbits of the solar planets are not exact ellipses, but slowly rotating ones. This phenomenon was interpreted in Newton's theory as a result of the influence of other planets. The perihelion precession of most planets could thus be accounted for—except for that nearest the Sun, Mercury, for which a precession of 42″ per century over the 5600″ observed (found by Leverrier in 1845) remained unexplained. The Einsteinian correction (1916) just filled the gap, which was a remarkable success for the new theory.

For a long time, it was thought that the not exactly symmetric shape of the Sun could also play a role but was too difficult to estimate,[8] but recent data have estimated the quadrupole moment of the Sun and shown its contribution to Mercury's perihelion advance to be quite small.

In 1974, a pulsar (a rotating neutron star, PSR 1913+16) was observed by Taylor and collaborators, orbiting around a companion that is most probably also a neutron star. Its orbit shows a precession of about 4.2 deg per year, which is about 27 100 times the precession rate of Mercury. This is believed to be an Einsteinian effect, but since the masses of the orbiting objects are not known, the precession cannot be used directly to test General Relativity; it is used instead to estimate these masses. To do this, one must generalize the above treatment of the dynamics of a small mass around a large one to the study of the relativistic dynamics of two comparable masses.

In order to compare the predictions of General Relativity with the observations of binary pulsars, it is necessary to go beyond the first post-Newtonian approximation presented in Section IV.5.3 in Chapter IV. First, as a neutron star is an object that generates at its surface a very strong deformation of the Minkowski metric ($g_{00}^{\text{surface}} \simeq -1 + h_{00}^{\text{surface}}$, with $h_{00}^{\text{surface}} \simeq 2\,GM/c^2R \simeq 0.4$ for a neutron star, to be compared with $h_{00}^{\text{surface}}(\text{Sun}) \sim 10^{-6}$ and $h_{00}^{\text{surface}}(\text{Earth}) \sim 10^{-8}$), one needs to develop approximation methods that go beyond the usual weak-field expansion and are able to deal with the motion of strongly self-gravitating bodies. Such methods, based on the *multichart* approach mentioned in Section IV.5.3, were developed by several authors in the 1970s and 1980s.[9] These approaches use the method of *matched asymptotic expansions* to transfer information between the tidal-like expansions of the metric $G_{\alpha\beta}^A(X_A^\gamma)$ considered in the local coordinate system X_A^α attached to each body (labelled by A) and the post-Minkowskian (or post-Newtonian) expansion of the metric $g_{\mu\nu}(x^\lambda)$ considered in the global coordinate system x^μ. In addition, it was necessary to push the accuracy of the post-Newtonian expansion used in the global chart x^μ much beyond the 1PN level. More precisely, it was necessary to go to the 2.5PN level, corresponding to keeping terms $(v/c)^5$ smaller than the Newtonian-level terms. The first complete results in the 2.5PN approximation were obtained in the early 1980s.[10] At this level of approximation, the fact that the gravitational interaction between the two bodies propagates (via a retarded Green function of the wave equation) at the velocity of light entails an observable effect: namely a slow decrease of the orbital period P_b given by[11]

$$\frac{dP_b}{dt} = -\frac{192\,\pi}{5\,c^5}\,\nu(GM\,\Omega)^{5/3}\,\frac{1 + \frac{73}{24}\,e^2 + \frac{37}{96}\,e^4}{(1 - e^2)^{7/2}}, \qquad (\text{V.6.16})$$

where $\nu \equiv m_A\,m_B/(m_A + m_B)^2$ is the symmetric mass ratio, $M \equiv m_A + m_B$ the total mass, $\Omega = 2\pi/P_b$ the orbital frequency, and e the eccentricity of the orbit. This effect has been observed in several

[8] In the case of satellites orbiting the Earth, the irregularities of their motions are used to determine the shape of the Earth.

[9] See, notably, D'Eath (1975) and Damour (1983a).

[10] See Damour and Deruelle (1981) and Damour (1982).

[11] See Damour (1983b).

binary pulsars, with a magnitude in precise agreement with the General Relativity prediction (6.16). As this effect is a *direct* consequence of the propagation of the gravitational interaction at the velocity of light, it constitutes an observational proof of the reality of gravitational radiation.

In addition, the development of the relativistic theory of the 'timing' of binary pulsars[12] has allowed several other comparisons between the predictions of Einstein's theory and binary pulsar data. More precisely, the measurement of n 'post-Keplerian' parameters appearing in the Damour–Deruelle timing formula allows one to extract $n - 2$ tests of General Relativity (or alternative theories of gravity). For instance, in the case of PSR 1913+16, one could use pulsar timing data to measure the following three 'relativistic' or 'post-Keplerian' observables: (i) advance of periastron, (ii) apparent modification of the proper time of the pulsar by combined Doppler and gravitational effects, and (iii) rate of change of the orbital period of the pulsar linked to gravitational radiation damping. Comparison of these three 'post Keplerian' observables with their General Relativity prediction leads to an accurate (10^{-3} level) confirmation of Einistein's theory in a regime that involves both strong-field effects (in the neutron star) and radiative effects. Other binary pulsars[13] have recently allowed further accurate confirmation of GR.

[12]See Damour and Deruelle (1985, 1986).

[13]For a review, see, for instance, Damour (2013b).

V.6.3   Escape velocity

As an exercise in the physical interpretation of General Relativity, we compute the radial velocity with respect to an observer at rest in the Schwarzschild metric that must be applied to a test object for it to escape the gravitational attraction.

Let r_0 be the r coordinate of the static observer in a Schwarzschild spacetime with mass $m < r_0/2$. Denote by $\dot{r}_0 = dr/ds(0)$ the proper-time initial velocity, supposed to be radial, of the launched rocket at parameter time $t = 0$. This rocket, supposed to be in free motion after its launch, follows a radial geodesic curve and hence satisfies, by (6.6), the equation

$$\left(1 - \frac{2m}{r}\right)\dot{t} = E = \left(1 - \frac{2m}{r_0}\right)\dot{t}_0. \qquad (V.6.17)$$

We also have, as a result of the definition of ds,

$$1 = \left(1 - \frac{2m}{r}\right)\dot{t}^2 - \left(1 - \frac{2m}{r}\right)^{-1}\dot{r}^2. \qquad (V.6.18)$$

Hence

$$\dot{r}^2 = E^2 - 1 + \frac{2m}{r}. \qquad (V.6.19)$$

The rocket can attain a maximum of the parameter r (and then turn back) when $\dot{r} = 0$, that is, when r takes the value

$$r_M = \frac{2m}{1 - E^2}. \tag{V.6.20}$$

The number r_M is an attained maximum of the parameter r if it is positive and finite, that is, if $E < 1$. The escape velocity for which r_M is infinite corresponds to $E = 1$, and hence, by (6.16), to

$$\dot{r}_0^2 = \frac{2m}{r_0}. \tag{V.6.21}$$

The relativistic escape velocity V for the observer at rest is given by the ratio of the space (radial) and time components V^1 and V^0 in the proper frame of this observer of the velocity vector, which has components $(\dot{r}_0, \dot{t}_0)$ in the natural frame of the coordinates t, r. The proper frame of the static observer is

$$\theta^0 = \left(1 - \frac{2m}{r_0}\right)^{\frac{1}{2}} dt, \quad \theta^1 = \left(1 - \frac{2m}{r_0}\right)^{-\frac{1}{2}} dr, \tag{V.6.22}$$

and therefore

$$V^1 = \left(1 - \frac{2m}{r_0}\right)^{-\frac{1}{2}} \dot{r}_0, \quad V^0 = \left(1 - \frac{2m}{r_0}\right)^{\frac{1}{2}} \dot{t}_0. \tag{V.6.23}$$

Hence, using (6.20) and (6.18) when $E = 1$, we obtain

$$V =: \frac{V^1}{V^0} = \sqrt{\frac{2m}{r_0}}. \tag{V.6.24}$$

This escape velocity coincides with its Newtonian value. It tends to 1, the velocity of light, when r_0 tends to $2m$, in agreement with interpretations that we will give in a forthcoming section.

V.7   Stability of circular orbits

To study the **stability of circular orbits**, we use the identity

$$g_{\alpha\beta}\dot{x}^\alpha \dot{x}^\beta = -1 \tag{V.7.1}$$

and (6.4) and (6.6) to obtain

$$\frac{1}{2}\dot{r}^2 + \frac{1}{2}\left(1 - \frac{2m}{r}\right)\left(\frac{\ell^2}{r^2} + 1\right) = \frac{1}{2}E^2. \tag{V.7.2}$$

This type of differential equation occurs in classical mechanics and governs the motion of a particle of unit mass and energy $\frac{1}{2}E^2$ in the potential[14]

$$V(r) \equiv \frac{1}{2}\left(1 - \frac{2m}{r}\right)\left(\frac{\ell^2}{r^2} + 1\right). \tag{V.7.3}$$

[14]Our consideration of the dynamics of a test particle by means of an effective potential can actually be generalized to the case of the motion of a binary system of comparable masses by using the 'effective one-body' approach to two-body systems (see Buonanno and Damour, 1999; Damour and Nagar, 2011).

By differentiating (7.3) with respect to s, we obtain the following second-order differential equation for r

$$\ddot{r} + \frac{dV}{dr} = 0. \tag{V.7.4}$$

A circular orbit $r = r_0$ is one for which $\dot{r} = 0$; hence $\ddot{r} = 0$ and $(dV/dr)(r_0) = 0$. That is, for such an orbit, the potential V has a critical point. Computation gives

$$\frac{dV}{dr} \equiv r^{-4} \left(mr^2 - \ell^2 r + 3m\ell^2 \right). \tag{V.7.5}$$

The critical points are therefore given by

$$R_{\pm} = \frac{\ell^2 \pm \sqrt{\ell^4 - 12\ell^2 m^2}}{2mr^2}. \tag{V.7.6}$$

For $\ell^2 < 12m^2$, there is no circular orbit. For $\ell^2 > 12m^2$, there are two possible circular orbits: $r_0 = R_+$ and $r_0 = R_-$. The circular orbit $r = r_0$ is stable if the critical value r_0 is a minimum of V; it is unstable if r_0 is a maximum. Indeed, linearization around r_0 of dV/dr shows that the equation (7.4) leads to oscillations of r around r_0 if $(d^2V/dr^2)(r_0) > 0$ and exponential growth of r if $(d^2V/dr^2)(r_0) < 0$. Elementary calculus shows that R_- is a maximum of V and hence the orbit $r_0 = R_-$ is unstable, while R_+ is a minimum and hence $r_0 = R_+$ is stable.

We see from (7.6) that, given m, the smallest possible value of R_+ is $6m$, while the smallest possible value of R_- (obtained when ℓ tends to infinity) is $3m$. We have proved the following theorem:

Theorem V.7.1 *In a Schwarzschild spacetime of mass m, there is no circular orbit with angular momentum ℓ less than $m\sqrt{12}$. The Schwarzschild coordinates r of circular orbits all satisfy $r > 3m$. The last stable circular orbit has $r \geq 6m$.*

V.8   Deflection of light rays

V.8.1   Theoretical prediction

Light rays are null geodesics, so they are bent by the curvature of the spacetime. The differential equations they satisfy are the same as the equations for timelike geodesics, except that the derivative denoted by a dot is now a derivative with respect to the affine parameter on the null geodesic and (7.1) is replaced by

$$g_{\alpha\beta}\dot{x}^{\alpha}\dot{x}^{\beta} = 0. \tag{V.8.1}$$

That is, using (6.4) and (6.6), we have

$$\frac{1}{2}\dot{r}^2 + \frac{1}{2}\left(1 - \frac{2m}{r}\right)\frac{\ell^2}{r^2} = \frac{1}{2}E^2. \tag{V.8.2}$$

We deduce from this equation and from (6.4) that on a null geodesic,

$$\frac{1}{r^4}\left(\frac{dr}{d\varphi}\right)^2 + \frac{1}{r^2} - \frac{2m}{r^3} = k^2, \tag{V.8.3}$$

with $k = E^2 \ell^{-2}$ a constant. Setting $u \equiv r^{-1}$ gives

$$\left(\frac{du}{d\varphi}\right)^2 + u^2 - 2mu^3 = k^2. \tag{V.8.4}$$

Hence, after differentiation, we obtain

$$\frac{d^2 u}{d\varphi^2} + u = 3mu^2. \tag{V.8.5}$$

The term $3mu^2$ in (8.5) is the Einsteinian correction to the following equation:

$$\frac{d^2 u}{d\varphi^2} + u = 0, \text{ with } u \equiv \frac{1}{r}. \tag{V.8.6}$$

This gives, in the absence of gravitation, straight lines as light rays, with equations in spherical coordinates taking the form

$$\frac{1}{r} = u_{\text{sphr}}(\varphi) \equiv \frac{1}{r_0}\cos(\varphi - \varphi_0), \tag{V.8.7}$$

where r_0 is the displacement from the centre. If we consider the Einstein correction $3mu^2$ as small, we obtain an approximation to Einstein's light rays by solving the differential equation

$$\frac{d^2 u}{d\varphi^2} + u = 3mu_{\text{sphr}}^2. \tag{V.8.8}$$

Setting $\varphi_0 = 0$ for simplicity, we find that the general solution to (8.8) is

$$\frac{1}{r} = \frac{1}{r_0}\cos\varphi + \frac{m}{r_0^2}(1 + \sin^2\varphi). \tag{V.8.9}$$

On the straight line given by (8.7), the coordinate r tends to infinity when ϕ tends to $-\pi/2$ or $\pi/2$. On the curve given by (8.9), r tends to infinity when ϕ tends to $\pm(\pi/2 + \alpha)$ with, by (8.9) and elementary trigonometry, α satisfying

$$-\sin\alpha + \frac{m}{r_0}(1 + \cos^2\alpha) = 0. \tag{V.8.10}$$

That is, if α is small,

$$\alpha \cong \frac{2}{r_0}. \tag{V.8.11}$$

The total deflection of a light ray is therefore estimated to be

$$\delta \cong \frac{4m}{r_0}. \tag{V.8.12}$$

For light rays grazing the solar surface, this angle is about $1.75''$.

V.8.2   Fermat's principle and light travel parameter time

We have said that light rays follow null geodesics of spacetime. The following property is a generalization to static spacetimes of a classical theorem of Fermat in Newton's space E^3.

Theorem V.8.1   *In a static spacetime with metric*

$$ds^2 = -g_{00}\, dt^2 + g_{ij}dx^i dx^j,$$

the projections on space of light rays are geodesics of the Riemannian metric

$$d\sigma^2 = \frac{g_{ij}dx^i dx^j}{g_{00}};$$

i.e. they are relative minima of the integral

$$\ell = \int_{t_1}^{t_2} \sqrt{\frac{g_{ij}dx^i dx^j}{g_{00}}}.$$

On a null curve, ℓ is the parameter-time duration.

Proof.   This is a straightforward computation, using the geodesic equations in ds and $d\sigma$. We leave it as an exercise.   ∎

In addition, we note that Fermat's principle can be generalized to arbitrary spacetimes as follows.[15]

Theorem V.8.2   *The 'proper time of arrival' of null curves to a given timelike curve admits a critical point (is a relative minimum) at a light ray.*

[15] Ferrarese (2004).

V.8.3   Results of observation

It was first necessary to wait for an eclipse to be able to observe the apparent displacement of the stars due to the deflection by the Sun of light coming from them. As early as 1919, an expedition was organized by Eddington to measure the bending of light by the Sun. The observed deflection was in reasonable agreement with the prediction but was not very precise, until data from the Hipparcos satellite verified the deflection of light to a 10^{-3} precision. More precise results have long since been obtained with the use of radio waves: it was not necessary in this case to wait for an eclipse. Measurements using several radio telescopes at intercontinental separations (very long baseline interferometry, VLBI) now give a precision of about 10^{-4}. They strongly support General Relativity, in particular in constrast to the Jordan–Brans–Dicke scalar–tensor theory.

V.9 Redshift and time dilation

The time dilation and redshift effects are not due to a change of the velocity of light in a gravitational field—in General as well as in Special Relativity, the speed of light is a universal constant. Rather, the effects are caused by dependence on the observer of the measure of the proper time (see Chapter III).

V.9.1 Redshift

We have already considered the gravitational redshift in Chapter III, in any static spacetime. Its measurement in a Schwarzschild spacetime is one of the classical tests of Einstein's equations.

Suppose for simplicity that the emitting atom and the observer are both at rest with the same angular coordinates, and with radial Schwarzschild coordinates r_A and r_O, respectively. The emitted period T_A and the observed period T_O are then linked by the relation (see (III.8.5))

$$T_O = \sqrt{\left(1 - \frac{2m}{r_O}\right)} \sqrt{\left(1 - \frac{2m}{r_A}\right)^{-1}} \, T_A. \qquad \text{(V.9.1)}$$

Hence $T_O > T_A$ if $r_O > r_A$, and so a redshift (smaller frequency ν) of spectral lines is then observed. If m/r_0 is small, then

$$\frac{T_A}{T_O} \equiv \frac{\nu_O}{\nu_A} \cong 1 - m\left(\frac{1}{r_A} - \frac{1}{r_O}\right). \qquad \text{(V.9.2)}$$

This formula shows that the time between signals emitted regularly in its own time by a source located at a point with parameter r_O tending to the horizon $2m$ seem progressively longer to a stationary faraway observer; it tends to infinity as the source tends to the horizon $r_O = 2m$. This effect, already noticed by Oppenheimer and Snyder, is called the **infinite-redshift effect**.

V.9.2 Time dilation

To state the elements of the theoretical prediction of time dilation, we treat the case of a rocket sent from the Earth in a radial direction with a velocity less than the escape velocity and moving then in free fall, i.e. following a timelike geodesic of the Schwarzschild metric representing the Earth's gravitational field. The parameter time t_M it takes the rocket to fall back to Earth is twice the parameter time it takes to attain its maximum r_M value (see Section V.6.3). We calculate the proper time using the formula

$$s_A = \int_0^{t_M} \frac{ds}{dt}\, dt = \int_0^{t_M} E^{-1} \left(1 - \frac{2m}{r}\right) dt$$

$$= \int_0^{t_M} \frac{1 - \dfrac{2m}{r}}{1 - \dfrac{2m}{r_0}} \frac{ds}{dt}(0)\, dt, \qquad (V.9.3)$$

with

$$\left[\left(\frac{ds}{dt}\right)(0)\right]^2 = \left(1 - \frac{2m}{r_0}\right) - \left(1 - \frac{2m}{r_0}\right)^{-1} \left[\frac{dr}{dt}(0)\right]^2. \qquad (V.9.4)$$

We set $(dr/dt)(0) = v$ and we obtain for s_A, when m/r_0 is small, the approximate expression

$$s_A \cong \int_0^{t_M} \left(1 - \frac{2m}{r} + \frac{m}{r_0} + \frac{mv^2}{r_0}\right) dt. \qquad (V.9.5)$$

The proper time observed on Earth at the fixed r_0 between the origin and the impact point is

$$s_O = \int_0^{t_M} \sqrt{1 - \frac{2m}{r_0}}\, dt \cong \int_0^{t_M} \left(1 - \frac{m}{r_0}\right) dt. \qquad (V.9.6)$$

Therefore, a standard clock carried by the rocket shows a delay over the same standard clock of the observer, given approximately by

$$s_A - s_O \cong \int_0^{t_M} \left(-\frac{2m}{r} + 2\frac{m}{r_0} + \frac{mv^2}{r_0}\right) dt > 0, \qquad (V.9.7)$$

since $r > r_0$ on the trajectory. Experiments made with caesium clocks have confirmed this time delay and its estimates with great precision.

Remark V.9.1 *On a Lorentzian manifold, timelike geodesics are, as in Minkowski spacetime, local maxima of the length of timelike curves joining two points. In the given example, it is the rocket that follows a geodesic, while in the twin paradox (see Chapter II), it is the travelling twin who is not in free fall—he has to use an engine to come back.*

V.10 Spherically symmetric interior solutions

A relativistic model for a spherically symmetric isolated star is a spherically symmetric spacetime $(V \equiv R^3 \times R, g)$ such that g satisfies the Einstein equations on V,

$$S_{\alpha\beta} \equiv R_{\alpha\beta} - \frac{1}{2} g_{\alpha\beta} R = T_{\alpha\beta},$$

where $T_{\alpha\beta}$, the source stress–energy tensor, is zero outside the star and depends inside it on the star's physical constitution—this is not well known and is difficult to model by a single formula. For this reason, there is no physically reliable exact solution for the considered problem. However, important qualitative features can be obtained from general considerations.

Remark V.10.1   *As in Newtonian theory, a spherically symmetric Einsteinian gravitational field vanishes in the interior of a hollow sphere—that is to say, namely a spherically symmetric vacuum solution defined for $r < a$ is necessarily flat in this domain. Indeed, the construction of Section V.3 shows that such a solution is a Schwarzschild metric (3.1), with m some arbitrary number. The only solution of this form that is continuous at $r = 0$ is the flat solution with $m = 0$.*

V.10.1   Static solutions. Upper limit on mass

We look for equilibrium configurations, i.e. static spacetimes. We use Schwarzschild coordinates. For $r > r_{\text{star}}$, with r_{star} the radial Schwarzschild coordinate of the star's boundary, we have vacuum; hence the solution is Schwarzschild. For $r < r_{\text{star}}$, we again look for a static spherically symmetric metric, of the form

$$g = -e^{\nu} dt^2 + e^{\lambda} dr^2 + r^2(d\theta^2 + \sin^2\theta \, d\varphi^2), \tag{V.10.1}$$

but now we solve the equations with a non-vanishing stress–energy tensor, which we take, in the absence of better modelling, to represent a perfect fluid; that is, the Einstein equations are

$$S_{\alpha\beta} = T_{\alpha\beta}, \quad \text{with} \ \ T_{\alpha\beta} \equiv (\mu + p)u_\alpha u_\beta + p g_{\alpha\beta}; \tag{V.10.2}$$

equivalently,

$$R_{\alpha\beta} = \rho_{\alpha\beta} \equiv (\mu + p)u_\alpha u_\beta + \frac{1}{2}(\mu - p)g_{\alpha\beta}. \tag{V.10.3}$$

For a static solution, the fluid unit velocity is $u^i = 0$, $u^0 = e^{-\nu/2}$, tangent to the timelines. Then $T_{0i} = R_{0i} = 0$,

$$T_{00} \equiv \mu e^{\nu}, \quad R_{00} = e^{\nu}\frac{1}{2}(\mu + 3p), \quad T_{ij} \equiv p g_{ij}, \quad R_{ij} = \frac{1}{2}(\mu - p)g_{ij}. \tag{V.10.4}$$

The scalar functions μ and p, like λ and ν, depend only on r.

The identities found in Section V.3 give (a prime denotes differentiation with respect to r)

$$R_{22} \equiv \frac{1}{\sin^2\theta}R_{33} \equiv -e^{-\lambda}\left[1 + \frac{r}{2}(\nu' - \lambda')\right] + 1,$$

$$r(e^{\lambda-\nu}R_{00} + R_{11}) \equiv \nu' + \lambda',$$

and for the spacetime scalar curvature

$$R \equiv -e^{-\nu}R_{00} + e^{-\lambda}R_{11} + r^{-2}(R_{22} + \sin^{-2}\theta \, R_{33}). \tag{V.10.5}$$

$$\equiv -e^{-\lambda}\left(\nu'' + \frac{\nu'^2}{2} - \frac{\nu'\lambda'}{2} + \frac{\nu'-\lambda'}{r} + \frac{2}{r^2}\right) + \frac{2}{r^2}. \qquad \text{(V.10.6)}$$

Elementary computations give

$$S_{00} \equiv R_{00} + \frac{1}{2}e^{\nu}R \equiv \frac{1}{2}[R_{00} + e^{\nu-\lambda}R_{11} + e^{\nu}r^{-2}[R_{22} + (\sin\theta)^{-2}R_{33}]\}.$$

The equation $S_{00} = T_{00} \equiv \mu e^{\nu}$ is then found to reduce[16] to a differential equation for λ that can be written as

$$1 - \frac{d}{dr}(e^{-\lambda}r) = r^2\mu, \quad \text{hence} \quad e^{-\lambda} = \frac{1}{r}\left\{\int_0^r \left[1 - \rho^2\mu(\rho)\right]d\rho + \text{constant}\right\}. \qquad \text{(V.10.7)}$$

Smoothness of the metric in a neighbourhood of the origin $r = 0$ of polar coordinates imposes the vanishing of the integration constant; hence

$$e^{-\lambda} = 1 - \frac{2M(r)}{r}, \quad \text{with} \quad M(r) \equiv \frac{1}{2}\int_0^r \rho^2\mu(\rho)\,d\rho. \qquad \text{(V.10.8)}$$

This solution defines a Lorentz metric in a domain $r \leq a$ only if

$$2M(a) \equiv \int_0^a \rho^2\mu(\rho)\,d\rho \leq a. \qquad \text{(V.10.9)}$$

Having computed λ, we use the equation

$$\nu' + \lambda' \equiv r(e^{\lambda-\nu}R_{00} + R_{11}) \equiv r(e^{\lambda-\nu}\rho_{00} + \rho_{11}) \equiv re^{\lambda}(\mu + p) \qquad \text{(V.10.10)}$$

to find that

$$\nu' = (e^{\lambda} - 1)r^{-1} + re^{\lambda}(p - \mu) = \frac{2M(r) + r^3(p-\mu)}{r[r - 2M(r)]}. \qquad \text{(V.10.11)}$$

On the other hand, the fluid equation (cf. Chapter IV) gives

$$(\mu + p)u^\alpha\nabla_\alpha u_1 + \partial_1 p \equiv (\mu + p)u^0 u_0\Gamma_{01}^1 + \partial_1 p = 0;$$

that is,

$$p' = -\frac{1}{2}(p + \mu)\nu',$$

and we find that

$$p' = -\frac{1}{2}(p + \mu)\frac{2M(r) + r^3(p-\mu)}{r[r - 2M(r)]}. \qquad \text{(V.10.12)}$$

This is known as the **Tolman–Oppenheimer–Volkov** equation of relativistic hydrostatic equilibrium. It can be integrated if one assumes some equation of state inside the star. In most situations, it is quite difficult to know with any reliability an equation of state, because of the complexity of the phenomena that occur. However, one can deduce some conclusions from this formula.

When the density μ is a constant, μ_0, the integral (10.8) gives

$$2M(r) = \mu_0\int_0^r \rho^2\,d\rho = \frac{1}{3}\mu_0 r^3. \qquad \text{(V.10.13)}$$

Equation (10.12) then reduces to a differential equation for p that can be integrated exactly. Assuming that p vanishes at the boundary $r = a$ of the star, one finds for the pressure at its centre

$$p(0) = \mu_0 \frac{1 - \left[1 - 2a^{-1}M(a)\right]^{\frac{1}{2}}}{3\left[1 - 2a^{-1}M(a)\right]^{\frac{1}{2}} - 1}. \qquad \text{(V.10.14)}$$

This pressure becomes infinite if

$$3\left[1 - 2a^{-1}M(a)\right]^{\frac{1}{2}} = 1, \quad \text{i.e.} \quad M(a) = \frac{4a}{9}. \qquad \text{(V.10.15)}$$

The pressure $p(0)$ becomes negative if $M(a) > 4a/9$.

Exercise V.10.1 *Prove these results.*

The classical conclusion is that stars with $M(a) \geq 4a/9$ cannot exist. The same result holds[17] if one supposes only that μ is a non-increasing function of r.

Theorem V.10.1 *There exist no equilibrium configuration of spherically symmetric stars filled with a perfect fluid with μ a non-increasing function of r and such that*

$$M(a) \geq \frac{4a}{9}, \qquad \text{(V.10.16)}$$

where a is the standard radius of the star and $M(a)$ is given by

$$M(a) = \frac{1}{2} \int_0^a \mu(\rho)\rho^2 \, d\rho.$$

V.10.2   Matching with an exterior solution

An Einsteinian model for the exterior and the interior of a spherically symmetric star is a manifold $R^3 \times R$ with a Lorentzian metric g that satisfies the Einstein equations on the whole manifold, induces an interior Schwarzschild metric in $B \times R$, with B a ball of R^3 filled with matter, and an exterior, vacuum, Schwarzschild metric in the complementary domain. The metric, spherically symmetric, reads on $R^3 \times R$

$$g = -e^\nu \, dt^2 + e^\lambda \, dr^2 + r^2(d\theta^2 + \sin^2\theta \, d\varphi^2),$$

with e^λ and e^ν equal to the coefficients of the interior metric g_{star} for $r < r_{\text{star}}$, denoted by a above, and to those of a vacuum Schwarzschild metric in the exterior. For such a metric to be a solution on $R^3 \times R$, some continuity[18] properties are required from the functions λ and ν at the boundary $r = a$ between the interior and exterior regions. Equation (10.8) in particular imposes that

$$e^{-\lambda_{\text{ext}}} = 1 - \frac{2m}{r} = e^{-\lambda_{\text{int}}} = 1 - \frac{2M(r)}{r} \quad \text{for} \quad r = a;$$

[17]The integration was first performed by Schwarzschild in 1916. For details, and a discussion of the case of a non-constant density, see Wald (1984), p. 129.

[18]If these quantities were not continuous, their differentiation would introduce a measure with support $r = r_{\text{star}}$ in the components of the Ricci tensor that are not present in the source.

that is,

$$m = M(a) \equiv \frac{1}{2} \int_0^a r^2 \mu(r) \, dr. \qquad (\text{V}.10.17)$$

The number $M(a)$, sometimes called the mass function, can be computed in terms of the density in classical units by reinserting in the above formula the Einstein gravitational constant $G_E = 8\pi G_N$. This leads to the integral of the density on the volume of a sphere B_a of radius a, namely

$$M^{\text{cgs}}(a) = 4\pi G_N^{\text{cgs}} \int_0^{a_{\text{cgs}}} r^2 \mu^{\text{cgs}}(r) \, dr.$$

Note, however that $M(a)$ is not the proper mass of a static spherical star of standard radius a, which should be computed with the space proper volume element

$$r^2 e^{\frac{1}{2}\lambda} dr \sin\theta \, d\theta \, d\phi,$$

resulting in a smaller quantity; the difference represents the gravitational binding energy of the star in equilibrium.

V.10.3   Non-static interior solutions

Birkhoff's theorem does not apply to interior solutions. There exist time-dependent, spherically symmetric solutions of the Einstein equations with sources in a domain $r < r_{\text{star}}$; the radius r_{star} of the star may be a function of t. The full solution is the considered interior solution g_{int} for $r \leq r_{\text{star}}$ and a Schwarzschild exterior solution with mass m for $r \geq r_{\text{star}}$, provided that $r_{\text{star}} > 2m$, the Schwarzschild radius of the star, which is constant for an exterior solution and is linked with the energy content of the star, which is considered as an isolated object. When the star contracts so much that r_{star} becomes smaller than $2m$, it is no longer observable from the region $r > 2m$ of the spacetime (see Chapter VI).

It is physically clear that there can be no static interior solution with zero pressure, because there is nothing in that case to resist the gravitational attraction. This can be checked mathematically by solving (10.12) with $p = p' = 0$, which gives

$$\mu(r) = 2r^{-3} M(r) \equiv 2r^{-3} \int_0^r \rho^2 \mu(\rho) \, d\rho, \qquad (\text{V}.10.18)$$

that is, setting $y(r) = \int_0^r \rho^2 \mu(\rho) \, d\rho$, the differential equation

$$y' = \frac{2}{r} y, \quad \text{hence} \quad y = C e^{r^2},$$

with C a constant. Then

$$\mu = C r^{-2} e^{r^2}, \qquad (\text{V}.10.19)$$

which is always infinite for $r = 0$.

V.11 Spherically symmetric gravitational collapse

It is believed that when a star has exhausted all thermonuclear sources of energy, it will collapse under its own gravitational field. Oppenheimer and Snyder[19] made a rigorous study of this process for a spherically symmetric dust cloud by using a special solution given by Tolman[20] in the restricted case of a spatially constant matter density starting from rest. This study was revisited by Gu[21] with the restriction to spatially constant matter lifted. Gu analysed shell-crossing singularities as well as the essential singularity at the centre of symmetry. He gave a detailed mathematical discussion of the formation of horizons and of what is now called the nakedness of the singularities. Hu[22] constructed the fully general spherically symmetric solutions for dust, not restricted to the Tolman class, and discussed their properties. The work done by Gu and Hu for non-homogeneous dust clouds was repeated, in ignorance of these papers published in China, by Müller zum Hagen, Seifert, and Yodzis[23], who proved the existence of shell-crossing singularities in the case of the Tolman class; later, Christodoulou proved the existence of central singularities,[24] starting from rest. Finally, Newman[25] obtained the general solution. These western papers were motivated by the **cosmic censorship conjecture**, which is violated both at shell-crossing singularities and at the centre, as was already shown in Gu's paper.

[19]Oppenheimer and Snyder (1939).

[20]Tolman (1934) and, Bondi (1947).

[21]Gu (1973).

[22]Hu (1974).

[23]Müller zum Hagen, Seifert, and Yodzis (1973).

[24]Christodoulou (1984).

[25]Newman (1986).

V.11.1 Tolman, Gu, Hu, and Claudel–Newman metrics

In the case of dust, the flow lines are the timelike geodesic trajectories of the particles; we take these geodesics as timelines (comoving coordinates). A spherically symmetric metric can then be written as

$$-dt^2 + e^{2\omega} dr^2 + R^2(d\theta^2 + \sin^2\theta \, d\phi^2), \qquad (V.11.1)$$

where ω and R are functions of r and t. The flow lines of the dust are the curves on which the coordinates r, θ, ϕ are constant; a so-called dust shell is labelled by its coordinate r. The metric is regular as long as $e^{2\omega}$ and R are smooth positive functions. R^2 is (up to multiplication by 4π) the area of the spherical dust shell of parameter r, at time parameter t. One fixes the parameter r by choosing it to be such that $r = R(r, 0)$.

The stress–energy tensor T of the dust reduces to

$$T_{\alpha\beta} = \mu u_\alpha u_\beta, \qquad (V.11.2)$$

with $u^\alpha = \delta_0^\alpha$, since the timelines are the matter flow lines. The only non-zero component of the stress energy tensor is

$$T_{00} = \mu, \qquad (V.11.3)$$

with

$$\mu = \mu(r, t) \text{ for } r \le a \text{ and } \mu(r, t) = 0 \text{ for } r > a, \qquad (V.11.4)$$

where a is the r parameter of the outermost dust shell[26] at time $t = 0$.

We denote by primes and dots respectively differentiation with respect to r and t. We compute the general solution in the chosen notation. The equation $R_{10} = 0$ is equivalent to the second-order equation

$$\dot{R}' - \dot{\omega} R' = 0, \qquad (V.11.5)$$

which integrates to the first-order equation

$$R' e^{-\omega} = f(r), \qquad (V.11.6)$$

with $f(r)$ an arbitrary function of r.

The conservation equations reduce to

$$\nabla_\alpha T^{\alpha 0} \equiv \dot{\mu} + \Gamma^\alpha_{\alpha 0} \mu \equiv \dot{\mu} + \mu(\dot{\omega} + 2R^{-1}\dot{R}) = 0; \qquad (V.11.7)$$

hence, with ϕ an arbitrary function of r, we obtain

$$\mu(t, r) = \frac{e^{-\omega}}{R^2} \phi(r). \qquad (V.11.8)$$

Using (12.6), this equation becomes

$$\mu(t, r) = \frac{r^2 \mu_0}{R^2 R'}. \qquad (V.11.9)$$

Here μ_0 is an arbitrary function of r that we identify with the initial density because we have chosen r such that $R(0, r) = r$ and hence $R'(0, r) = 1$. It has been shown by R. Newman that the Einstein equations then admit the first integral

$$\frac{1}{2} \dot{R}^2 - \frac{M(r)}{R} = \frac{1}{2}[f^2(r) - 1], \qquad (V.11.10)$$

with

$$M(r) = \int_0^r f(\rho) \mu(t, \rho) R^2(t, \rho) e^\omega \, d\rho. \qquad (V.11.11)$$

The function $M(r)$ is the integral over the volume occupied by the dust shells $0 \leq \rho \leq r$ of the density weighted by the factor $f(\rho)$.[27] It is independent of t, as can be verified by using (12.6) and (12.9), which show that

$$M(r) = \int_0^r \mu_0(\rho) \rho^2 \, d\rho. \qquad (V.11.12)$$

The explicit formulas for $R(r, t)$ and $\omega(r, t)$ depend on the sign of $f^2(r) - 1$; they have been computed and studied in the general case by Hu and Newman, to whom we refer the reader.[28] The case where the dust cloud starts from rest, i.e.

$$\dot{R}(0, r) = 2\frac{M(r)}{r} + f^2(r) - 1 = 0, \qquad (V.11.13)$$

has been studied by Christodoulou, using parametric equations for R and t.

Here, following Gu[29] we study the case $f^2(r) = 1$, in which exact integration permits a clear and comparatively short discussion. In the case $f^2(r) = 1$, (11.10) reads

$$R^{\frac{1}{2}}\dot{R} = \pm\sqrt{2M(r)}. \qquad (V.11.14)$$

We suppose that the star starts contracting; then the minus sign must be chosen when t increases, and the equation integrates to

$$R(r,t)^{\frac{3}{2}} = \Phi(r) - \frac{3}{2}\sqrt{2M(r)}\,t,$$

with Φ an arbitrary function. Since we have normalized the radial parameter r by the condition $R(r,0) = r$, it holds that $\Phi(r) = r^{\frac{3}{2}}$. We use the notation

$$\frac{3}{2}\sqrt{2M(r)} =: h^{\frac{1}{2}}(r), \qquad (V.11.15)$$

and then

$$R(r,t) = [r^{\frac{3}{2}} - h^{\frac{1}{2}}(r)t]^{\frac{2}{3}}. \qquad (V.11.16)$$

Using (11.6), we then find

$$e^{\omega} = R' = [r^{\frac{3}{2}} - h^{\frac{1}{2}}(r)t]^{-\frac{1}{3}}\left[r^{\frac{1}{2}} - \frac{1}{3}h^{-\frac{1}{2}}h'(r)t\right].$$

Here, as a consequence of the definition (11.15), we have

$$h'(r) = \frac{9}{2}r^2\mu_0(r).$$

On the other hand, the equation for μ gives

$$\mu(t,r) = \frac{r^2\mu_0}{[r^{\frac{3}{2}} - h^{\frac{1}{2}}(r)t][r^{\frac{1}{2}} - \frac{3}{2}h^{-\frac{1}{2}}h'(r)t]}. \qquad (V.11.17)$$

We make a change of coordinates: instead of choosing as radial parameter the number characterizing a dust shell, we take the area R of a dust shell at time t. Since we have chosen $f(r) = 1$, i.e. $R' = e^{\omega}$, it holds that

$$dR = e^{\omega}\,dr + \dot{R}\,dt = e^{\omega}\,dr - R^{-\frac{1}{2}}\sqrt{2M(r)}\,dt. \qquad (V.11.18)$$

The change of coordinates from r, t to R, t is admissible if R' does not vanish. In the new coordinates, the spacetime metric becomes, for $0 \leq r \leq a$,

$$-\left[1 - \frac{\sqrt{2M(r)}}{R}\right]dt^2 + 2R^{-\frac{1}{2}}\sqrt{2M(r)}\,dR\,dt + dR^2 + R^2(d\theta^2 + \sin^2\theta\,d\phi^2).$$

$$(V.11.19)$$

We see that the metric of the space sections $t = $ constant reduces to the Euclidean metric in polar coordinates. If the space manifold is, for every t, homeomorphic to a ball of three-dimensional Euclidean space and R is an admissible polar coordinate, then $R(t,r)$ is also the distance at time t between the centre and the dust shell with parameter r.

[29]Gu (1973).

V.11.2   Monotonically decreasing density

Collapse of dust shells

In this subsection, we suppose that μ_0 is a monotonically decreasing function of the parameter r, as r increases from 0, the centre of the star, to a, the value of r at the surface boundary of the star at time $t = 0$.

To study the possible collapse, we study the evolution in proper time t of the function $R(t, r)$ for a given dust shell, i.e. for a given value of r. We see from (11.16) that the shell collapses at the centre of symmetry at the time $t_1(r)$ where $R(t_1, r) = 0$; that is,

$$t_1(r) = h^{-\frac{1}{2}}(r) r^{\frac{3}{2}}.$$

It follows that

$$\frac{dt_1(r)}{dr} = \frac{1}{2} r^{\frac{1}{2}} h^{-\frac{3}{2}} \left[3h(r) - rh'(r) \right] = \frac{9}{4} r^{\frac{1}{2}} h^{-\frac{3}{2}} \left[3M(r) - rM'(r) \right]. \tag{V.11.20}$$

If μ_0 is monotonically decreasing, then, using the definition (11.11) of $M(r)$, we find that

$$3M(r) - rM'(r) \equiv 3 \int_0^r \mu_0(\rho)\rho^2 \, d\rho - \mu_0(r)r^3 \geq 0. \tag{V.11.21}$$

Hence the shells with increasing parameter r arrive successively at the centre, and there is no shell crossings.[30] If the density is uniform (the Oppenheimer–Snyder case), then the dust shells all arrive at the same time at the centre.

The metric (11.19) is a regular Lorentzian metric if the linear form $dR - \dot{R}\, dt$ does not vanish, i.e. if $R' = e^\omega > 0$. We deduce from the expression (11.16) for R that

$$\frac{3}{2} R^{\frac{3}{2}} R' = \left[\frac{3}{2} r^{\frac{1}{2}} - \frac{1}{2} h^{-\frac{1}{2}}(r) h'(r) t \right] = 0. \tag{V.11.22}$$

Hence $R'(t, r)$ vanishes[31] at a time t_2 given by

$$t_2 = \frac{3h^{\frac{1}{2}}(r) r^{\frac{1}{2}}}{h'(r)} = \frac{2h^{\frac{1}{2}}(r)}{3r^{\frac{3}{2}}\mu_0(r)} \equiv \frac{\sqrt{2M(r)}}{r^{\frac{3}{2}}\mu_0(r)}. \tag{V.11.23}$$

We note that

$$\frac{t_1}{t_2} = \frac{h'(r)r}{3h(r)}. \tag{V.11.24}$$

The monotonicity of μ_0 and the mean value theorem led Gu to the conclusion that

$$t_1 \leq t_2; \tag{V.11.25}$$

that is, the dust evolution does not induce a singularity in the metric before a dust shell arrives at the centre, in agreement with the previous

[30] Shell crossing, leading to non-central singularities, exists when the density is not monotonically decreasing. See Gu (1973), Hu (1974), Müller zum Hagen et al. (1984), and Newman (1986).

[31] Vanishing of R' signals a shell crossing.

conclusion of the absence of shell crossing. The first occurence of the singularity is at a time t_0, with

$$t_0 = \lim_{r=0} \left[\frac{r^3}{h(r)} \right]^{\frac{1}{2}} = \frac{1}{\sqrt{\frac{3}{2}\mu_0(0)}} ; \qquad (\text{V.11.26})$$

the greater $\mu_0(0)$ is, the sooner the singularity appears.

Matching with an exterior metric

The interior metric is given by (11.19) for $r \leq a$, i.e.

$$0 \leq R \leq \left[a^{\frac{3}{2}} - h^{\frac{1}{2}}(a)t \right]^{\frac{2}{3}} . \qquad (\text{V.11.27})$$

The computations that we have made are still valid outside the star, where $r > a$ and $\mu_0(r) = 0$, but for $r > a$ it holds that

$$M(r) \equiv M_a := \int_0^a \mu_0(\rho)\rho^2 \, d\rho, \quad r > a, \qquad (\text{V.11.28})$$

and the exterior metric reads

$$-\left(1 - \frac{2M_a}{R} \right) dt^2 + 2\sqrt{\frac{2M_a}{R}} \, dR \, dt + dR^2 + R^2(d\theta^2 + \sin^2\theta \, d\phi^2). \qquad (\text{V.11.29})$$

It is a boosted Schwarzschild metric with horizon

$$R = 2M_a . \qquad (\text{V.11.30})$$

This solution takes the usual Schwarzschild form

$$-\left(1 - \frac{2M_a}{R} \right) dt^2 + \left(1 - \frac{2M_a}{R} \right)^{-1} dR^2 + R^2(d\theta^2 + \sin^2\theta \, d\phi^2) \qquad (\text{V.11.31})$$

if we make the change of time coordinate

$$d\tau = dt - \left(1 - \frac{2M_a}{R} \right)^{-1} \sqrt{\frac{2M_a}{R}} \, dR. \qquad (\text{V.11.32})$$

The full spacetime metric g is defined on $R^3 \times R^+$, with R, θ, ϕ polar coordinates on R^3 and $t \in R^+$ (i.e. $t \geq 0$) by

$$g = g_{\text{dust}}, \quad 0 \leq R \leq \left[a^{\frac{3}{2}} - h^{\frac{1}{2}}(a)t \right]^{\frac{2}{3}}, \qquad (\text{V.11.33})$$

$$g = g_{\text{ext}}, \quad R \geq \left[a^{\frac{3}{2}} - h^{\frac{1}{2}}(a)t \right]^{\frac{2}{3}} . \qquad (\text{V.11.34})$$

The interior dust solution is hidden behind the horizon when

$$\left[a^{\frac{3}{2}} - h^{\frac{1}{2}}(a)t \right]^{\frac{2}{3}} < 2M_a . \qquad (\text{V.11.35})$$

This begins at a positive time t_3 given by

$$t_3 = h^{-\frac{1}{2}}(a)[a^{\frac{3}{2}} - (2M_a)^{\frac{3}{2}}] \qquad (\text{V.11.36})$$

if $a > 2M_a$, that is, if the star is initially visible. We have seen that the outer shell of the star collapses into a singularity when $t_1(a) = h^{-\frac{1}{2}}(a)a^{\frac{3}{2}}$, hence after the star has ceased to be visible. This fact was an inspiration for the formulation of the **cosmic censorship conjecture** by Penrose (see Chapter VIII).

V.12   Problems

V.12.1   Relativistic and Newtonian gravitational masses

Show that the relativistic gravitational proper mass of a spherical star with radius a and density $\mu(r)$ is larger that its Newtonian gravitational mass.

Solution

The Newtonian gravitational mass is, with G_N the Newtonian gravitational constant,

$$M(a) = G_N \int_{S^2} \int_0^a r^2 \mu(r)\, dr \sin\theta\, dr\, d\theta\, d\phi \equiv 4\pi G_N \int_0^a r^2 \mu(r)\, dr.$$

The relativistic gravitational mass, computed with the proper volume element, is

$$2M_a := G_E \int_0^a r^2 \mu(r) \left[1 - \frac{2M(r)}{r}\right]^{-\frac{1}{2}} dr. \qquad \text{(V.12.1)}$$

The proper mass M_a is greater than $M(a)$ because $[1 - 2M(r)/r]^{-\frac{1}{2}} > 1$. The difference represents the gravitational binding energy of the star in equilibrium.

V.12.2   The Reissner–Nordström solution

Show that the **Reissner–Nordström** metric (found in 1916 by Reissner) and given in standard coordinates by

$$-\left(1 - \frac{2m}{r} + \frac{Q^2}{r^2}\right) dt^2 + \left(1 - \frac{2m}{r} + \frac{Q^2}{r^2}\right)^{-1} dr^2 + r^2(d\theta^2 + \sin^2\theta\, d\varphi^2)$$
$$\text{(V.12.2)}$$

is a spherically symmetric solution of the Einstein–Maxwell equations with electromagnetic potential

$$A_i = 0, \quad A_0 = -\frac{Q}{r}, \qquad \text{(V.12.3)}$$

identical with the classical electrostatic potential of a spherical body with charge Q.

Show that the metric (12.1) is smooth and Lorentzian with t a time variable, provided that

$$\frac{2m}{r} - \frac{Q^2}{r^2} < 1. \qquad (V.12.4)$$

Remark V.12.1 *For large r, the term Q/r^2 is small in comparison with m/r. Since the total charge of celestial bodies is observed to be negligible in comparison with their mass, the Reissner–Nordström solution has little application in astrophysics. It was also abandoned as a possible model for the electron, for which $m^{-1}Q^2 = 2.8 \times 10^{-13}$ cm, and so the influence of the term $r^{-2}Q^2$ would be important only at distances where quantum effects cannot be neglected.*

The Reissner–Nordström solution has attracted interest in connection with modern aspects of mathematical General Relativity. Indeed, if $m^2 > Q^2$, it possesses two event horizons (see Chapter VI) given in standard coordinates by

$$r_\pm = m \pm \sqrt{m^2 - Q^2}. \qquad (V.12.5)$$

It would invalidate a censorship conjecture if the word 'generic' was not included in the hypotheses (see Chapter VIII).

Reissner–Nordström-type solutions have recently become important in string theories.

V.12.3   Schwarzschild spacetime in dimension $n + 1$

Construct by reasoning and computations analogous to those for $n + 1 = 4$ the $(n + 1)$-dimensional Schwarzschild metric in spherical standard coordinates, $r \in R^+$,

$$g_{\text{Schw}} = -\left(1 - \frac{2m}{r^{n-2}}\right) dt^2 + \left(1 - \frac{2m}{r^{n-2}}\right)^{-1} dr^2 + r^2\, d\omega^2, \quad (V.12.6)$$

where $d\omega^2$ is the metric of the sphere S^{n-1}. The Schwarzschild spacetime is defined as for $n = 3$ by this metric supported by the manifold $M \times R$, with M the exterior of the ball $r^{n-2} = 2m$.

V.12.4   Schwarzschild metric in isotropic coordinates, $n = 3$

Show that in isotropic coordinates the four-dimensional Schwarzschild metric reads, with $R^2 = X^2 + Y^2 + Z^2$,

$$g_{\text{Schw}} = -\left(\frac{2R - m}{2R + m}\right)^2 dt^2 + \left(1 + \frac{m}{2R}\right)^4 (dX^2 + dY^2 + dZ^2). \quad (V.12.7)$$

Solution

Recall that, in standard coordinates,

$$g_{\text{Schw}} = -\left(1 - \frac{2m}{r}\right) dt^2 + \left(1 - \frac{2m}{r}\right)^{-1} dr^2 + r^2(d\theta^2 + \sin^2\theta\, d\phi^2).$$

$$(V.12.8)$$

The relation between r and R is

$$r = R\left(1 + \frac{m}{2R}\right)^2 = \frac{4R^2 + 4Rm + m^2}{4R}, \qquad (V.12.9)$$

and hence the coefficient of dt^2 becomes

$$1 - \frac{2m}{r} = 1 - \frac{8Rm}{(2R+m)^2} = \left(\frac{2R-m}{2R+m}\right)^2.$$

To compute the space metric in isotropic coordinates, we first remark that

$$\left(1 - \frac{2m}{r}\right)^{-1} = \frac{r}{r - 2m} = \left(\frac{2R+m}{2R-m}\right)^2$$

and

$$r = R\left(1 + \frac{m}{2R}\right)^2 = R + m + \frac{m^2}{4R}, \qquad (V.12.10)$$

$$dr = dR\left(1 - \frac{m^2}{4R^2}\right).$$

We then proceed as follows. From the definitions

$$R^{-1}X = r^{-1}x, \quad R^{-1}Y = r^{-1}y, \quad Z = Rr^{-1}z, \text{ and}$$
$$R^{-2}(X^2 + Y^2 + Z^2) = r^{-2}(x^2 + y^2 + z),$$

$$(V.12.11)$$

it results that,

$$d(r^{-1}x) = r^{-1}\, dx - r^{-2}x\, dr = R^{-1}\, dX - R^{-2}X\, dR$$

with analogous relations for y and z, and hence

$$r^{-2}(dx)^2 - 2r^{-3}x\, dx\, dr + r^{-4}x^2(dr)^2 = R^{-2}(dX)^2$$
$$- 2R^{-3}X\, dX\, dR + R^{-4}X^2(dR)^2,$$

and, by summation and simplification,

$$r^{-2}(dx^2 + dy^2 + dz^2 - dr^2) = R^{-2}(dX^2 + dY^2 + dZ^2 - dR^2).$$

Using

$$r^2(d\theta^2 + \sin^2\theta\, d\phi^2) = dx^2 + dy^2 + dz^2 - dr^2$$
$$= r^2 R^{-2}(dX^2 + dY^2 + dZ^2 - dR^2)$$

shows that the space part of g_{Schw} can be written as

$$\left(\frac{2R+m}{2R-m}\right)^2 \left(1 - \frac{m^2}{4R^2}\right)^2 dR^2 + \left(1 + \frac{m}{2R}\right)^4 (dX^2 + dY^2 + dZ^2 - dR^2),$$

and, after simplification, the Schwarzschild metric is obtained in isotropic coordinates.

V.12.5   Wave coordinates for the Schwarzschild metric in dimension $n + 1$

Show that the requirement that $x^\mu = (t, x^i)$ be wave coordinates, $\Box_g x^\mu = 0$, with $x^i = \bar{r}(r)n^i$, $n^i \in S^n$, reduces to the equation

$$-\triangle^* \bar{r} - (n-1)\bar{r} = \frac{d}{dr}\left[\frac{d\bar{r}}{dr} r^{n-1}(1 - 2mr^{2-n})\right] - (n-1)\bar{r} = 0, \quad \text{(V.12.12)}$$

with $\triangle^*$ the Laplacian on the sphere S^{n-1}.

Show that setting $s = 1/r$ gives as equation singular at $s = 0$:

$$\frac{d}{ds}\left[s^{3-n}(1 - 2ms^{n-2})\frac{d\bar{r}}{ds}\right] = (n-1)s^{1-n}\bar{r}.$$

Show the asymptotic expansions

$$\bar{r} = r + \frac{m}{(n-2)r^{n-3}} + \begin{cases} \frac{m^2}{4}r^{-3}\ln r + O(r^{-5}\ln r), & n = 4, \\ O(r^{5-2n}), & n \geq 5 \end{cases}.$$

Solution

See Choquet-Bruhat, Chruściel, and Loiselet (2006).

Black holes

VI.1   Introduction

Laplace and Michell had already foreseen that light can become trapped by a massive body, so that the latter becomes black to observers and is perceptible only through its gravitational field. However, the apparent 'Schwarzschild singularity' $r = 2m$ is a phenomenon that has no analogue in classical mechanics, and the completed spacetimes are indeed very strange. It was long believed that these extensions have no physical reality, i.e. that matter cannot be compressed so that it is included in the region $r < 2m$. It was mainly through the vision of Robert Oppenheimer and John Archibald Wheeler that the reality of black holes was seriously considered. They cannot be seen directly, but astronomical observations, in particular perturbations of motions of various stars, reveal gravitational fields too strong to be explained by an invisible massive companion other than a black hole. Some X-ray sources and active galactic nuclei are interpreted in terms of black holes. Also, some γ-ray bursts are thought to be due ultimately to the explosion of matter accreted near the boundary of a black hole by the gravitational field it generates. There is now a large consensus among astrophysicists on the existence of many black holes in the universe, even at the centre of our own galaxy.

VI.2   The Schwarzschild black hole

The first model of a black hole appears in the first exact solution of the Einstein equations constructed by Schwarzschild in 1916, the spherically symmetric one that we have studied in Chapter V (Fig. VI.1). We have defined the Schwarzschild spacetime as the manifold $r > 2m$ in $R^3 \times R$, with r a polar coordinate in R^3, with Lorentzian metric given in standard coordinates by

$$-\left(\frac{2m}{t} - 1\right) dt^2 + \left(\frac{2m}{t} - 1\right) dr^2 + r^2(d\theta^2 + \sin^2 \theta \, d\phi^2), \quad \text{(VI.2.1)}$$

and we have called the submanifold $r = 2m$ of $R^3 \times R$, diffeomorphic to the product $S^2 \times R$, the event horizon.

The Schwarzschild metric in standard coordinates with $m > 0$ ceases to be a smooth Lorentzian metric for $r = 2m$; at this value of r, the coefficient g_{00} vanishes and g_{11} becomes infinite.

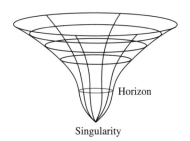

Fig. VI.1 Schwarzschild black hole.

For $0 < r < 2m$, the Schwarzschild metric in standard coordinates is again a smooth Lorentzian metric, *but t is a space coordinate while r is a time coordinate.*

Exercise VI.2.1   *Show that the Schwarzschild metric with $0 < r < 2m$ is neither spherically symmetric nor static.*

The volume element of g_{Schw} in standard coordinates, $r^2 \sin\theta \, dr \, d\theta \, d\varphi \, dt$, is smooth[1] and non-vanishing. More importantly, a straightforward computation shows that the coordinate-independent scalar $R^{\alpha\beta\gamma\delta} R_{\alpha\beta\gamma\delta}$, called the **Kretschman scalar**, is finite at $r = 2m$, in fact for all $r > 0$, being equal to $48m^2/r^6$. This property led to the belief that the Schwarzschild spacetime $(S^2 \times \{r > 2m\}) \times R$ is extendible, in the sense that it can be immersed in a larger Einsteinian spacetime, whose manifold is not covered by the Schwarzschild standard coordinates with $r > 2m$.

[1] If r, θ, ϕ are interpreted as polar coordinates on R^3, then θ, ϕ are only locally coordinates in the usual sense; the sphere S^2 is not diffeomorphic to R^2 and hence cannot be covered by a single coordinate patch.

VI.3   Eddington–Finkelstein extensions

Let us consider (following Eddington in 1924, Lemaître in 1926, and Finkelstein in 1958) the change of coordinates, defined for $r > 2m$, obtained by replacing the canonical Schwarzschild time t by the 'retarded time' v given by,[2]

$$v = t + r + 2m \log\left(\frac{r}{2m} - 1\right). \qquad \text{(VI.3.1)}$$

[2] This change of coordinates from t, r, θ, ϕ to v, r, θ, ϕ is singular for $r = 2m$.

Using this new coordinate v together with r, θ, φ, the Schwarzschild metric can be expressed as the so-called **Eddington–Finkelstein** (EF) metric

$$-\left(1 - \frac{2m}{r}\right) dv^2 + 2\, dr \, dv + r^2(\sin^2\theta \, d\varphi^2 + d\theta^2). \qquad \text{(VI.3.2)}$$

The EF metric is a smooth metric of Lorentzian signature[3] on the manifold $S^2 \times R^+ \times R$, $(\theta, \varphi) \in S^2$, $r \in R^+$, $v \in R$ defining a vacuum Einsteinian spacetime called the Eddington–Finkelstein black hole. By its construction, the Schwarzschild spacetime is isometric with the domain $r > 2m$ of the EF black hole.

[3] The vanishing of the coefficient g_{vv} at $r = 2m$ does not correspond to a singularity of the metric, because its determinant, $\det g \equiv 1$, does not vanish there.

Exercise VI.3.1   *Prove these statements.*

The submanifold $r = 2m$ is null (isotropic), since $g^{rr} = 0$ for $r = 2m$.

One family of radial (i.e. $\theta = \text{constant}$, $\varphi = \text{constant}$) light rays is represented by straight lines $v = \text{constant}$; the other family is given by

$$-\left(1 - \frac{2m}{r}\right) dv + 2\, dr = 0;$$

that is,

$$dv = -2\frac{r}{2m - r}\, dr = \left(-\frac{4m}{2m - r} + 2\right) dr,$$

which integrates for $r < 2m$ to

$$v = 2r + 4m \log(2m - r) + \text{constant}. \qquad (\text{VI.3.3})$$

Under the change of coordinates

$$t = v - r - 2m \log\left(\frac{2m}{r} - 1\right),$$

the EF metric takes, in the domain $r < 2m$, the Schwarzschild form

$$\left(\frac{2m}{r} - 1\right) dt^2 - \left(\frac{2m}{r} - 1\right)^{-1} dr^2 + r^2(d\theta^2 + \sin^2\theta \, d\phi^2),$$

but t is now spacelike, and r timelike. The metric is no longer static. It is singular for $r = 0$—in general interpreted as a limiting spacelike 3-surface. The following theorem justifies the name of black hole given to the EF spacetime: no future light ray issuing from a point where $r < 2m$ crosses the **event horizon** $r = 2m$, as can be seen from the light cones and null geodesics in Fig. VI.2. More generally, the theorem is as follows:

Theorem VI.3.1   *On a timelike line issuing from a point with $r < 2m$, in standard coordinates, the variable r is always decreasing and tends to zero in a finite proper time. Hence any observer crossing the Schwarzschild radius $r = 2m$ attains the singularity $r = 0$ in a finite proper time.*

Proof. This is conveniently done in terms of the t, r coordinates, notwithstanding the fact that t is now a space coordinate and r a time one.

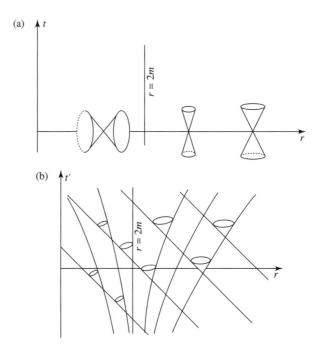

Fig. VI.2 (a) Orientation of the light cones in the standard Schwarzschild coordinates. (b) Null geodesics and light cones in the extension. Here $t' = v - r$, where v is the 'retarded' time defined in (3.1).

(1) Radial geodesics: we have found in Chapter V that on such geodesics, with s the proper time,[4]

$$\dot{r}^2 = E^2 - (1 - 2mr^{-1}), \quad \dot{r} := \frac{dr}{ds}. \qquad (VI.3.4)$$

Hence the point $r = 0$ is attained from $r = r_0 < 2m$ after the finite proper time

$$\int_0^{r_0} \frac{ds}{dr} dr = \frac{r^{\frac{1}{2}} dr}{\sqrt{2m - (1 - E^2)r}}. \qquad (VI.3.5)$$

(2) Non-radial geodesics and non-geodesic[5] motions lead to a smaller lapse of proper time between points with coordinates $r_0 < 2m$ and 0. ∎

[4]We are in the region $r < 2m$, and hence

$$E^2 - (1 - 2mr^{-1})$$
$$\equiv r^{-1} \left[2m - r(1 - E^2) \right] > 0.$$

[5]Recall that in a Lorentzian manifold causal geodesics realize a local maximum of length between two causally related points.

The amount of proper time taken for an astronaut entering the black hole to reach the singularity is indeed very short—it is of the order of the Schwarzschild radius estimated in geometric units. For a black hole with mass of the order of the solar mass, this gives a time of the order of 10^{-5} s.

Remark VI.3.1 *Because of the singularity for $r = 0$, the Finkelstein black hole cannot be considered as a generalized solution of the Einstein equations, in contrast to the $1/r$ singularity in the potential of Newtonian gravitation, which is solution of the Poisson equation with right-hand side a delta measure mass located at the origin.[6] The idea now is rather to consider that, near this singularity, the gravitational field is so strong that it becomes a quantum field or even a string field, for which the Einstein equations are no longer applicable.*

[6]Remember also that for $r < 2m$, the coordinate r is timelike. In a two-dimensional, necessarily somewhat misleading, conformal diagram, $r = 0$ is represented by a spacelike line.

VI.3.1 Eddington–Finkelstein white hole

By time reversal, one obtains manifestly another extension of the Schwarzschild spacetime. The manifold is again $S^2 \times (r > 0) \times R$ and the metric is

$$- \left(1 - \frac{2m}{r} \right) dv^2 - 2\, dr\, dv + r^2 (\sin^2 \theta\, d\varphi^2 + d\theta^2).$$

The extension to $r < 2m$ now appears to observers in the Schwarzschild spacetime as a white hole: nothing can penetrate into it, but every past inextendible light ray or timeline in the Schwarzschild spacetime emanates from this white hole, as can be seen from the radial null geodesics and light cones in Fig. VI.3.

VI.3.2 Kruskal spacetime

It is possible to embed the Schwarzschild spacetime and both of its extensions in a larger spacetime containing an additional copy of the

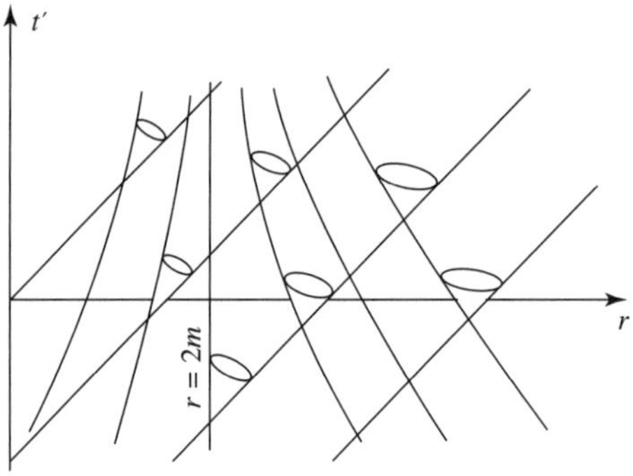

Fig. VI.3 Radial null geodesics and light cones of the Eddington–Finkelstein white hole. Here $t' = v - r$, where v is the 'retarded' time defined in (3.1).

Schwarzschild spacetime for which the previous black hole extension now plays the role of a white hole and vice versa. The support of the obtained spacetime is the manifold $S^2 \times R^2$.

The metric of the Kruskal spacetime reads, in **Kruskal coordinates**,

$$\frac{32m^3}{r}e^{-r/2m}(dz^2 - dw^2) + r^2(d\theta^2 + \sin^2\theta \, d\phi^2),$$

where θ, ϕ are coordinates on S^2 while z, w are coordinates on the open set diffeomorphic to R^2 defined by

$$z^2 - w^2 > -1,$$

and r is the function of z and w defined by

$$z^2 - w^2 = \frac{1}{2m}(r - 2m)e^{r/2m}.$$

In Kruskal coordinates, the radial light rays are represented by straight lines.

Exercise VI.3.2 *Obtain the portion of the Kruskal spacetime isometric to the Schwarzschild spacetime by the change of coordinates*

$$z = e^{v/4m} + e^{-u/4m}, \qquad w = e^{v/4m} - e^{-u/4m}, \tag{VI.3.6}$$

with

$$u = t - r - 2m\log\left(\frac{r}{2m} - 1\right). \tag{VI.3.7}$$

Obtain other portions by analogous changes of coordinates.

The Kruskal spacetime has two asymptotically flat regions (I and I′ in Fig. VI.4(a)), each of which is isometric to the Schwarzschild spacetime. A section through the Kruskal spacetime connecting these two regions, for instance $w = 0$, is called the Schwarzschild throat (Fig. VI.4(b)). No non-tachyonic[7] signal can travel from one of the asymptotically flat regions to the other.

The Kruskal spacetime cannot be embedded in a larger spacetime.

[7]Tachyon is the name given to a (so-far unobserved) particle travelling faster than light.

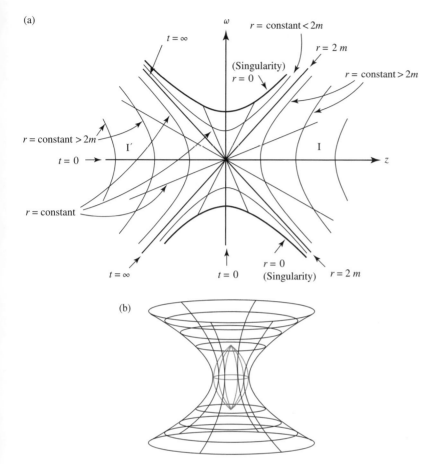

(a)

$t = \infty$

$r = \text{constant} < 2m$

$r = 2m$

(Singularity)
$r = 0$

$r = \text{constant} > 2m$

$r = \text{constant} > 2m$

$t = 0 \rightarrow$

I'

I

z

$r = \text{constant}$

$t = \infty$

$t = 0$

$r = 0$
(Singularity) $r = 2m$

(b)

Fig. VI.4 (a) Kruskal spacetime. (b) Schwarzschild throat.

VI.4   Stationary black holes

Most stars are not static[8] with respect to their local inertial reference frame, nor are they exactly spherically symmetric, so they cannot be modelled by a Schwarzschild spacetime or, after their collapse, by a Schwarzschild black hole. The simplest physical situations after static are the stationary ones (see Chapter IV), where the spacetime is of the type $(M \times R, g)$ with g invariant under translations along R.

VI.4.1   Axisymmetric and stationary spacetimes

In General Relativity, physical considerations[9] led to the conjecture[10] that a stationary black hole is necessarily axisymmetric, a spacetime (V, g) being said to be **axisymmetric** if the underlying manifold is of the type $M \times R$, with $M_t \equiv M \times \{t\}$ spacelike and $\{x\} \times R$ timelike, and the spacetime metric is such that M_t is diffeomorphic to the exterior of a subset of R^3, g_t admits a group S^1 of isometries that acts on R^3 like

[8]In particular, their motion is not invariant under time reversal.

[9]A non-axisymmetric rotating body would lose energy by gravitational radiation and hence cannot be stationary.

[10]The conjecture was formulated by Hawking in the 1970s and proved by him under an analyticity hypothesis and some restrictive geometric assumptions (see Hawking and Ellis, 1973). The analyticity hypothesis and some of the geometric assumptions have recently been removed by Alexakis, Ionescu, and Klainerman (2013).

a rotation group with given origin O_t, and the lapse and shift of g are also invariant under this group.

A spacetime is called **stationary axisymmetric** if, the origins O_t being derived from each other by time translations, the two Killing fields $\partial/\partial\phi$ and $\partial/\partial t$ commute. Moreover, for such a spacetime, it is required that the metric be invariant under simultaneous reversals of time $t \mapsto -t$ and angle of rotation $x^1 \equiv \phi \mapsto -\phi$, as is consistent with physical interpretation. It holds then that

$$g_{02} = g_{03} = g_{12} = g_{13} = 0,$$

since the signs of these coefficients change under such reversals. Using the fact that a two-dimensional metric can always be put in diagonal form, we write the metric of an arbitrary stationary axisymmetric spacetime as follows:

$$g = g_{00}\,dt^2 + 2g_{01}\,dt\,d\phi + g_{11}\,d\phi^2 + g_{22}(dx^2)^2 + g_{33}(dx^3)^2.$$

VI.5   The Kerr spacetime and black hole

In 1963, Roy Kerr found a stationary axisymmetric metric, an exact solution of the Einstein vacuum equations, which was later interpreted as a black hole and became very important in General Relativity (see the following sections).

VI.5.1   Boyer–Lindquist coordinates

The Kerr metric, written in Boyer–Lindquist coordinates (found by Boyer and Lindquist in 1967), reads

$$-\left(1 - \frac{2mr}{A}\right) dt^2 - 4mra\frac{\sin^2\theta}{A}\,dt\,d\phi + \frac{A}{B}\,dr^2 + A\,d\theta^2$$

$$+ \left(r^2 + a^2 + \frac{2mra^2\sin^2\theta}{A}\right)\sin^2\theta\,d\phi^2, \qquad \text{(VI.5.1)}$$

with a and m constants and

$$A \equiv r^2 + a^2\cos^2\theta, \quad B \equiv r^2 - 2mr + a^2. \qquad \text{(VI.5.2)}$$

The coordinate r appearing in this metric can be considered as defining the manifold structure of the support of this metric by interpreting it together with θ and ϕ as polar coordinates on the outside of the subset of R^3 such that $A > 2mr$ and $B > 0$; $t \in R$ is a time coordinate.

The Kerr metric (written here in Boyer–Lindquist coordinates) is axisymmetric (it does not depend on ϕ) and stationary but not static (it does not depend on t, but it is not invariant under time reversal). It is invariant under simultaneous reversals of time $t \mapsto -t$ and angle of rotation $\phi \mapsto -\phi$, as required. The Kerr metric is interpreted as the gravitational field outside a rotating black hole. It has not been found to be the gravitational field outside any realistic rotating star.

To interpret the constants a and m, we go to Cartesian-type coordinates

$$r^2 = x^2 + y^2 + z^2, \quad \theta = \text{Arctan} \frac{\sqrt{x^2 + y^2}}{z}. \tag{VI.5.3}$$

We then see that for large r, the Kerr metric tends to the Minkowski metric and we have the following equivalence for its deviation from Minkowski:

$$g_{\text{Kerr}} \cong - \left(1 - \frac{2m}{r}\right) dt^2 + \left(1 + \frac{2m}{r}\right) dr^2 + r^2(d\theta^2 + \sin^2\theta\, d\phi^2) \tag{VI.5.4}$$

$$- \frac{4ma}{r^3}(x\,dy - y\,dx)\,dt. \tag{VI.5.5}$$

The first line is equivalent for large r to the Schwarzschild metric: we identify m with the mass of the system. The last term leads to the identification of a as the angular momentum per unit mass of the system. The Kerr metric is invariant under the change $(a, t) \to (-a, -t)$, in agreement with this interpretation of a.

VI.5.2   The Kerr–Schild spacetime

The Kerr metric in Boyer–Lindquist coordinates becomes singular on various surfaces. As with the Schwarzschild metric, it is possible to transform it to another metric admissible in a larger domain. The following coordinates adapted to light propagation and redefinition of angular variable adapted to the rotation were introduced by Kerr and Schild in 1965:

$$dv = dt + \frac{r^2 + a^2}{B}\, dr$$
$$d\Phi = d\phi + \frac{a}{B}\, dr.$$

The Boyer–Lindquist metric then takes the Kerr–Schild form

$$g_{\text{KS}} \equiv - \left(1 - \frac{2mr}{A}\right) dv^2 + 2\,dr\,dv + A\,d\theta^2 + \frac{\Sigma^2}{A}\sin^2\theta\, d\Phi^2$$
$$+ 4amr\frac{\sin^2\theta}{A}\, dv\,d\Phi + 2a\sin^2\theta\, dr\,d\Phi,$$

with

$$\Sigma^2 \equiv (r^2 + a^2)^2 - Ba^2\sin^2\theta.$$

This metric extends to a regular Lorentzian metric for all $A > 0$. It reduces to the Minkowski metric in retarded time and polar coordinates of R^3 if $m = a = 0$, and it reduces to the Eddington–Finkelstein metric when $a = 0$.

VI.5.3   Essential singularity

The Kerr metric becomes singular for $A \equiv r^2 + a^2 \cos^2 \theta = 0$. This is a genuine singularity: the **Kretschman** scalar tends to infinity when A tends to zero.

In order to have the Kerr metric reducing to a form of the Minkowski metric when $m = 0$ and $a \neq 0$, the coordinates are reinterpreted as follows: θ and ϕ are coordinates on S^2, but $r = 0$ is not a single point of R^3; it is assumed instead that the manifold R^3 defined by the Cartesian coordinates x, y, z is represented in the variables r, θ, ϕ through the mapping (oblate polar 'coordinates')

$$x = (r^2 + a^2)^{\frac{1}{2}} \sin \Phi \sin \theta, \quad y = (r^2 + a^2)^{\frac{1}{2}} \cos \Phi \sin \theta,$$
$$z = r \cos \theta.$$

The singularity $A = 0$ is then interpreted as the circle $r = 0, \theta = \pi/2$, i.e.

$$z = 0, \quad x^2 + y^2 = a^2$$

VI.5.4   Horizons

The case $|a| > m$

This would apply to very rapidly rotating bodies. We then have $B > 0$ for all r, no surface $r = $ constant is a null surface, and the essential singularity is naked. This case is presently considered as physically unrealistic.

The case $|a| < m$

The Boyer–Lindquist metric then appears as singular for $r = r_+$ or $r = r_-$, solutions of $B = 0$ given by

$$r_\pm = m \pm \sqrt{m^2 - a^2}.$$

When a tends to zero, the Kerr metric tends to the Schwarzschild metric, r_+ tends to $2m$, and r_- tends to zero. The surfaces $r = r_\pm$ are not singular in the Kerr–Schild spacetime, but they are null surfaces—the contravariant component g^{rr} of the Kerr–Schild metric vanishes for $B = 0$, as can be foreseen and checked by direct calculus.

The surface $r = r_+$ is the event horizon: no particle entering the region $r < r_+$ can escape from it: the future light cone at points where $r = r_+$ points entirely towards the interior. The surface $r = r_-$ has no obvious physical meaning.

VI.5.5   Limit of stationarity. The ergosphere

The variable r becomes a time variable when

$$\frac{2mr}{A} > 1.$$

The metric is then no longer stationary. The surface $r = r_{\text{stat}}$, with r_{stat} the largest root of $A - 2mr = 0$, is called the **limit of stationarity**. It does not coincide with the horizon when $a \neq 0$: it holds that

$$r_{\text{stat}} = m + \sqrt{m^2 - a^2 \cos^2 \theta} \geq r_+.$$

The domain between the limit of stationarity and the horizon r_+ is called the **ergosphere**.

The **Penrose process** is the extraction of energy from a Kerr black hole by dropping a particle into the ergosphere, which will emit a particle with a greater energy. The process can be described as follows. Consider the Killing vector $X \equiv \partial/\partial v$, with components $(X^\alpha) = (1, 0, 0, 0)$ in Kerr–Schild coordinates. This vector is timelike outside the ergosphere $r > r_{\text{stat}}$,

$$g_{\alpha\beta} X^\alpha X^\beta \equiv - \left(1 - \frac{2mr}{A} \right) < 0 \quad \text{when} \quad r > r_{\text{Stat}},$$

and spacelike when $r_+ < r < r_{\text{stat}}$. A particle with timelike 4-momentum p has an energy $E = -p^\alpha X_\alpha$ with respect to the vector field X; this scalar is constant along the trajectory (a geodesic) because of the law of dynamics and the fact that X is a Killing vector. It holds that $E > 0$ outside the ergosphere. Suppose that the particle enters the ergosphere and splits there into two particles with momenta p_1 and p_2. By the conservation of 4-momentum, it holds that

$$p = p_1 + p_2, \quad E = E_1 + E_2,$$

with $E_i = -p_i^\alpha X_\alpha$. Since X is spacelike inside the ergosphere, it is possible to have the timelike vector p_1 such that $E_1 < 0$; hence $E_2 = E - E_1 > E$. Both scalars E_1 and E_2 are conserved along the trajectories of the respective fragments. If the second fragment returns to the outside of the ergosphere, it will be with a greater energy (with respect to the X observer) than the whole piece which left it. It can be considered that it has extracted energy from the rotating black hole.

VI.5.6   Extended Kerr spacetime

The Kerr–Schild spacetime that we have defined is not complete. It is possible to extend it (even to negative r), but the results[11] are much more complicated than in the Schwarzschild case and their physical interpretation very unclear.

[11]Cf. for instance, Wald (1984) and Chandrasekhar (1983).

VI.5.7   Absence of realistic interior solutions or models of collapse

Despite considerable effort, no realistic source has been found to make an exterior domain of a Kerr metric a complete admissible Einsteinian spacetime.

No construction of gravitational collapse analogous to those due to Oppenheimer–Snyder, Gu–Hu, and Claudel–Newman in the spherically symmetric case has been made for axisymmetric gravitational collapse.

VI.6   Uniqueness theorems for stationary black holes

We have seen in Chapter IV that there are no gravitational solitons with support compact in space and sources with non-negative energy. There is no theorem on the non-existence of spacetimes with compact space outside the black hole region—but nor are there any examples of such spacetimes. Unless otherwise specified, a black hole is considered as an isolated object in spacetime; the black hole region in space is an asymptotically Euclidean manifold with boundary its event horizon.

VI.6.1   The Israel uniqueness theorem

The uniqueness of a static black hole in vacuum, the Schwarzschild black hole, had been conjectured in the early 1950s by G. Darmois. It was proved with simple and precise assumptions by W. Israel.

VI.6.2   Uniqueness of the Kerr black hole

The following uniqueness theorem was proved in 1975 by Robinson, improving previous results of Carter from 1972.

Theorem VI.6.1 *Stationary, axisymmetric spacetimes that are solutions of the vacuum Einstein equations are uniquely specified by two parameters—the mass m and the angular momentum a—if they have a regular event horizon, are smooth outside the horizon, and are asymptotically flat.*

Since the Kerr metric provides a solution of the vacuum Einstein equations satisfying the hypothesis of the theorem for any given m and a, the Kerr metric represents the only such axisymmetric stationary black hole.

The Hawking–Klainerman theorem on the axisymmetric property of a stationary black hole completes, under its hypotheses,[12] the 'no-hair' theorem[13] for a stationary black hole in a vacuum spacetime of dimension 4.

In higher dimensions, other stationary black holes have been constructed by Emparan and Reall.[14]

VI.6.3   Stability of the Kerr black hole

The intuitive definition of the stability of a dynamical system is that its evolution does not change much under a change in its state at an initial time. A stronger requirement if the system admits an asymptotic

[12]For details on the state of these hypotheses—in particular on the horizon—see Chapter XIV, Section 11 of YCB-OUP 2009 (contributed by P.Chruściel), Bray and Chruściel (2004), and Alexakis et al. (2013).

[13]This uniqueness was predicted in the early 1960s by J. A. Wheeler and described by the picturesque statement 'Black holes have no hair'.

[14]See again Chapter XIV, Section 11 by Chruściel in YCB-OUP 2009.

state for given initial data is that the system admits the same asymptotic state for data that are perturbations of these data. If the dynamics obeys nonlinear differential equations

$$N(u) \equiv f(D^k u, \ldots, Du, u) = 0, \qquad \text{(VI.6.1)}$$

one is tempted to study the stability of a solution $\underline{u}$, at least in a first step, to look for the evolution of a perturbation δu that satisfies the system linearized at u:

$$\frac{\partial N}{\partial (D^k u)}(\underline{u}) D^k \delta u + \ldots + \frac{\partial N}{\partial u}(\underline{u}) \delta u = 0. \qquad \text{(VI.6.2)}$$

If δu remains small for all times for small initial data, the solution $\underline{u}$ is called linearization-stable. If δu tends to zero when the time tends to infinity, $\underline{u}$ is called asymptotically linearization-stable. Linearization stability is often treated by physicists by considering perturbations given by Fourier series. The linearization stability of the Schwarzschild spacetime was thus found by Zerilli and Wheeler in the early 1960s. Results on the linearized stability of the Kerr spacetime have been obtained.[15]

Nonlinear stability is a much more difficult problem, requiring global existence of solutions of nonlinear differential equations. The nonlinear stability of the Kerr spacetime is a subject of active research, in particular by S. Klainerman and collaborators, with conjectures and partial results.

[15]See, in particular, Dafermos, Holzegel, and Rodnianski (2013).

VI.7   General black holes

An intuitive physical definition of a black hole is easy to give: a black hole is a region of spacetime in which the gravitational field is so strong that no signal, in particular light, can get out. A black hole may reveal its presence by the motion of surrounding stars that indicates the presence of an enormous mass confined in a region too small for any known matter to be contained in it. The presence of a black hole can also sometimes be deduced from the very short and bright emission[16] of γ rays (a γ-ray burst) due to the explosion of an accretion ring formed around the black hole by its tremendous gravitational pull. Black holes are the subject of active observational research. The black holes interior is physically terra incognita, and very strange indeed. It is the subject of fascinating advanced and varied mathematical studies and conjectures.

[16]Although it is rather thought now that this emission is in general quasi-stationary.

VI.7.1   Definitions

A mathematical definition of an isolated black hole given by Penrose through a conformal diagram of an asymptotically Euclidean spacetime is often used. A definition that does not appeal to conformal compactification can be given for isolated black holes, or for a group of black holes and material bodies that is isolated in the sense that it is so far from other bodies that it may be considered to be embedded in an asymptotically Euclidean spacetime (V, g). The following is a possible definition:

Definition VI.7.1   *The **black hole region** $\mathcal{B}$ of the spacetime (V, g) is the complement of the past of the set covered by the null geodesics that have an infinite future canonical parameter.*

*The **event horizon** H is the boundary $\partial \mathcal{B}$ of the black hole region $\mathcal{B}$ in the spacetime V. Under appropriate reasonable assumptions, the event horizon H is a null hypersurface (C^0 but not necessarily C^1), generated by null geodesics.*

VI.7.2   Weak cosmic censorship conjecture

[17]See, for instance, Penrose (1979).

The original idea of Penrose,[17] coming from the study of spherical gravitational collapse where a black hole forms and hides the singularity from timelike observers, is the conjecture that generic Einsteinian spacetimes with physically reasonable sources do not admit any **naked singularity**, that is, a singularity visible by an observer. With the definition of singularity by incompleteness, trivial counterexamples can be obtained by cutting out regions of any spacetime, for example Minkowski spacetime. Therefore, the mathematics must be more precise to grasp the physical content. We give the following definition of a nakedly singular spacetime.

Definition VI.7.2   *An inextendible spacetime (V, g) is said to be future nakedly singular if it admits a future inextendible causal curve that lies entirely in the past of some point $x \in V$.*

We formulate the conjecture as follows:

Definition (No naked singularity conjecture).   *An inextendible, generic,[18] Einsteinian spacetime with physically reasonable sources admits no naked singularity.*

[18]Generic is usually taken to mean without exceptional properties or stable under small perturbations.

Note that the big bang is not a counterexample to this conjecture—it has no past, and hence does not correspond to any future inextendible causal curve.

A generic spacetime can be understood as a spacetime that is stable, in some sense to be defined, under small perturbations. Reasonable sources are physical sources that have a hyperbolic, causal evolution and do not have their own singularities (shocks, shell crossings, etc.).

We will briefly return to cosmic censorship in Chapter VIII.

VI.7.3   Thermodynamics of black holes

There are many theorems and conjectures about the thermodynamics and quantum properties of black holes—difficult to prove and even often to state in a precise way. Several properties have analogies with the fundamental laws of thermodynamics. They have led physicists to attribute to general black holes both entropy and temperature.

The first black hole 'thermodynamic' property is the **Hawking area theorem**, which says that the surface area A of the horizon H at a given time cannot decrease with time: it led to the identification of A, up to a constant factor, with the **entropy** of the black hole.

The temperature of a black hole has been defined through its **surface gravity**,[19] often denoted by κ (not to be mistaken with the gravitational constant G_N, which is also often denoted by κ). The usual definition of the surface gravity does not apply to the horizon of a black hole. It is defined for stationary black holes by the formula, with k the Killing vector tangent to the horizon,

$$k^\alpha \nabla_\alpha k^\beta = \kappa k^\beta.$$

This definition has been challenged by many authors and is still the subject of active discussion and possible generalizations to non-stationary black holes.

Hawking radiation from black holes is a quantum effect. It is well known that there are no absolute barriers in quantum theory. Hawking introduced quantum waves to study radiation from black holes and provide a consistent definition of their entropy and its variation (i.e. the decrease of their horizon area) with 'black hole evaporation'. Such radiation could be due to splitting of virtual pairs of particles emerging from the quantum vacuum outside the black hole, with one particle falling into the black hole and the other escaping from it. Hawking radiation is very weak and has not been observed.

These properties of black holes are fascinating and the subject of active study and discussion, involving also quantum gravity, a largely open subject. They are at present beyond any possible physical confirmation and mostly also any rigorous mathematical proof, although progress in the construction of a quantum theory of gravitation, in particular through string theory, opens new roads to the modelling of black hole interiors.

An exception, as far as mathematical proofs are concerned, is the proof through sophisticated mathematics by Huisken and Ilmanen and independently by Bray[20] of the **Riemannian Penrose inequality**, which admits a purely geometric formulation. It is linked with several theorems or conjectures due to Penrose (see Problem VI.10.5).

VI.8   Conclusions

VI.8.1   Observations

The conjectured existence of the first black hole, Cygnus X1 in 1972, was deduced from the study of a star's orbital parameters, the values of which implied the existence of an invisible companion too massive to be thought to be even a neutron star. An enormous number of candidate

---

[19] For material bodies, the surface gravity in Newtonian theory is the gravitational acceleration at points of its surface, about $9.81 \, \text{m s}^{-2}$ in the case of the Earth.

[20] See, for instance, Bray and Chruściel (2004).

black holes have now been detected. Astronomers usually classify black holes as follows.

Stellar-Mass black holes with masses between 10 and 24 solar masses reveal their presence by the motion of companion stars due to the existence of a strong gravitational field that cannot be explained otherwise. Another indication of the presence of a black hole is radiation emitted from matter falling into the black hole under its tremendous gravitational pull.

Supermassive black holes, millions (or billions) times more massive than the Sun, are thought to be in the centres of many galaxies, including our own Milky Way. They are detected by the motion of nearby stars and gas.

Recently, evidence for intermediate-mass black holes has been obtained by powerful new telescopes, such as Chandra, XMM-Newton, and Hubble.

There is an abundant literature on the subject in relevant scientific journals and on the Internet.

VI.8.2   The interiors of black holes

We have seen that the representation of the interior of a Schwarzschild black hole by a Lorentzian manifold ceases to be meaningful at the singularity $r = 0$, attained in a very short time[21] by any infalling object. The representation of the singularity in the Kruskal diagram by a spacelike submanifold is misleading. The interior of the Kerr black hole and its singularity are very strange. They cannot be considered as Einsteinian spacetimes in any classical sense. The general belief is that a new theory of gravitation must be constructed to represent them. This is a subject of very active research, linked with the old problem of the quantization of the gravitational field. Previous research in this difficult field has known dramatic transformations through the introduction of new and fascinating ideas, in particular string theory. However, further discussions of black holes and quantum gravity are outside the scope of this book.

[21] Although not so short for a very large mass M, since it is estimated as $10^{-5}\,\mathrm{s} \times M/M_{\mathrm{Sun}} \cong 10^3\,\mathrm{s}$ for $M = 10^8 M_{\mathrm{Sun}}$.

VI.9   Solution of Exercise VI.3.1

(1) With t now denoted by ρ and r by τ, the Schwarzschild metric reads

$$-\left(\frac{2m}{\tau} - 1\right)^{-1} d\tau^2 + \left(\frac{2m}{\tau} - 1\right) d\rho^2 + \tau^2(d\theta^2 + \sin^2\theta\, d\phi^2),$$

It has Lorentzian signature for $0 < \tau < 2m$, with τ a timelike coordinate. For any such given τ, the space metric is defined on the product $R \times S^2$; it is degenerate at $\tau = 2m$ and singular for $\tau = 0$.

(2) The definition

$$v = t + r + 2m \log\left(\frac{r}{2m} - 1\right), \qquad (VI.9.1)$$

i.e.

$$dv = dt + dr + \frac{2m\,dr}{r - 2m} = dt + \frac{r\,dr}{r - 2m},$$

implies by a simple computation the announced result:

$$-\left(1 - \frac{2m}{r}\right)dt^2 + \left(1 - \frac{2m}{r}\right)^{-1}dr^2 = \left(1 - \frac{2m}{r}\right)(dv^2 + 2\,dv\,dr).$$

(3) The metric is Lorentzian because

$$2\,dv\,dr = \frac{1}{2}\left[(dv + dr)^2 - (dv - dr)^2\right].$$

VI.10   Problems

VI.10.1   Lemaître coordinates

As was done by Lemaître in 1933, consider coordinates τ, ρ, θ, ϕ given in terms of the standard Schwarzschild coordinates t, r, θ, ϕ on R^4 by the functions

$$\tau := t + 2\sqrt{2mr} + 2m \log\left|\frac{\sqrt{2m} - \sqrt{r}}{\sqrt{2m} + \sqrt{r}}\right|,$$

$$\rho - \tau := r^{3/2}\frac{2}{3\sqrt{2m}}.$$

1. Show that the coordinates change is singular for $r = 2m$.
2. Show that in the Lemaître coordinates, the Schwarzschild metric takes a form that is singular only for $r = 0$, but non-static:

$$-d\tau^2 + \frac{2m}{r}d\rho^2 + r^2(d\theta^2 + \sin^2\theta\,d\phi^2).$$

3. Determine the radial geodesics.

Solution

1. The logarithm tends to $-\infty$ when r tends to $2m$.
2. This is a straightforward computation.
3. The radial geodesics are lines where only τ varies.

VI.10.2   Reissner–Nordström black hole

Discuss for which values of the constants m and Q the **Reissner–Nordström** metric (see Chapter V)

$$g_{\mathrm{RN}} := -\left(1 - \frac{2m}{r} + \frac{Q^2}{r^2}\right) dt^2 + \left(1 - \frac{2m}{r} + \frac{Q^2}{r^2}\right)^{-1} dr^2 \qquad \text{(VI.10.1)}$$
$$+ r^2(d\theta^2 + \sin^2 \theta \, d\varphi^2),$$

with electromagnetic potential $A_i = 0$, $A_0 = -Q/r$ describes a black hole and study its event horizon.

Solution

The metric g_{RN} is a smooth Lorentzian metric for all $r \neq 0$ if

$$r^2 - 2mr + Q^2 > 0,$$

that is, $m^2 < Q^2$, since the roots of this polynomial in r,

$$r_\pm = m \pm \sqrt{m^2 - Q^2}, \qquad \text{(VI.10.2)}$$

are then imaginary. For $Q^2 < m^2$, the metric becomes singular in the standard coordinates on the two manifolds $r = r_+$ and $r = r_-$. The manifold $r = r_+$ is the event horizon.

VI.10.3   Kerr–Newman metric

The Kerr–Newman metric reads

$$g_{\mathrm{KN}} = -\left(1 - \frac{2mr}{A}\right) dt^2 - 4mra\frac{\sin^2 \theta}{A} dt \, d\phi + \frac{A}{B + Q^2} dr^2$$
$$+ A \, d\theta^2 + \left(r^2 + a^2 + \frac{2mra^2 \sin^2 \theta}{A}\right) \sin^2 \theta \, d\phi^2, \qquad \text{(VI.1.2)}$$

with a and m constants and

$$A \equiv r^2 + a^2 \cos^2 \theta, \qquad B \equiv r^2 - 2mr + a^2. \qquad \text{(VI.1.3)}$$

Check that g_{KN} coincides with the Kerr metric in Boyer–Lindquist coordinates if $Q = 0$, and with the Reissner–Nordström metric if $a = 0$.

Show that g_{KN} satisfies the Einstein equations with electromagnetic source.

[22] Christodoulou and Ruffini had shown in 1971 (see Ohanian and Ruffini, 2013) that the irreducible mass increases in any Penrose problem, proving thus in a particular case the Hawking area theorem. For the general case, see Hawking and Ellis (1973).

VI.10.4   Irreducible mass (Christodoulou–Ruffini)

Assume the Hawking area theorem,[22] which says that no physical process can make the area of the event horizon of a black hole decrease. Show that the energy loss in a Penrose process in Kerr spacetime is at best 50%.

Solution

The radius of the event horizon of a Kerr spacetime with parameters m and a is

$$r_+ = m + m\sqrt{1 - |J|^2},$$

where J denotes the spin of the black hole.

The event horizon of the static spacetime resulting from an optimal (i.e. absorbing all the rotational energy) Christodoulou–Ruffini–Penrose process with mass m' is $r = 2m'$. The Hawking area theorem implies

$$m' \geq \frac{1}{2}m(1 + \sqrt{1 - |J|^2}).$$

The gain in mass is

$$m - m' \leq \frac{1}{2}m\left(1 - \frac{1}{\sqrt{\sqrt{1 - |J|^2}}}\right) < \frac{1}{2}m.$$

It tends to $m/2$ when $|J|$ tends to its maximum value 1.

VI.10.5   The Riemannian Penrose inequality

We have given in Chapter V the Schwarzschild metric in isotropic coordinates,

$$g_{\mathrm{Schw}} = -\left(\frac{2R - m}{2R + m}\right)^2 dt^2 + \left(1 + \frac{m}{2R}\right)^4 (dX^2 + dY^2 + dZ^2) \quad (\text{VI.10.3})$$

and have shown in Chapter IV that it has ADM mass $m_{\mathrm{ADM}} = m$.

Prove that

$$m = \sqrt{\frac{A}{16\pi}},$$

where A is the area of the horizon.

Solution

The horizon, deduced from the expression for the spacetime metric, is

$$R = \frac{m}{2}.$$

Straightforward classical calculus gives the area of this 2-sphere:

$$A = \int_{B_{m/2}} 2^4 \, dX \, dY \, dZ = \frac{m^2}{4} 2^4 4\pi = 16\pi m^2.$$

Comment. The above equalities and the Beckenstein–Hawking area theorem led Penrose to conjecture that general black holes should satisfy an inequality of the form

$$m \leq \sqrt{\frac{A}{16\pi}}.$$

This inequality considering only the space manifold, then called the 'Riemannian Penrose inequality', has been proved by Huisken and Ilmanen under the hypotheses that the space manifold is a Riemannian asymptotically Euclidean manifold with non-negative scalar curvature that admits a compact boundary composed of minimal surfaces but no other minimal surface.[23] The considered area A is the area of any disconnected such surface. The relevance to physics is that such boundary surfaces are 'trapped surfaces' conjectured to be linked with black holes containing a singularity of the corresponding spacetime.

[23]For details, see Bray and Chruściel (2004) and references therein.

Introduction to cosmology

VII

VII.1 Introduction

A cosmological model is a spacetime that is supposed to represent the whole past, present, and future of our universe. It has long been argued that in the framework of General Relativity, gravitation should model the geometry of the cosmos because its forces are long range (in contrast to nuclear forces) and non-compensated (there are no negative masses—in contrast to the existence of positive and negative electric charges). Since Einstein's formulation of his equations, it was assumed by most cosmologists that the cosmos in which we live is a Lorentzian 4-manifold (V, g) that satisfies the Einstein equations with the non-gravitational energies as sources. In the beginning, cosmological models relied essentially on a priori conjectures based on philosophical prejudices, the aim of simplicity, and the desire for unification.

Because of considerable progress in astronomical observations by both Earth-based telescopes and satellites,[1] cosmology has become a fully fledged part of physics. There is a wealth of information that has been accumulated in recent years from observations of the cosmos and that needs to be analysed. There have been remarkable advances in our knowledge, but also new and puzzling questions. We cannot in this book enter into a detailed exposition of the amazing number of observations made since the beginning of this century, and we can only give some indication of how remarkably precise conclusions, often convincing but sometimes speculative, have been deduced from these observations.

We will in this chapter give the main facts that led a majority of cosmologists (although there are still dissidents) to believe that our cosmos is represented by a four-dimensional Robertson–Walker spacetime (see Section VII.4) solution of the Einstein equations with source all the non-gravitational energies. Radiation energy is today negligible with respect to matter energy, but is thought to have been dominant at early stages. The matter source was considered until fairly recently as a perfect fluid, whose particles would be galaxies or clusters of galaxies. Observations made at various scales have now shown that the usual matter sources represent only a quite small fraction, an estimated 4.5%, of the energy content of the universe.[2] Modern cosmologists have introduced new types of sources: dark matter (about 25%) and dark energy (about 70%). There are numerous physical conjectures regarding the nature of these sources. A possible interpretation of dark matter is the existence of WIMPS (weakly interacting massive particles), particles that interact

[1] Such as COBE (Cosmic Background Explorer), and WMAP (Wilkinson Microwave Anisotropy Probe), and the information gathered by the Planck satellite (analysed and published in March 2013).

[2] Dynamical properties of celestial bodies, for instance the flat aspect of some galaxies, cannot be explained by their 'normal' mass content.

[3]Though not quite by all of them—there are still adherents of the steady-state and continuous-creation theory of Gold, Hoyle, and Narlikar and also of I. E. Segal's theory of chronogeometry (Segal, 1976).

[4]See, for instance, for results and references prior to 2002, Cotsakis and Papantonopoulos (2002) and for references prior to 2010, Cheng (2010). Most recent observations can be found in papers on the arXiv (gr-qc or astro-ph).

only through the weak field and gravitation. Dark energy is often interpreted as an energy of the vacuum, represented by a cosmological constant, or possibly a scalar field, called quintessence.

In Section VII.7, we give a brief description of what is currently thought by most cosmologists[3] to be the birth and infancy of the cosmos in which we live.[4]

VII.2   The first cosmological models

Soon after the Einstein equations were first framed, models were proposed for the whole cosmos.

VII.2.1   Einstein static universe

Historically, the first cosmological model was found in 1917 by Einstein himself, looking for a solution that was homogeneous and eternal, would solve Olbers' paradox,[5] and would be in agreement with Mach's principle. He was thus led to the **Einstein static universe** (see Problem IV.11.1) with closed space manifold S^3, and spacetime manifold $S^3 \times R$ endowed with the static metric

$$- dt^2 + a_0^2 \gamma_+, \qquad (VII.2.1)$$

[5]Olbers pointed out in 1823 that if space was infinite and filled with a homogeneous distribution of stars, the sky would appear uniformly bright—in clear contradiction with the dark night sky. Historians of astronomy have found that Olbers was not the first to make this remark, but it is his name that is still attached to it.

with a_0 a constant and γ_+ the metric of the unit 3-sphere S^3, that is, in the usual angular coordinates,

$$\gamma_+ := d\theta^2 + \sin^2 \theta \, (d\phi^2 + \sin^2 \phi \, (d\psi^2)$$

Setting $\sin \theta = r$ gives the form familiar to geometers:

$$\gamma_+ = \frac{dr^2}{1 - r^2} + r^2 (d\phi^2 + \sin^2 \phi \, d\psi^2). \qquad (VII.2.2)$$

We have seen that the Einstein static universe is a solution of the Einstein equations with static perfect fluid source of constant density and pressure given by

$$\mu_0 = \frac{3}{2} a_0^{-2}, \quad p_0 = -a_0^{-2} < 0. \qquad (VII.2.3)$$

[6]Another obstacle to the acceptance of the Einstein static universe as a model for our cosmos is that it is unstable.

[7]An idea Einstein did not like, because it introduces a new parameter, a priori not geometrically defined.

A negative pressure is unacceptable on classical physical grounds.[6] This difficulty was remedied by Einstein by introducing a positive cosmological constant.[7] The formulas (2.3) are then replaced by

$$\mu_0 = \frac{3}{2} a_0^{-2} - \Lambda, \quad p_0 = -a_0^{-2} + \Lambda; \qquad (VII.2.4)$$

Einstein choose $\Lambda = a_0^{-2}$ so that $p_0 = 0$.

[8]These properties have been used by Penrose to give a geometric definition of asymptotic flatness; see, for instance, Hawking and Ellis (1973) or Section 6 in Appendix VI of YCB-OUP2009.

The Einstein static spacetime, also called the **Einstein cylinder**, is (but with a different interpretation) the arena of the **Segal cosmos**. The manifold $S^3 \times R$ is the universal cover of compactified Minkowski spacetime; it is conformally locally flat,[8] and the conformal group acts on this geometric, non-metric, structure. The fact that representations of the conformal group govern the world of elementary particles seemed to I. E. Segal a justification of his cosmological model. He explains the cosmological redshift by the existence of a local time different from the cosmic time.

A few years after introducing the cosmological constant (in order to obtain a physically realistic time-independent cosmos), Einstein abandoned it (calling it the greatest blunder of his life) and accepted time-dependent models, motivated by the observation by astronomers of the redshift of light coming from distant stars, interpreted as being due to the expansion of the universe.[9]

[9]Recent observations indicate that the cosmological expansion is accelerating. This has led cosmologists to reintroduce a cosmological term.

VII.2.2   de Sitter spacetime

The de Sitter spacetime (see Problem IV.11.2) was constructed in 1917 by Willem de Sitter. It is a solution of the four-dimensional Einstein equations with positive cosmological constant. It is not realistic, because it is a vacuum solution, but was the first example of a time-dependent, expanding, Einsteinian spacetime.

Einstein and de Sitter had long discussions about their respective models. These discussions, and the discovery of the expansion of the universe, led Einstein to abandon the cosmological constant, but accept time-dependent spacetimes.

VII.2.3   General models

In the classical models, still basic for cosmological studies, the cosmos is modelled by a four-dimensional Einsteinian spacetime. Nowadays, all cosmologists consider, as it is experienced by everybody in daily life, that there is a flow of time—that is, a past and a future. They speak of the universe as the cosmos at a given time. Note that we do not observe the present state of the cosmos, since information comes to us at most with the speed of light. Most cosmologists now accept that the universe is expanding—indeed that the expansion is accelerating, according to recent observational data.

VII.3   Cosmological principle

VII.3.1   Assumptions

The **standard cosmological models** are based on the so called **cosmological principle,** which is composed of two assumptions.

The first assumption is that the cosmos is a manifold $M \times R$ endowed with a Lorentzian metric, four-dimensional in the usual case, such that the lines $x \times R$, $x \in M$, called trajectories of the '**fundamental observers**', are timelike geodesics orthogonal to the manifold $M \times \{t\}$ at each point $(x, t) \in M \times R$. The Lorentzian metric of the cosmos can therefore be written as

$$^{(4)}g := -dt^2 + {}^{(3)}g.$$

The *Riemannian manifold* $(M, {}^{(3)}g)$ *is the universe at time t.* The proper time of fundamental observers is called **cosmic time.**

The second assumption of the cosmological principle is that the universe at each time should look the same in all directions and also to any fundamental observer. This assumption is called the Copernican principle—in the name of Copernicus, who deprived us of a central position in the Solar System. We also know that the Sun does not occupy a remarkable place in our galaxy. The mathematical content of the second assumption, modulo the first one, is as follows:

(1) **Isotropy**: the Riemann tensor of the space metric $^{(3)}g$ is at each point x invariant under rotation in the tangent space to M centred at x.

(2) The space metric is **homogeneous**, i.e. it admits a transitive[10] group of (global) isometries.

[10]That is, for each pair of points x and y in M_t, there exists an isometry of (M, g) that brings x to y.

VII.3.2 Observational support

We see the stars very unevenly distributed in the night sky. Galaxies, and even clusters of galaxies, are also observed in our telescopes as being very anisotropically and inhomogeneously distributed. An argument in favour of adopting the cosmological principle is that, at a still larger scale, isotropy and homogeneity seem to be attained.

For most astrophysicists, the strongest evidence for the validity of the cosmological principle is the isotropy of the **CMB (cosmic microwave background) radiation**. This is the faint background glow[11] that sensitive radio telescopes detect in the sky, which otherwise appears dark between stars and galaxies to traditional optical telescopes. The CMB radiation is measured to be very nearly isotropic, with a temperature of about $2.725\,\mathrm{K}$ and a black body[12] spectrum. An observed discrepancy with isotropy of order 10^{-3} is interpreted as being due to our own motion relative to fundamental observers. There are further anisotropies of the order of 10^{-5}: the most recently observed[13] anisotropic structures at this scale seem to be best explained by random inhomogeneities in the big bang.

An argument in favour of homogeneity is that the fundamental physical constants seem to be and have been the same throughout the cosmos with a remarkable accuracy.

[11]First found by Penzias and Wilson in 1964, but predicted earlier by Gamov from the big bang theory.

[12]A black body is one that absorbs all radiation. Contrary to its name, a black body emits radiation when heated, but the graph of the emitted energy in terms of wavelength (the black body spectrum) is a roughly Gaussian-type curve with a peak depending on the temperature.

[13]Planck satellite in 2013.

[14]There is no real difficulty in extending the definition to $n + 1$ dimensions, although the space Riemann tensor is then no longer equivalent to the Ricci tensor. Note also that the topological classification of three-dimensional manifolds does not extend to higher dimensions.

VII.4   Robertson–Walker spacetimes

The **Robertson–Walker** spacetimes are $(3 + 1)$-dimensional[14] models satisfying the assumptions of the cosmological principle; that is, they are represented by metrics that read, on a product $M \times R$ with M a three-dimensional manifold,

$$^{(4)}g \equiv -dt^2 + {}^{(3)}g, \qquad (VII.4.1)$$

where $^{(3)}g$ is a t-dependent Riemannian metric on M whose Riemannian curvature is isotropic at each point of M. We will see that this implies that the metric $^{(3)}g$ is homogeneous.

VII.4.1   Robertson–Walker universes, metric at given t

The curvature tensor of a Riemannian metric g at a point $x \in M$ is **isotropic** at x, that is, invariant under rotations in the tangent space to M at x, if and only if it is of the form

$$R_{ij,hk}(x) = K(x)(g_{ih}\ g_{jk} - g_{jh}\ g_{ik})(x). \qquad (VII.4.2)$$

Exercise VII.4.1   *Prove the 'if' part of this property.*

Since in dimension 3 the Riemann tensor is equivalent to the Ricci tensor (4.2) can be replaced in that case by (recall that $S_{ij} \equiv R_{ij} - \frac{1}{2}g_{ij}R$ denotes the Einstein tensor)

$$^{(3)}R_{ij} \equiv 2K\ {}^{(3)}g_{ij}, \quad \text{hence} \quad {}^{(3)}S_{ij} \equiv -K {}^{(3)}g_{ij}. \qquad (VII.4.3)$$

The contracted Bianchi identity implies

$$^{(3)}\nabla_j^{(3)}S^{ij} \equiv 0, \quad \text{i.e.} \quad \partial_i K = 0.$$

Hence, K is a constant; the isotropic metric $^{(3)}g$ is also homogeneous. Riemannian spaces with curvature of the form (4.2) with K a constant are called **spaces of constant curvature**.

To determine the general Riemannian spacetimes $(M, {}^{(3)}g)$ of constant curvature, it is convenient to use, in a neighbourhood of an arbitrarily chosen point, polar pseudo-coordinates centred at that point, in which the spherical symmetry resulting from the cosmological principle is manifest. In these coordinates, the metric (see Chapter V) takes the form

$$^{(3)}g \equiv e^\mu\, dr^2 + r^2(d\theta^2 + \sin^2\theta\, d\phi^2), \quad \text{with} \quad \mu = \mu(r). \qquad (VII.4.4)$$

Equations (4.2) then read (the prime denotes the derivative with respect to r)

$$^{(3)}R_{ik} \equiv 0,\ i \neq k, \qquad {}^{(3)}R_{11} \equiv r^{-1}\mu' = 2Ke^\mu, \qquad (VII.4.4)$$

$$\frac{1}{\sin^2\theta}\,{}^{(3)}R_{33} \equiv {}^{(3)}R_{22} \equiv -e^{-\mu} + 1 + \frac{r}{2}e^{-\mu}\mu' = 2Kr^2. \qquad (VII.4.5)$$

The general solution of (4.4) is trivially found to be $e^{-\mu} = -Kr^2 +$ constant, and (4.5) then gives

$$e^{-\mu} = 1 - Kr^2. \qquad (VII.4.6)$$

We already know that a metric of constant curvature $K = 0$ is locally flat.

A metric with constant curvature $K > 0$ [respectively $K < 0$] is locally a 3-sphere of radius $K^{\frac{1}{2}}$ [respectively locally a hyperbolic 3-space of radius $|K|^{-\frac{1}{2}}$]. In these cases, one classically scales r by setting $r = |K|^{-\frac{1}{2}}\bar{r}$ and relabels $\bar{r}$ as r. In the new coordinate r, the metric takes one or other of the standard forms according to the sign of K:

$$^{(3)}g \equiv |K|^{-1}\gamma_\varepsilon, \quad \gamma_\varepsilon \equiv \frac{dr^2}{1 - \varepsilon r^2} + r^2(\sin^2\theta\, d\phi^2 + d\theta^2), \quad \varepsilon = \operatorname{sign} K.$$
(VII.4.7)

The metric γ_ε can be transformed into the familiar forms of the unit sphere or pseudosphere metrics as follows:

when $\varepsilon = 0$, $r \equiv \chi$, we have

$$\gamma_0 = d\chi^2 + \chi^2(\sin^2\theta\, d\phi^2 + d\theta^2);$$
(VII.4.8)

when $\varepsilon = 1$, $r \equiv \sin\chi$, we have

$$\gamma_+ = d\chi^2 + \sin^2\chi\,(\sin^2\theta\, d\phi^2 + d\theta^2);$$
(VII.4.9)

when $\varepsilon = -1$, $r = \sinh\chi$, we have

$$\gamma_- = d\chi^2 + \sinh^2\chi\,(\sin^2\theta\, d\phi^2 + d\theta^2).$$
(VII.4.10)

The isotropic homogeneous simply connected Riemannian manifold corresponding to $\varepsilon = 1$ with $0 \leq \chi < \pi$ and θ, ϕ angular coordinates on S^2, is the sphere S^3, a compact Riemannian manifold.[15]

The metrics with $\varepsilon = 0$ or $\varepsilon = -1$ can both be supported by the non-compact manifold R^3, with coordinate $0 \leq \chi < \infty$. The corresponding isotropic, homogeneous, simply connected Riemannian manifolds are the Euclidean space E^3 or the hyperbolic 3-space $\mathcal{H}^3$.

We have computed Robertson–Walker spaces supported by simply connected manifolds by looking for globally spherically symmetric metrics around one point. There are other metrics that are isotropic and homogeneous but not globally spherically symmetric, which can be obtained as quotients of the previous metrics by an isometry group. They are supported by manifolds with different topologies. An example in the case $\varepsilon = 0$ is the flat 3-torus.

VII.4.2   Robertson–Walker cosmologies

From Section VII.4.1, there are three types of Robertson–Walker space-time metrics:

$$-dt^2 + a^2(t)\gamma_\varepsilon, \quad \varepsilon = 1, -1, \text{ or } 0,$$
(VII.4.11)

where a is an arbitrary function[16] of t. We see that $a(t)$ is a scaling factor of local spatial distances of fundamental observers. In an expanding universe, (i.e. if $\dot{a} := da/dt > 0$), this distance increases proportionally to a and to the original distance, as in an inflated balloon. Indeed, on

[15] In an old terminology, universes supported by compact manifolds were called 'closed' (recall that, unless otherwise specified, manifolds are without boundary). Those with non-compact support were called 'open'.

[16] When $\varepsilon \neq 0$, $a(t)$ is equal to $|K|^{-\frac{1}{2}}$, K being the constant-in-space, time-dependent, scalar Riemann curvature of $^{(3)}g$.

the trajectory of a fundamental observer, only t varies; the space distance at time t of two fundamental observers moving respectively on the timelines $x = x_0$ and $x = x_1$ is the product by $a(t)$ of the distance in the metric γ_ε between these two points of M.

VII.5   General properties of Robertson–Walker spacetimes

VII.5.1   Cosmological redshift

The first essential cosmological data came from the observation of the redshifts of stars and galaxies. The **redshift parameter** is defined to be

$$z \equiv \frac{\nu_S}{\nu_0} - 1, \tag{VII.5.1}$$

where ν_0 is the observed frequency and ν_S the emitted frequency. There is a shift towards the red [respectively towards the blue] if $z > 0$ [respectively $z < 0$]. We now prove the following proposition.

Proposition VII.5.1   *In an expanding Robertson–Walker spacetime, there is a cosmological redshift with parameter given approximately by*

$$1 + z \equiv \frac{\nu_S}{\nu_0} \simeq \frac{a(t_0)}{a(t_S)}, \tag{VII.5.2}$$

if we assume that the variation of a is negligible during a period of the emitted and a period of the received light signal.

Proof.   Recall that a Robertson–Walker metric is of the form

$$- dt^2 + a^2(t) \left[d\chi^2 + f^2(\chi)(d\theta^2 + \sin^2 \theta \, d\phi^2) \right]. \tag{VII.5.3}$$

Let O_0 and O_S be two fundamental observers. Take coordinates such that they have the same θ and ϕ and have χ coordinates 0 and χ_S, respectively. An observer O with space coordinate $\chi_O = 0$ (this is no restriction) and $t = t_O$ receives at time t_O light emitted at time $t_S < t_O$ by a source S with space coordinate $\chi(t_S) = \chi_S > 0$ if it is on the light ray joining these two spacetime points. This light ray is the solution of the differential equation

$$\frac{dt}{d\chi} = a(t) \tag{VII.5.4}$$

given by

$$\chi_S = \int_{t_S}^{t_O} \frac{dt}{a(t)}. \tag{VII.5.5a}$$

Photons emitted by O_S situated at χ_s at times t_S and $t_S + T_S$ follow null rays and arrive at O_0 situated at $\chi_O = 0$ at times t_0 and $t_0 + T_0$

that satisfy

$$\int_{t_S}^{t_O} \frac{dt}{a(t)} = \int_{t_S+T_S}^{t_O+T_O} \frac{dt}{a(t)}. \qquad \text{(VII.5.6)}$$

Elementary calculus shows that the times T_S and T_O are therefore related by

$$\int_{t_S}^{t_S+T_S} \frac{dt}{a(t)} = \int_{t_O}^{t_O+T_O} \frac{dt}{a(t)}. \qquad \text{(VII.5.7)}$$

Hence, if a is considered as constant during the small amounts of time T_S (period of the emission) and T_O (period of the reception), we obtain

$$\frac{T_S}{T_O} = \frac{a(t_S)}{a(t_O)}. \qquad \text{(VII.5.8)}$$

Since the frequency is the inverse of the period, this result leads to

$$\frac{\nu_S}{\nu_0} = \frac{a(t_0)}{a(t_S)}, \qquad \text{(VII.5.9)}$$

which is the desired relation (5.2).   ∎

A positive cosmological redshift, $\nu_0 < \nu_S$ signals an expansion of the universe, $a(t_0) > a(t_S)$.

Statistical observations[17] of distant galaxies confirm, if we analyse them in the framework of a Robertson–Walker spacetime, that our universe is at present expanding, $a'(t_0) > 0$. Over the last few years, it has been observed that the expansion is accelerating: $a''(t_0) > 0$.

VII.5.2   The Hubble law

In a Robertson–Walker spacetime, the distance between two given fundamental observers at some cosmic time t is proportional to $a(t)$. The **Hubble parameter** is defined to be the constant-in-space, t-dependent scalar

$$H \equiv a^{-1}a'. \qquad \text{(VII.5.10)}$$

It measures the rate of expansion (or possibly contraction) of the universe. It has dimension $(\text{time})^{-1}$. An expanding Robertson–Walker universe has a positive Hubble parameter.

The **Hubble law** says that *the observed redshift is proportional to the distance* of the source.[18] It is true as consequence of the theory only in *first approximation* for not too distant sources, as we now show.

Approximate computation of the redshift gives

$$1 + z := \frac{\nu_S}{\nu_0} \doteq \frac{a(t_0)}{a(t_S)}, \qquad \text{(VII.5.11)}$$

$$z := \frac{\nu_S}{\nu_0} - 1 = \frac{a(t_O)}{a(t_S)} - 1 \doteq \frac{a'(t_O)(t_O - t_S)}{a(t_S)}.$$

Recall that

$$\chi_S = \int_{t_S}^{t_O} \frac{dt}{a(t)} \doteq (t_0 - t_S)a^{-1}(t_S), \qquad (VII.5.12a)$$

and hence, with $d_S = a(t_0)\chi_S$,

$$z \doteq H(t_O)\frac{a(t_0)}{a(t_S)}(t_0 - t_S) \doteq H(t_0)d_S.$$

For a given observer at a given time, the redshift is, in rough approximation, proportional to the distance of the source.

The Hubble law has long been used by astronomers to estimate distances from the Earth of various astronomical objects that are too far away for these distances to be measured by triangulation. It is then not logical to use these redshifts to estimate distances. Fortunately, distances can also be estimated by comparing the luminosity of a star as observed from Earth (the 'apparent luminosity') with the emitted luminosity (the 'absolute luminosity') as calculated from the physical nature of the star. In the past, cepheid variables (whose absolute luminosity is related to their period of variation) were used as such 'standard candles', but it was discovered that there are two kinds of cepheids. More reliable 'standard candles' have been found by astronomers, particular so-called type Ia supernovae.[19]

[19]Supernovae of type Ia have weak hydrogen lines and a strong Si line at 6150 Å.

VII.5.3 Deceleration parameter

A second approximation of the relation between redshift and distance is obtained by introducing the **deceleration parameter** linked with the second derivative of the scale factor a. This is a dimensionless scalar that measures the rate of variation with time of the expansion (or contraction), and is defined by

$$q = -\frac{aa''}{a'^2}. \qquad (VII.5.13)$$

An expanding accelerated universe has a negative deceleration parameter.

Taylor formula gives

$$a(t_S) = a(t_0)\left[1 + (t_S - t_0)H(t_0) - \frac{1}{2}qH^2(t_S - t_0)^2 + \ldots\right] \qquad (VII.5.14)$$

and

$$\chi_S = a^{-1}(t_0)\left[t_0 - t_S + \frac{H}{2}(t_O - t_S)^2 + \ldots\right], \qquad (VII.5.15)$$

which gives the next approximation,

$$z = H(t_0)d_S + \frac{1}{2}(q+1)H(t_0)d_S^2 + \ldots \ . \qquad (VII.5.16)$$

VII.5.4   Age and future of the universe

It is generally believed that the universe started with a big bang,[20] the structure of which is a subject of active debate (see Section VII.7). We denote by $t = 0$ the time at which it is believed to have started expanding as a Robertson–Walker universe. We set $a(0) = a_0$. We denote by t_0 the present time, that is, the age of the universe. By definition, $a(t_0) > 0$; also, $a'(t_0) > 0$, since we observe (statistically) redshifts, not blueshifts—the universe is presently expanding.

Until near the end of the twentieth centuries, it was thought for classical physical reasons that the expansion of the universe has been, is, and will be always slowing down—that is, $a''(t) < 0$ for all $t > 0$, and the curve is concave-downwards. The universe would end in a big crunch at a time $t = T < \infty$. In fact, more recent observations have led to the belief that the expansion of the universe is now accelerating $a''(t_0) > 0$. Estimates of t_0 and the future of the universe depend on the properties of the function $a(t)$, that is, of the qualitative and quantitative properties of sources.

VII.6   Friedmann–Lemaître universes

VII.6.1   Equations

A Friedmann–Lemaître cosmos is a Robertson–Walker spacetime whose metric satisfies the equations

$$S_{\alpha\beta} = -\Lambda g_{\alpha\beta} + T_{\alpha\beta}, \quad \text{with} \ \ S_{\alpha\beta} \equiv R_{\alpha\beta} - \frac{1}{2} g_{\alpha\beta} R, \qquad \text{(VII.6.1)}$$

with $T_{\alpha\beta}$ such that

$$T_{00} := \mu, \quad T_{ij} = pa^2 \gamma_{ij}, \quad T_{i0} = 0.$$

The non-zero Christoffel symbols of the spacetime metric (4.11), $-dt^2 + a^2 \gamma_\varepsilon$, are computed to be

$$\Gamma^0_{ij} \equiv aa' \gamma_{ij}, \quad \Gamma^j_{0i} \equiv a^{-1} a' \delta^j_i, \quad \Gamma^i_{jh} = \gamma^i_{jh}, \quad \text{with} \ \ a' := \frac{da}{dt}, \quad \text{(VII.6.2)}$$

where γ^i_{jh} are the Christoffel symbols of the relevant metric γ_ε.

The non-zero components of the Ricci tensor of the spacetime metric are, using the value $r_{ij} = 2\varepsilon \gamma_{ij}$ for the Ricci tensor of the spatial metric γ_ε,

$$R_{00} \equiv -3a^{-1} a'', \quad R_{ij} \equiv \left(2\varepsilon + aa'' + 2a'^2 \right) \gamma_{ij} \quad a'' := \frac{d^2 a}{dt^2} \quad \text{(VII.6.3)}$$

The scalar curvature is

$$R \equiv -R_{00} + a^{-2} \gamma^{ij} R_{ij} \equiv 3a^{-2} \left(2\varepsilon + 2aa'' + 2a'^2 \right).$$

Hence the component S_{00} of the Einstein tensor reads as a first-order differential operator on a:

$$S_{00} \equiv R_{00} - \frac{1}{2} g_{00} R \equiv 3a^{-2} \left(\varepsilon + a'^2 \right)$$

The corresponding Einstein equation[21] gives μ in terms of a, a', and Λ; it is called the **Friedmann equation**:

$$\mu = S_{00} - \Lambda \equiv 3a^{-2} \left(\varepsilon + a'^2 \right) - \Lambda. \qquad \text{(VII.6.4)}$$

The equation

$$\gamma^{ij} S_{ij} \equiv \gamma^{ij} T_{ij} - 3\Lambda \equiv 3pa^2$$

gives p in terms of a, a', and a'', namely

$$p = -a^{-2} \left(\varepsilon + a'^2 + 2aa'' \right) + \Lambda. \qquad \text{(VII.6.5)}$$

Exercise VII.6.1 *Prove (6.4) and (6.5).*

Exercise VII.6.2 *Using the formulas obtained for μ and p, prove that the stress–energy tensor satisfies the conservation laws, which read, since Λ is a constant,*

$$\nabla_\alpha T^{\alpha\beta} = 0.$$

VII.6.2 Density parameter

In discussing cosmological observations, it is useful to replace the density function μ by a dimensionless quantity Ω, called the '**density parameter**', which is defined by

$$\Omega \equiv \frac{\mu + \Lambda}{3H^2}, \qquad \text{(VII.6.6)}$$

where $H := a^{-1} a'$ is the Hubble parameter. Note that Ω depends only on cosmic time. The Friedmann equation (6.4) reads

$$3 \left(a^{-2} \varepsilon + H^2 \right) = \mu + \Lambda, \qquad \text{(VII.6.7)}$$

that is,

$$\Omega \equiv 1 + \frac{\varepsilon}{a^2 H^2}. \qquad \text{(VII.6.8)}$$

If our universe is modelled by a Robertson–Walker cosmology, then the value of Ω determines the type of Robertson–Walker spacetime in which we live: $\Omega > 1$ implies that $\varepsilon = 1$, with closed space sections; $\Omega < 1$ implies $\varepsilon = -1$, with the spatial sections being open if they are simply connected. The critical case $\Omega = 1$ implies $\varepsilon = 0$, and hence locally flat space (Euclidean space if simply connected). Physicists argue that in the models we are studying, if $\Omega > 1$ by an appreciable amount at an early time, the universe will subsequently collapse in an extremely short time. If $\Omega < 1$, it will expand so fast that no stars could form. Thus one must

[21] It is a constraint on a and a', as predicted from a general analysis of the Einstein equations.

suppose that Ω was initially very close to 1; this is called the **flatness problem**.

VII.6.3   Einstein–de Sitter universe

At the present cosmic time, p is negligible with respect to μ.

The **Einstein–de Sitter universe** (found by Einstein and de Sitter in 1932) is a Friedmann–Lemaître universe such that $p = 0$ (dust model), $\Lambda = 0$, and $\varepsilon = 0$ (flat case). Then a satisfies the second-order differential equation

$$a'^2 + 2aa'' = 0, \tag{VII.6.9}$$

which can be integrated easily. Indeed, it can be written as

$$2aa'' + a'^2 \equiv 2aa' \left(\frac{a''}{a'} + \frac{1}{2} \frac{a'}{a} \right) = 0,$$

which is equivalent to

$$\frac{d}{dt} \ln(a' a^{\frac{1}{2}}) = 0. \tag{VII.6.10}$$

Hence, if $a(0) = 0$, it holds that

$$a^{\frac{3}{2}} = kt, \ k \text{ a constant.}$$

By setting $\chi := kr$, the metric reads

$$- dt^2 + k^2 t^{\frac{4}{3}} \left[dr^2 + r^2(d\theta^2 + \sin^2 \theta \, d\phi^2) \right]. \tag{VII.6.11}$$

Exercise VII.6.3   *Find this metric by using (6.5) and (6.6).*

VII.6.4   General models with $p = 0$

In the general case with $p = 0$, one can use the Friedmann equation

$$\mu = 3a^{-2} \left(\varepsilon + a'^2 \right) - \Lambda \tag{VII.6.12}$$

and the conservation law

$$\frac{d}{dt} (\mu a^3) = 0$$

to obtain

$$a'^2 = f(a) := \frac{a^{-1}C + a^2\Lambda}{3} - \varepsilon, \quad C = \mu a^3 \text{ a constant.} \tag{VII.6.13}$$

Hence one has the differential equation

$$\frac{dt}{da} = \sqrt{\frac{1}{f(a)}}.$$

for the function $a \mapsto t(a)$, whose inverse gives the variation of a in function of t.

In fact, it is now possible with a good approximation to deduce density μ from observations by a dynamical analysis, and the distance $a\chi$ by using standard candles. Analysis of the redshift then gives the Hubble parameter H and the deceleration parameter q, and hence a' and a''. It is found that today a' is positive (expansion) and also $a'' > 0$ (acceleration). The belief among most cosmologists is that at present the universe is in accelerated expansion.

VII.6.5   ΛCDM cosmological model

The standard model of the universe adopted at present by most cosmologists is, up to small corrections, a **Friedmann–Lemaître universe**; that is, a Robertson–Walker spacetime solution of the Einstein equations with cosmological constant and source a stress–energy tensor of matter, with negligible contribution from radiation and neutrinos (even if the latter have a tiny non-zero mass). Observations of the motions of stars and galaxies using powerful Earth-based and satellite telescopes seem to imply that at a cosmological scale classical baryonic matter[22] is a very small part of the energy content of the cosmos, around 4.5%. The observations indicate the existence of **cold dark matter**, estimated at about 25% of the universe energy content and probably composed of WIMPS (weakly interacting massive patricles) with only weak and gravitational interactions. The acceleration of the universe is explained in the ΛCDM model by the existence of dark energy represented by the term Λg, usually considered as an energy of the vacuum, of quantum origin and constant in spacetime, in agreement with the observational result of the constancy throughout spacetime of the fundamental dimensionless parameters, in particular the fine-structure constant (see Chapter IV).

Remark VII.6.1   *The expansion history for the ΛCDM cosmological model is presently obtained from the supernova distance–redshift relation deduced from observation of type Ia supernovae. It assumes the validity of the Friedmann equation. However, as pointed out by some cosmologists, the Friedmann equation itself has not been independently tested.*

VII.7   Primordial cosmology

The general belief is that the cosmos started with a singularity[23] about 14×10^9 years ago.[24] There was a primeval phase[25] for which the physics was very different from anything we have ever investigated experimentally. This physics, linked with the search for quantum gravity, is the subject of intensive investigation, mainly in the context of string theories, with amazing results. In the most popular string theory, the universe is an $(n+1)$-dimensional manifold that has $n - 3$ of its spacelike dimensions compactified or a very small size (possibly of the order of the Planck length) and that is filled with one-dimensional strings. In some theories, space and time began to appear only after the Planck time.[26]

---

[22] Protons, neutrons, and electrons.

[23] The structure of the conjectured initial singularity (big bang or oscillatory) has led to interesting mathematical work but remains a mystery.

[24] This estimated age was augmented by a few million years after analysis in 2013 of data from the Planck satellite, which led to a reduction of the value of the Hubble parameter (see Section VII.5) by about 10%.

[25] The duration of this primeval phase is estimated as a Planck time, that is, about 10^{-44} s (see Chapter III). Note that giving an estimate for the duration of this preliminary stage is somewhat contradictory to the fact that time and space perhaps did not exist at this stage.

[26]Or rather no time less than the Planck time 10^{-44} s exists.

[27]The transition phase of inflation could have been a weakening of the inflaton called the slow roll. It is believed that damping oscilations were probably present (see Damour and Mukhanov, 1998).

[28]Protons and neutrons.

[29]By a process that is not yet fully understood, as a result of the existence at some time of a slightly greater number of baryons than antibaryons, most of the latter were annihilated, leaving behind an overwhelming predominance of matter over antimatter in the universe.

[30]In the radiation epoch, the photons lost energy by interations with electrons (Thomson scattering) and the universe was opaque. After recombination and the capture of electrons by atoms, the universe became transparent.

For the next earliest phase, the favoured paradigm is that the universe (that is, a section $t = $ constant of the Friedmann–Lemaître model), entered a phase of quasi-exponential growth, called **inflation**, driven by an energy content approximately equivalent to a large cosmological term, a constant or a function of a scalar field, called the inflaton. Inflation is generally considered by cosmologists as solving the flatness problem, every region of space being stretched almost flat by the expansion. Inflation is also considered as solving the '**horizon problem**', that is, the observed similarity in properties (cosmological principle) in regions of space so far apart now that no signal emitted near the big bang could have reached them both without inflation.

The inflationary stage (during which the universe is similar to a de Sitter spacetime) was followed by a transition[27] to a radiation-dominated era, during which the universe was filled by an extremely hot plasma of quarks, gluons, leptons, photons, and neutrinos. The evolution of the universe was governed until a time of the order of 10^{-6} seconds by quantum mechanics, and the physics was more like that which is or soon will be accessible to laboratory experiments, albeit under extreme conditions.

The universe was expanding and cooling; there came periods of so-called **recombinations** and the formation of baryons[28] and antibaryons.[29] A later recombination between ions and electrons led to the formation of atoms and molecules. The universe became **transparent**.[30] The cosmic microwave background (CMB) radiation observed now is thought to be made of the free photons surviving from this epoch. Finally, stars and galaxies appeared, and the universe as we know it was born at a time estimated at about 10^8 years.

The modelling of these fascinating early phases is, however, outside the scope of this book.

VII.8   Solution of Exercises VII.6.1 and VII.6.2

We need to show that in a Friedmann–Lemaître universe

$$\mu = 3a^{-2}\left(\varepsilon + a'^2\right) - \Lambda, \tag{VII.8.1}$$

$$p = -a^{-2}\left(\varepsilon + a'^2 + 2aa''\right) + \Lambda \tag{VII.8.2}$$

$$\nabla_\alpha T^{\alpha\beta} = 0 \tag{VII.8.3}$$

Solution

(VII.8.1):

$$S_{00} = T_{00} - g_{00}\Lambda$$

gives

$$\mu = S_{00} - \Lambda \equiv 3a^{-2}\left(\varepsilon + a'^2\right) - \Lambda.$$

(VII.8.2):

$$S_{ij} \equiv R_{ij} - \frac{a^2}{2}\gamma_{ij}R \equiv \left[2\varepsilon + aa'' + 2a'^2 - 3(\varepsilon + aa'' + a'^2)\right]\gamma_{ij}$$

When $n = 3$, $\gamma^{ij}\gamma_{ij} = 3$, and hence

$$\gamma^{ij}S_{ij} \equiv 3\left(-\varepsilon - 2aa'' - a'^2\right) = \gamma^{ij}T_{ij} - 3\Lambda = 3(pa^2 - \Lambda)$$

implies

$$p = -a^{-2}\left(\varepsilon + a'^2 + 2aa''\right) + \Lambda.$$

(VII.8.3): The stress–energy tensor is such that $T^{i0} = 0$, and hence

$$\nabla_\alpha T^{ai} \equiv \partial_j T^{ji} + \Gamma^h_{hj}T^{ji} + \Gamma^i_{hj}T^{hj}.$$

The non-zero Christoffel symbols of the spacetime metric are $\Gamma^0_{ij} \equiv aa'\gamma_{ij}$, $\Gamma^j_{0i} \equiv a^{-1}a'\delta^j_i$, and $\Gamma^i_{jh} = \gamma^i_{jh}$. Denoting by $^{(3)}\nabla$ the covariant derivative in the metric γ, we have

$$\nabla_\alpha T^{ai} \equiv \partial_j T^{ji} + \Gamma^h_{hj}T^{ji} + \Gamma^i_{hj}T^{hj} \equiv {}^{(\gamma)}\nabla_j T^{ji} \equiv \gamma^{ij}\partial_j(pa^2) = 0,$$

because a and p depend only on t. We also have

$$\nabla_\alpha T^{a0} \equiv \partial_0 T^{00} + 3a^{-1}a'T^{00} + aa'\gamma_{ij}T^{ij} \equiv \mu' + 3a^{-1}a'(\mu + p).$$

Using the previous expressions for μ and p, we find

$$\mu' = -6a^{-3}a'\left(\varepsilon + a'^2\right) + 6a^{-2}a'a'',$$

$$\mu + p = 3a^{-2}\left(\varepsilon + a'^2\right) - a^{-2}\left(\varepsilon + a'^2 + 2aa''\right)$$

$$\equiv 2a^{-2}\left(\varepsilon + a'^2 - aa''\right),$$

and hence

$$\mu' + 3a^{-1}a'(\mu + p) = 0.$$

Thus

$$\nabla_\alpha T^{a0} = 0.$$

VII.9   Problems

VII.9.1   Isotropic and homogeneous Riemannian manifolds

Isotropy of the space section M_t at a point x means that there is no privileged direction in the tangent space T_x to M_t at x.

1. Show that if the Riemann tensor of the metric g takes at x the form, with $K(x)$ some number,

$$R_{ij,hk}(x) = K(x)(g_{ih}g_{jk} - g_{jh}g_{ik})(x), \qquad \text{(VII.9.1)}$$

it is invariant on the tangent plane at x under rotations centred at x.

2. Show that the tensor (9.1) is locally homogeneous.

Solution

1. The tangent space to an n-dimensional Riemannian manifold at x is an n-dimensional Euclidean vector space. A rotation around its origin x is represented by an $n \times n$ matrix A (called an orthogonal matrix) whose elements A_p^m acting on tangent vectors X at x, $Y_p = A_p^m X_m$, are such that

$$\sum_p Y_p^2 := \sum_p A_p^m X_m A_p^q X_q \equiv \sum_q X_q^2;$$

that is,

$$\begin{aligned} \sum_p A_p^m A_p^q = 0, \quad m \neq q, \\ \sum_p A_p^m A_p^q = 1, \quad m = q. \end{aligned} \qquad \text{(VII.9.2)}$$

In an orthonormal frame at x, the components of the Riemann tensor (9.1) read

$$R_{ij,hk}(x) = K(x)(\delta_{ih}\delta_{jk} - \delta_{jh}\delta_{ik})(x).$$

Under a rotation A, the transformed curvature tensor R' has components

$$R'_{i'j',h'k'}(x) = K(x)A_{i'}^i A_{j'}^j A_{h'}^h A_{k'}^k (\delta_{ih}\delta_{jk} - \delta_{jh}\delta_{ik})(x);$$

that is, using the formulas (9.2),

$$R'_{i'j',h'k'}(x) = K(x)(\delta_{i'h'}\delta_{j'k'} - \delta_{j'h'}\delta_{i'k'})(x).$$

VII.9.2   Age of the universe

1. Use the Taylor formula[31]

$$a(t) = a(t_0) + a'(t_0)(t - t_0) + (t - t_0)^2 \int_0^1 (1 - \lambda)a''(t_0 + \lambda(t - t_0))\, d\lambda$$

to find estimates of the age t_0 and the time T of the end of an expanding Robertson–Walker universe when we assume that $a''(t) < 0$ for $0 \leq t \leq T$.

2. What can we say if $a''(t_0) > 0$.

Solution

1. The Taylor formula implies

$$a(0) = a(t_0) - t_0 a'(t_0) + t_0^2 \int a''(t_0(1-\lambda)) \, d\lambda, \quad 0 \le \lambda \le 1.$$

Hence, if $a(0) = 0$, $a'(t_0) > 0$, and $a''(t) \le 0$ for $t \le t_0$, then

$$t_0 \ge \frac{a(t_0)}{a'(t_0)} = H(t_0).$$

Suppose that $a''(t) \le -C$, some positive constant for $t \ge t_0$; then

$$a(t) \le a(t_0) + a'(t_0)(t - t_0) - \frac{1}{2}C(t - t_0)^2,$$

and hence $a(T) = 0$ (big crunch) as soon as

$$a(t_0) + a'(t_0)(t - t_0) - \frac{1}{2}C(t - t_0)^2 = 0;$$

that is,

$$T - t_0 = \frac{-a'(t_0) + \sqrt{a'^2(t_0) + 2Ca(t_0)}}{C}.$$

2. The age t_0 depends on the behaviour of $a'(t)$, and hence of $a''(t)$ for $0 < t < t_0$. Assume $a'(t) \ge 0$ for $0 \le t \le t_1$; hence $a(t_1) \ge 0$ and (inflation) $a''(t) \ge C > 0$ for $t_1 \le t \le t_0$. Then the Taylor formula reads

$$a(t_0) = a(t_1) + a'(t_1)(t_0 - t_1) + (t_0 - t_1)^2 \int_0^1 (1-\lambda)a''(t_1 + \lambda(t_0 - t_1)) \, d\lambda,$$

and implies

$$a(t_0) \ge \frac{C}{2}(t_0 - t_1)^2.$$

VII.9.3   Classical Friedmann–Lemaître universes

Consider a Friedmann–Lemaître universe with $\Lambda = 0$ filled with a perfect fluid with equation of state $p = (\gamma - 1)\mu$, with γ a constant. Note that $\gamma = 1$ corresponds to dust and $\gamma = \frac{4}{3}$ to pure radiation.

1. Show that the velocity of acoustic waves is less than the speed of light if $\gamma \le 2$ (see Chapter IX).

2. Show that in such a Friedmann–Lemaître universe, it holds that, with C a constant,

$$\mu a^{3\gamma} = C.$$

Therefore, the density μ must decrease with increasing a.

Solution

If $p = (\gamma - 1)\mu$, the equation (6.5) reads

$$\frac{d}{dt}(a^3\mu) + 3a^{-1}a'(\gamma - 1)\mu = 0$$

and simplifies to

$$\mu' + 3a^{-1}a'\gamma\mu = 0.$$

Hence, by integration, with C a constant,

$$\mu a^{3\gamma} = C.$$

VII.9.4   Milne universe

The **Milne universe** is a Robertson–Walker spacetime with $\varepsilon = -1$, $\Lambda = 0$, space support R^3, and is a vacuum ($\mu = p = 0$). The Milne universe, or analogous spacetimes with compact (non-simply connected) space sections, play an important role as asymptotic states in cosmology.

Write the spacetime metric of the Milne universe. Show that it is locally flat, isometric to a wedge of Minkowski spacetime.

Solution

The Friedmann equation implies

$$a'^2 = 1.$$

Hence $a' = 1$, and, up to the choice of label of the time origin, $a(t) = t$. The spacetime metric on the manifold $R^3 \times (0, \infty)$ then takes the form

$$-dt^2 + t^2[dr^2 + \sinh^2 r(d\theta^2 + \sin^2\theta\,(d\phi^2)],$$

where r, θ, ϕ are polar (pseudo-)coordinates on R^3. The space metric collapses for $t = 0$ and expands indefinitely when t tends to infinity. The spacetime metric is locally flat, as can be seen by computing its Riemann tensor, for instance by using the $3 + 1$ decomposition given in Chapter VIII.

Part B

Advanced topics

This part provides a deeper study of the properties of general solutions of the Einstein equations either in vacuum or with stress–energy–momentum sources. Particular attention is paid to relativistic fluids and kinetic models, both of which have become important in astrophysics and cosmology.

General Einsteinian spacetimes. The Cauchy problem

VIII.1   Introduction

The Cauchy problem for a partial differential equation of order m on R^{n+1} with unknown a function f is the search for a solution such that f and its derivatives of order less than m take given values on an n-dimensional submanifold M. In coordinates x^0, x^i, $i = 1, 2, \ldots, n$, with M represented by $x^0 = 0$, the independent data are

$$f(0, x^i) \text{ and } \frac{\partial^p f}{(\partial x^0)^p}(0, x^i), \ p = 1, \ldots, m-1.$$

In the regular[1] analytic case, this problem has one and only one analytic solution in a neighbourhood of M.

The stability[2] of the solution of the Cauchy problem, and its existence in the non-analytic case, depend crucially on its characteristic determinant, the polynomial obtained in replacing in the terms of order m the derivative ∂f by a vector X. For a linear second-order equation

$$a^{\alpha\beta} \frac{\partial^2 f}{\partial x^\alpha \partial x^\beta} + b^\alpha \frac{\partial f}{\partial x^\alpha} + cf = h, \qquad \text{(VIII.1.1)}$$

the **characteristic polynomial** is the scalar function

$$P(X) := a^{\alpha\beta} X_\alpha X_\beta. \qquad \text{(VIII.1.2)}$$

The properties of solutions depend essentially on the signature of this quadratic form. In a domain U of R^{n+1}, the equation is **elliptic** if $P(X)$ is positive- (or negative-) definite; it is **hyperbolic** if it is of Lorentzian signature.

For *elliptic equations*, the solution of the Cauchy problem exists in the analytic case, but is not stable: one says that the Cauchy problem is not **well posed**.[3] Possibly well-posed problems for elliptic equations are global problems, for instance the so-called **Dirichlet problem**, which is the data of the unknown on the boundary of the domain.[4]

For *hyperbolic equations*, the Cauchy problem is well posed in the analytic case, but also for more general functional spaces,[5] in particular smooth functions. For linear equations, the solution is global, i.e. it

[1] The regularity condition for a quasilinear equation with analytic coefficients and analytic data on M, represented by $x^0 = 0$, is that the coefficient of the derivative $\partial^m f/(\partial x^0)^m$ does not vanish for $x^0 = 0$.

[2] That is, contiguous dependence on initial data.

[3] A problem is said to be well posed if it has one and only one solution and this solution depends continuously on the data.

[4] In this case, the well posedness depends also on the lower-order terms.

[5] An important property for General Relativity, since the relativistic causality imposes that a solution of a relativistic equation depends only on the past of this point, while analytic functions are entirely determined by their values in any open set.

exists on the whole of R^{n+1}, although it may grow, even exponentially, with $t \equiv x^0$. Results extend to some (see J. Leray (1953)) quasidiagonal[6] systems and to systems on manifolds. That is, for linear second-order systems on a Lorentzian manifold (V^{n+1}, g) that read

$$g^{\alpha\beta}\nabla^2_{\alpha\beta}f^I + b_J^{\alpha,I}\nabla_\alpha f^J + c_J^I f^J = h^I, \tag{VIII.1.3}$$

the Cauchy problem with data on a spacelike section M^n is globally well posed in relevant functional spaces, with relevant hypotheses on the coefficients and Cauchy data. In particular, the solution f^I is smooth if (V^{n+1}, g) is a smooth and regularly sliced Lorentzian manifold and the Cauchy data are smooth. An important property for Relativity is that *the solution at a point depends only on the values of the initial data in the past of this point.*

A quasidiagonal, quasilinear system of second-order differential equations on a manifold V^{n+1} reads, in a domain U of local coordinates,

$$g^{\alpha\beta}\underline{\nabla}^2_{\alpha\beta}f^I + b_J^{\alpha,I}\underline{\nabla}_\alpha f^J + c_J^I f^J = h^I, \tag{VIII.1.4}$$

[7]The introduction of $\underline{\nabla}$ permits a global geometric formulation, impossible with ∇ for the Einstein equations, where g itself is the unknown and satisfies $\nabla g \equiv 0$.

where g, b, and c depend on the unknown f and its first derivative. The covariant derivative $\underline{\nabla}$ is taken in an a priori given[7] metric $\underline{g}$ in U. It is always possible to take for $\underline{g}$ in U the flat metric, i.e. $\underline{\nabla}_\alpha := \partial/\partial x^\alpha$. The Cauchy problem for the system (1.4) is well posed for some initial data if the linear system obtained by replacing in the coefficients g, b, and c the quantities f and ∂f by their initial data is a well-posed system for these initial data. Note that the existence of the solution (which can be obtained by iteration), is in general for nonlinear equations only *local in time.*

The *Einstein equations in vacuum are a geometric system* for the pair (V, g), invariant under diffeomorphisms of the manifold V and the associated isometries of the Lorentzian metric g. These equations constitute, from the analyst's point of view, a system of $(n + 1)(n + 2)/2$ second-order quasilinear partial differential equations for the $(n + 1)(n + 2)/2$ coefficients[8] of the metric in local coordinates. However, the equations are not independent, because of the contracted Bianchi identities, a consequence of the invariance under diffeomorphisms. We will see that the *Einstein equations have both hyperbolic and elliptic aspects.*

[8]In the classical physical case $n+1 = 4$, the number of independent coefficients $g_{\alpha\beta}$ of a general metric g is 10.

VIII.2   Wave coordinates

[9]Einstein used them in the Minkowskian approximation. It seems that de Donder was the first to introduce them for the full Einstein equations, shortly before G. Darmois.

The Einstein equations are invariant under diffeomorphisms; to construct generic Einsteinian spacetimes, one fixes the local coordinates by a general condition, called a gauge choice. One then has to prove that the constructed spacetime satisfies the gauge condition.

The results of the Minkowskian approximation have led various authors, over many years,[9] to select in various problems what are now called harmonic or wave coordinates, that is, to impose that the scalar functions $x \mapsto x^\alpha$ defining local coordinates satisfy the wave equations[10]

[10]In computations done in local coordinates (x^α), it holds that $\partial_\alpha \equiv \partial/\partial x^\alpha$.

$$F^\alpha := \Box_g x^\alpha \equiv g^{\lambda\mu}\nabla_\lambda\partial_\mu x^\alpha = 0. \qquad \text{(VIII.2.1)}$$

In terms of the Christoffel symbols associated with the coordinates x^α, it holds that

$$F^\alpha \equiv g^{\lambda\mu}\Gamma^\alpha_{\lambda\mu}. \qquad \text{(VIII.2.2)}$$

The following identity results from a straightforward computation:

$$R_{\alpha\beta} \equiv R^{(h)}_{\alpha\beta} + L_{\alpha\beta}, \qquad \text{(VIII.2.3)}$$

where

$$L_{\alpha\beta} \equiv \frac{1}{2}\left(g_{\alpha\lambda}\partial_\beta F^\lambda + g_{\beta\lambda}\partial_\alpha F^\lambda\right) \qquad \text{(VIII.2.4)}$$

vanish in wave coordinates while the $R^{(h)}_{\alpha\beta}$ are a system of **quasilinear**, **quasidiagonal** (i.e. linear and diagonal in the principal, second-order, terms) wave operators

$$R^{(h)}_{\alpha\beta} \equiv -\frac{1}{2}g^{\lambda\mu}\partial^2_{\lambda\mu}g_{\alpha\beta} + P_{\alpha\beta}(g,\partial g), \qquad \text{(VIII.2.5)}$$

where P is a quadratic form in the components of ∂g, with coefficients that are polynomials in the components of g and its contravariant associate.

The system of partial differential equations

$$R^{(h)}_{\alpha\beta} = \rho_{\alpha\beta} \qquad \text{(VIII.2.6)}$$

are called the (harmonically) **reduced Einstein equations** (the vacuum reduced Einstein equations if $\rho_{\alpha\beta} \equiv 0$).

VIII.2.1   Generalized wave coordinates

It is clear that one can also deduce from the Ricci tensor a quasidiagonal operator on the components of the metric, namely

$$R^{(h,H)}_{\alpha\beta} \equiv R^{(h)}_{\alpha\beta} + \frac{1}{2}\left(g_{\alpha\lambda}\partial_\beta H^\lambda + g_{\beta\lambda}\partial_\alpha H^\lambda\right), \qquad \text{(VIII.2.7)}$$

if the wave coordinate conditions are replaced by the more general ones[11]   [11] Friedrich (1986).

$$F^\alpha_H := F^\alpha - H^\alpha = 0, \quad \text{with} \quad F^\alpha \equiv g^{\lambda\mu}\Gamma^\alpha_{\lambda\mu}, \qquad \text{(VIII.2.8)}$$

where the H^α are known scalar functions. Their advantage is that they contain a new set of functions H^α, which can be freely specified, or eventually chosen to satisfy ad hoc equations.

VIII.2.2   Damped wave coordinates

[12]Gundlach, Calabrese, Hinder, and Martin-Garcia (2005), Pretorius (2005b), and Lindblom, Scheel, Kidder, Owen, and Rinne (2006).

Recently, numerical analysts[12] have been led to introduce reduced Einstein equations in a wave-related gauge, called **damped wave coordinates**. The operator $R_{\alpha\beta}^{(h,H)}$ is replaced by

$$R_{\alpha\beta}^{(h,H,\gamma_0)} \equiv R_{\alpha\beta}^{(h)} + \frac{1}{2}\left(g_{\alpha\lambda}\partial_\beta H^\lambda + g_{\beta\lambda}\partial_\alpha H^\lambda\right) + \frac{1}{2}\gamma_0 F_H^\lambda(g_{\beta\lambda}n_\alpha$$
$$+ g_{\alpha\lambda}n_\beta - g_{\alpha\beta}n_\lambda),$$

[13]For the Einstein equations with non-zero sources, one must solve the coupled system of these equations and the equations satisfied by the sources, which depend on their nature.

with n the unit vector normal to spacelike slices and γ_0 a constant. The presence of this non-zero constant seems to stabilize the results of numerical calculations of the evolution system; it determines the time rate at which the gauge conditions $F_H^\lambda = 0$ are damped under evolution.

The use of a wave gauge gives the basic elements for the proof of fundamental properties of Einstein gravity using known results for partial differential equations.

[14]A constructive method inspired by representation formulas for solutions of linear equations given in dimension 4 by Sobolev and in even dimension $n + 1 > 4$ by de Rham (restricted to constant coefficients) was used for the first proof of existence of solutions of the Cauchy problem for quasilinear, quasidiagonal systems of wave equations by Fourès (Choquet)-Bruhat (1952) for the case $n + 1 = 4$ and by Fourès (Choquet)-Bruhat (1953) for the case $n + 1 > 4$. The Leray theory of general hyperbolic systems with its energy method can also be used. The multiplicity of roots of the characteristic determinant is irrelevant for the application of Leray theory because the characteristic matrix is diagonal. This multiplicity destroys the Leray hyperbolicity when the characteristic matrix cannot be diagonalized into blocks with determinants with distinct roots. K. O. Friedrichs's theory of first-order symmetric hyperbolic systems has no difficulty with multiple characteristics, but such first-order systems destroy the spacetime appearance of the problem and hide the propagation properties of the solution. Friedrichs hyperbolic systems have no analogue for higher-order Leray hyperbolic systems. Moreover the weak hyperbolicity of Leray–Ohya (see Chapter IX), important for relativistic causality, has no analogue in Friedrichs's theory of symmetric hyperbolic first-order systems.

VIII.3   Evolution in wave gauge

VIII.3.1   Solution of the reduced equations in vacuum

We consider the vacuum case,[13] that is, the search for solutions of a system of quasilinear wave equations. The natural well-posed problem, called the Cauchy problem, is the construction of a solution taking together with its first derivatives given values on a submanifold M of V assumed to be spacelike for these initial values. If M has equation $x^0 = 0$ in a local coordinate system and the analytic data are $g_{\alpha\beta}(0, x^i)$ (a quadratic form of Lorentzian signature) and $(\partial g_{\alpha\beta}/\partial x^0)(0, x^i)$, then the manifold M is spacelike for these data if the initial data $\bar{g}$ (the induced metric on M with components $g_{ij}(0, x^h)$), is positive-definite. Known theorems for quasidiagonal quasilinear systems of wave equations[14] lead to the following theorem, valid in relevant functional spaces,[15] in particular spaces of smooth functions with some finite number of derivatives.

Theorem VIII.3.1 *The Cauchy problem for the vacuum Einstein equations in wave gauge is well posed; that is, it has one and only one local solution depending continuously on the initial data. This solution has the physically important property of exhibiting **propagation** of the gravitational field with the speed of light: its value at a spacetime point depends only on the data in the past of this point.*

[15]For results in classical Sobolev spaces, see YCB-OUP2009, Appendix III.

The equations being nonlinear, the solution in general is defined only in a neighbourhood of the initial space manifold. The appearance of singularities and their nature is a fundamental field of research in General Relativity.

VIII.3.2   Equations with sources

In the presence of sources, the reduced Einstein equations contain other unknowns in addition to the metric g, namely fields or matter. The necessary conservation laws satisfied by the sources, $\nabla_\alpha T^{\alpha\beta} = 0$, are equations that contain both g and these other unknowns; they must be coupled with the Einstein equations. The hyperbolicity of this coupled system with a dependance domain of a solution of the Cauchy problem determined by the null cones of the metric g is important for the satisfaction of the Einstein principle of causality (see Problems VIII.12.6 and VIII.12.7 and Chapters IX and X).

VIII.4   Preservation of the wave gauges

A solution of the reduced Einstein equations that is a metric in wave gauge satisfies the full Einstein equations, and, indeed, for any solution of the reduced equations to be a solution of the full equations, it has to be in wave gauge, that is satisfy the equations

$$F^\alpha \equiv g^{\lambda\mu}\Gamma^\alpha_{\lambda\mu} = 0. \qquad (VIII.4.1)$$

We use the Bianchi identities to show that if g is a solution of the reduced Einstein equations[16] in a wave (harmonic) gauge, then the gauge conditions satisfy a second-order linear quasidiagonal homogeneous differential system, because of the Bianchi identities, which imply the following identities for the Einstein tensor S:

$$\nabla_\alpha S^{\alpha\beta} \equiv 0. \qquad (VIII.4.2)$$

Indeed, if g is a solution of the equations in wave gauge

$$R^{\alpha\beta}_{(h)} = \rho^{\alpha\beta}, \qquad (VIII.4.3)$$

then it holds that, with $T^{\alpha\beta} := \rho^{\alpha\beta} - \frac{1}{2}g^{\alpha\beta}\rho$ the stress–energy tensor of sources,[17]

$$S^{\alpha\beta} - T^{\alpha\beta} = -\frac{1}{2}(g^{\alpha\lambda}\partial_\lambda F^\beta + g^{\beta\lambda}\partial_\lambda F^\alpha - g^{\alpha\beta}\partial_\lambda F^\lambda). \qquad (VIII.4.4)$$

An elementary computation using the Bianchi identities and the conservation laws for the source T shows that, the functions F^α then satisfy a linear homogeneous system of wave equations

$$g^{\alpha\lambda}\partial^2_{\alpha\lambda}F^\beta + A^{\beta\lambda}_\alpha\partial_\lambda F^\alpha = 0, \qquad (VIII.4.5)$$

where the A's are linear functions in the Christoffel symbols of g.

Exercise VIII.4.1 *Compute the coefficients A. Write similar equations in the cases of generalized and damped generalized wave coordinates.*

[16]This preservation of the wave gauge property holds for Einstein equations with sources satisfying the conservation laws.

[17]Recall that we use geometric units where $G_E = 1$; equivalently, in the case $n + 1 = 4$, $8\pi G_N = 1$.

The homogeneous wave equations (4.5). imply $F^\alpha = 0$ if the initial data for g are such that the F^α and their first derivatives vanish on the initial manifold M; that is,

$$F^\alpha|_M = 0 \text{ and } \partial_0 F^\alpha|_M = 0. \tag{VIII.4.6}$$

Corresponding properties hold for the generalized and damped wave gauges.

VIII.4.1   Wave gauge constraints

We will deduce from (4.6) relations that must hold between initial data on the submanifold $x^0 = 0$ for the solution of the reduced Einstein equations to be a solution of the full Einstein equations, that is, to be a metric in wave gauge. The simplest way to find these relations is to use the contravariant tensor densities

$$\mathcal{G}^{\alpha\beta} := |\mathrm{Det}\, g|^{\frac{1}{2}} g^{\alpha\beta}. \tag{VIII.4.7}$$

It holds that

$$F^\alpha := g^{\lambda\mu}\Gamma^\alpha_{\lambda\mu} \equiv -|\mathrm{Det}\, g|^{-\frac{1}{2}}\partial_\beta \mathcal{G}^{\alpha\beta}. \tag{VIII.4.8}$$

Exercise VIII.4.2   *Prove this identity.*

Hint: The derivatives of the determinant of g are $\partial_\alpha(\mathrm{Det}\, g) \equiv (\mathrm{Det}\, g)g^{\lambda\mu}\partial_\alpha g_{\lambda\mu}$.

Straightforward computation using the identity (2.3) satisfied by the Ricci tensor shows that the Einstein tensor S satisfies the identity

$$S^{\alpha\beta} := R^{\alpha\beta} - \frac{1}{2}Rg^{\alpha\beta} \equiv S^{\alpha\beta}_{(h)} + \frac{1}{2}\left(g^{\alpha\lambda}\partial_\lambda F^\beta + g^{\beta\lambda}\partial_\lambda F^\alpha - g^{\alpha\beta}\partial_\lambda F^\lambda\right), \tag{VIII.4.9}$$

where the $S^{\alpha\beta}_{(h)}$ are a quasidiagonal system of wave operators in the metric g for the $\mathcal{G}^{\alpha\beta}$.

Suppose that $F^\alpha|_{t\equiv x^0=t_0} = 0$. Then also $\partial_i F^\alpha|_{x^0=t_0} = 0$, and the identity (4.9) implies the equality

$$S^{\alpha0}|_M = S^{\alpha0}_{(h)}|_M + \frac{1}{2}(g^{00}\partial_0 F^\alpha)|_M. \tag{VIII.4.10}$$

The expressions for $S^{\alpha0}_{(h)}$ and F^α show that $S^{\alpha0}|_M$ does not contain any second time derivative of g; that is, the equations

$$\mathcal{C}^\alpha := S^{\alpha0}|_M - T^{\alpha0}|_M = 0 \tag{VIII.4.11}$$

are **constraints on the initial data**. If they are satisfied, then a solution of the reduced Einstein equations with these initial data, i.e. such that $S^{\alpha0}_{(h)}|_M - T^{\alpha0}|_M = 0$, satisfies also $\partial_0 F^\alpha|_M$. We have proved the following theorem:

Theorem VIII.4.1   *A solution of the reduced Einstein equations with initial data such that $F^\alpha|_M = 0$ satisfies the full Einstein equations if and only if the initial data satisfy the constraints $\mathcal{C}^\alpha = 0$.*

An analogous theorem holds for the other types of wave gauges.

VIII.5   Local existence and uniqueness

An Einsteinian spacetime is constructed in local coordinates,[18] in a neighbourhood of a hypersurface M with equation $x^0 \equiv t = t_0$, from initial data $g_{\alpha\beta}|_M$, a Lorentzian metric for which M is spacelike, and $(\partial_0 g_{\alpha\beta})|_M$.

[18]See the geometric intrinsic formulations in Section VIII.7.

Theorem VIII.5.1 (*Fourès (Choquet)-Bruhat, 1952*)

1. *The Cauchy problem for the vacuum Einstein equations with initial data on a spacelike hypersurface $g_{\alpha\beta}|_M$ and $(\partial_t g_{\alpha\beta})|_M$ satisfying the wave gauge constraints has a local-in-time solution, a Lorentzian metric in wave gauge depending continuously on the initial data. Its value at a given point depends only on the past of this point.*

2. *The solution is locally geometrically unique.*

Part 1 of this theorem is a consequence of Theorems VIII.3.1 and VIII.4.1.

For Part 2, we first recall that the solution constructed in wave gauge from given initial data $g_{\alpha\beta}|_M$, $(\partial_0 g_{\alpha\beta})|_M$ satisfying the constraints is unique. To show local geometric uniqueness, we show that any given vacuum Einsteinian spacetime defined in a neighbourhood of M can be put in wave gauge in a possibly smaller neighbourhood of M, taking initial data that depend only on the original data. We will state a more general, global, uniqueness property after defining geometric initial data.

VIII.6   Solution of the wave gauge constraints

The expressions for $S^{\alpha 0}$ and F^α show that the wave gauge constraints are of the form, for the Einstein equations with source the stress–energy tensor T,

$$\mathcal{C}^\alpha \equiv \frac{1}{2}\mathcal{G}^{ij}\partial^2_{ij}\mathcal{G}^{\alpha 0} - \frac{1}{2}\mathcal{G}^{00}\partial^2_{it}\mathcal{G}^{\alpha i} + \mathcal{G}^{i0}\partial^2_{it}\mathcal{G}^{\alpha 0} + \mathcal{K}^{\alpha 0} - \mathcal{T}^{\alpha 0} = 0, \quad \text{(VIII.6.1)}$$

where $\mathcal{K}^{\alpha 0}$ depends only on $\mathcal{G}$ and its first derivatives.

Remark VIII.6.1   *Giving $\mathcal{G}^{\alpha\beta}, \partial_t \mathcal{G}^{\alpha\beta}$ for $t = t_0$ is equivalent to giving $g_{\alpha\beta}, \partial_t g_{\alpha\beta}$, for $t = t_0$.*

The constraints appear as a system of $n+1$ equations for $(n+2)(n+1)$ unknowns, $\mathcal{G}^{\alpha\beta}, \partial_t \mathcal{G}^{\alpha\beta}$, on the manifold $x^0 = 0$. This system, with more unknowns than equations, is undetermined. It is natural to try to split the initial data into specified quantities and $n+1$ unknowns, for which we wish to find a well-posed system of partial differential equations, which we expect to be elliptic because this is a pure, non-evolutionary, space problem. This was done for the first time in the general case[19] by using the equations (6.1). Two choices were suggested for splitting the initial Cauchy data between unknowns and given data:

[19]Bruhat (Choquet-Bruhat) (1962).

(1) Give arbitrarily on the initial manifold M the quantities $\mathcal{G}^{\alpha i}$. The $\partial_j \mathcal{G}^{\alpha i}$ are then known and the $\partial_t \mathcal{G}^{\alpha 0}$ are determined by the harmonicity conditions. The remaining unknowns, in this first step, are $u := \mathcal{G}^{00}$ and $v^{ij} := \partial_t \mathcal{G}^{ij}$. Equations (6.1) can be written in the following form:

- For $\alpha = 0$,

$$\mathcal{C}^0 \equiv \frac{1}{2} \mathcal{G}^{ij} \partial^2_{ij} \mathcal{G}^{00} - A = 0, \qquad \text{(VIII.6.2)}$$

where

$$A := \frac{1}{2} \mathcal{G}^{00} \partial^2_{it} \mathcal{G}^{0i} + \mathcal{G}^{i0} \partial^2_{it} \mathcal{G}^{00} + \mathcal{K}^{00} - \mathcal{T}^{00} \qquad \text{(VIII.6.3)}$$

depends only on the specified quantities and on the space derivatives of the unknowns of order at most 1

- For $\alpha = h$,

$$\mathcal{C}^h \equiv -\frac{1}{2} \mathcal{G}^{00} \partial_i (\partial_t \mathcal{G}^{hi}) + \mathcal{B}^h = 0, \qquad \text{(VIII.6.4)}$$

where

$$\mathcal{B}^h := \frac{1}{2} \mathcal{G}^{ij} \partial^2_{ij} \mathcal{G}^{h0} + g^{i0} \partial^2_{it} \mathcal{G}^{h0} + \mathcal{K}^{\alpha 0} - \mathcal{G}^{\alpha 0}. \qquad \text{(VIII.6.5)}$$

Equation (6.2) is a semilinear second-order equation for $\mathcal{G}^{00}$ with **principal symbol**[20] the positive-definite quadratic form $\mathcal{G}^{ij} \xi_i \xi_j$; it is an **elliptic**[21] equation.

To write the system (6.4) of n equations with the $n(n+1)/2$ unknowns $X^{hi} := \partial_t \mathcal{G}^{hi}$ as an elliptic system for n unknowns and $n(n-1)/2$ specified functions, one can use the **Berger–Ebin decomposition theorem** for a functional space E as

$$E = \ker D + \operatorname{range} D^*,$$

where D and D^* are a differential operator and its adjoint, one of which has injective symbol; the product DD^* is then elliptic. Since in this section we work in local coordinates, we simply recall here the classical procedure for decomposing a symmetric unknown X^{ij} on R^n. We set

$$X^{ij} \equiv Y^{ij} + Z^{ij}, \quad \text{with } \partial_i Y^{ij} = 0 \quad \text{and} \quad Z^{ij} = \partial_i U^j + \partial_j U^i.$$

Equations (6.4) take the form

$$\sum_i \partial^2_{ii} U^j + \partial_i \partial_j U^i = f^j.$$

They imply

$$\sum_i \partial^2_{ii} (\partial_j U^j) = \frac{1}{2} \partial_j f^j,$$

which is an elliptic equation for $\partial_j U^j$ and an elliptic system for U^j when $\partial_i U^i$ is known.

[20]The principal symbol of a scalar partial differential operator is the polynomial obtained by replacing in the highest-order terms the partial derivative ∂_i by the component ξ_i of a vector. In the case of a system, the principal symbol is a matrix (see the example below).

[21]A system of partial differential operators is elliptic if its principal symbol is an isomorphism (i.e. an invertible square matrix) for each non-zero vector ξ. For one scalar equation, this means that the principal symbol does not vanish for any non-zero set of real numbers ξ_i.

(2) In the spirit of Wheeler's 'thin-sandwich conjecture', we give arbitrarily on M the quantities $\mathcal{G}^{ij}$ and $\partial_t \mathcal{G}^{ij}$. The harmonicity conditions then determine $\partial_t \mathcal{G}^{0i} = -\partial_j \mathcal{G}^{ji}$ and imply the relation

$$\partial_t \mathcal{G}^{00} = -\partial_i \mathcal{G}^{0i}. \qquad \text{(VIII.6.6)}$$

The constraints now read, without needing the introduction of a further splitting, as a system of $n+1$ equations for the $n+1$ unknowns $\mathcal{G}^{00}$, $\mathcal{G}^{0i}$:

$$\mathcal{C}^h \equiv \frac{1}{2}\mathcal{G}^{ij}\partial^2_{ij}\mathcal{G}^{0h} - \mathcal{F}^h = 0, \qquad \text{(VIII.6.7)}$$

$$\mathcal{C}^0 \equiv \frac{1}{2}\mathcal{G}^{ij}\partial^2_{ij}\mathcal{G}^{00} - \mathcal{G}^{0i}\partial^2_{ij}\mathcal{G}^{j0} - \mathcal{F}^0 = 0, \qquad \text{(VIII.6.8)}$$

where $\mathcal{F}^0$ and $\mathcal{F}^h$ depend only on the given quantities and on the first space derivatives of the unknowns.

The system is elliptic because of the expression of its characteristic determinant, the determinant of its principal symbol.

Exercise VIII.6.1 *Show that this determinant is proportional to a power of the determinant of a Laplace operator.*

Hint: The principal part in the equation $C^h = 0$ contains only $\mathcal{G}^{0h}$.

The solution of an elliptic system on a manifold is a global problem. Two cases are of particular interest in General Relativity: the case of asymptotically Euclidean manifolds and the case of compact manifolds.

VIII.6.1   Asymptotically Euclidean manifolds

A particularly physically relevant case is the study in Einsteinian gravitation of isolated systems, composed of a few bodies far from any other source of gravitation, so that the metric tends to be flat far away from the studied system. Such systems are modelled in general by asymptotically Euclidean manifolds, defined in Chapter IV. For mathematical study of solutions of elliptic differential equations on asymptotically Euclidean manifolds, one uses either weighted Hölder spaces[22] or weighted Sobolev spaces.[23]

The linearization of the nonlinear constraints (6.7) and (6.8) for given Euclidean data reduces to the Euclidean Laplace equations, for which solutions are known in the mentioned functional spaces. The nonlinear constraints have been shown to have a solution[24] for given data near Euclidean data.

Various numerical methods are available for solution of the relevant equations. It would be interesting to solve them numerically, their solution giving initial data directly for the harmonically reduced evolution equations, which form a well-posed system.

[22]Choquet-Bruhat (1974), Chaljub-Simon and Choquet-Bruhat (1979).

[23]Cantor (1979), Choquet-Bruhat and Christodoulou (1981).

[24]Vaillant-Simon (1969).

VIII.6.2   Compact manifolds

Mathematicians are especially interested in the case of compact space manifolds without boundary, where topology plays an important role. Solutions of the constraints on compact manifolds have been extensively studied, but after writing them as elliptic equations for geometric unknowns—the induced metric $\bar{g}$ and extrinsic curvature K.

Compact space manifolds are of interest as cosmological models. However, the choice of a model for our whole cosmos is very speculative.

VIII.7   Geometric $n + 1$ splitting

We will in the following sections formulate geometric, i.e. coordinate independent, theorems about the existence and uniqueness of solutions of the Cauchy problem. We first give the following decomposition,[25] often called the ADM decomposition, although Arnowitt, Deser, and Misner[26] are only accountable for its Hamiltonian interpretation. This decomposition has been for many years intensively used for numerical computation, in spite of the fact that the evolution equations that it gives are not a hyperbolic system.

[25] Fourès (Choquet)-Bruhat (1956).
[26] Arnowitt et al. (1962).

VIII.7.1   Adapted frame and coframe

We consider a spacetime with manifold $V = M \times R$ and hyperbolic metric g such that the submanifolds $M_t \equiv M \times \{t\}$ are spacelike. We take a frame with space axes e_i tangent to the space slice M_t and time axis e_0 orthogonal to it. Such a frame, particularly adapted to the solution of the Cauchy problem, is called a **Cauchy adapted frame**. We take local coordinates adapted to the product structure, $(x^\alpha) = (x^i, x^0 = t)$, and we choose for e_i the vectors $\partial/\partial x^i$ of a natural frame on M_t, i.e.

$$\partial_i = \partial/\partial x^i. \qquad \text{(VIII.7.1)}$$

The dual coframe is found to be such that, with β a time-dependent vector tangent to M_t called the **shift**,

$$\theta^i = dx^i + \beta^i dt, \qquad \text{(VIII.7.2)}$$

while the 1-form θ^0 does not contain dx^i. We choose

$$\theta^0 = dt. \qquad \text{(VIII.7.3)}$$

The vector e_0, i.e. the Pfaffian derivative ∂_0, is then

$$\partial_0 \equiv \partial_t - \beta^j \partial_j, \quad \text{with} \quad \partial_t := \partial/\partial t. \qquad \text{(VIII.7.4)}$$

The vector e_0 is timelike since it is orthogonal to spacelike surfaces. We suppose that it defines the positive time orientation. In the coframe θ^α, one has $g_{i0} = 0$, and the metric reads

$$ds^2 = -N^2(\theta^0)^2 + g_{ij}\theta^i\theta^j, \quad \theta^0 = dt, \quad \theta^i = dx^i + \beta^i dt. \qquad \text{(VIII.7.5)}$$

The function N is called the **lapse**. We shall assume throughout that $N > 0$. The time-dependent, properly Riemannian, space metric induced by g on M_t is denoted by either g_t or $\bar{g}$. An overbar denotes a spatial tensor or operator, i.e. a t-dependent tensor or operator on M. Note that in our frame, $\bar{g}_{ij} = g_{ij}$ and $\bar{g}^{ij} = g^{ij}$.

A spacetime (V, g) with $V = M \times R$ and metric (7.5) is called a **sliced spacetime**.

The non-zero structure coefficients of a Cauchy adapted frame are found to be

$$C^i_{0j} = -C^i_{j0} = \partial_j \beta^i. \qquad \text{(VIII.7.6)}$$

We denote by $\bar{\nabla}$ the covariant derivative corresponding to the space metric $\bar{g}$. Using the general formulas from Chapter I, we find that

$$\omega^i_{jk} \equiv \Gamma^i_{jk} \equiv \bar{\Gamma}^i_{jk}, \qquad \text{(VIII.7.7)}$$

and

$$\omega^i_{00} \equiv N g^{ij} \partial_j N, \quad \omega^0_{0i} \equiv \omega^0_{i0} \equiv N^{-1} \partial_i N, \quad \omega^0_{00} \equiv N^{-1} \partial_0 N,$$

and

$$\omega^0_{ij} \equiv \frac{1}{2} N^{-2} \left(\partial_0 g_{ij} + g_{hj} C^h_{i0} + g_{ih} C^h_{j0} \right), \qquad \text{(VIII.7.8)}$$

from which we obtain

$$\omega^0_{ij} \equiv \frac{1}{2} N^{-2} \bar{\partial}_0 g_{ij}, \qquad \text{(VIII.7.9)}$$

where the operator $\bar{\partial}_0$ is defined on any t-dependent space tensor T by the formula

$$\bar{\partial}_0 := \frac{\partial}{\partial t} - \bar{L}_\beta, \qquad \text{(VIII.7.10)}$$

where $\bar{L}_\beta$ is the Lie derivative on M_t with respect to the spatial vector β. Note that $\bar{\partial}_0 T$ is a t-dependent space tensor of the same type as T.

The **extrinsic curvature** (also called the **second fundamental form**) of M_t is the t-dependent symmetric space tensor K given by

$$K_{ij} \equiv -\frac{1}{2} N^{-1} \bar{\partial}_0 g_{ij} \equiv -\omega^0_{ij} n_0 \equiv -N \omega^0_{ij}. \qquad \text{(VIII.7.11)}$$

The remaining connection coefficients are found to be (indices are raised with $\bar{g}$)

$$\omega^i_{0j} \equiv -N K^i_j + \partial_j \beta^i, \quad \omega^i_{j0} \equiv -N K^i_j. \qquad \text{(VIII.7.12)}$$

The trace in the metric $\bar{g}$ of the extrinsic curvature, often denoted by τ, is called the **mean curvature** of M_t :

$$\tau \equiv \mathrm{tr}_{\bar{g}} K \equiv \bar{g}^{ij} K_{ij}. \qquad \text{(VIII.7.13)}$$

K and τ play an important role in the initial-value formulation and other geometric problems of General Relativity.

Exercise VIII.7.1   *Show that the extrinsic curvature K of M_t is the projection on M_t of the spacetime gradient of the past-oriented unit normal n to M_t (some authors take the opposite orientation of n to define K). Show that $\tau := \mathrm{Tr}_{\bar g} K$ of K is equal to the spacetime divergence of n.*

Remark VIII.7.1   *The tensor K is symmetric. It can equivalently be defined, up to a factor $1/2$, as the Lie derivative of the spacetime metric in the direction of n. It does not depend on the value of n outside M_t.*

Remark VIII.7.2   *A positive value of τ signals a positive divergence of the past-directed normals. A negative value of τ corresponds to convergence of the past-directed normals, and hence future expansion of the submanifolds M_t.*

We deduce from the general formula giving the Riemann tensor and the splitting of the connection the following identities:

$$R_{ij,kl} \equiv \bar{R}_{ij,kl} + K_{ik}K_{lj} - K_{il}K_{kj}, \tag{VIII.7.14}$$

$$R_{0i,jk} \equiv N(\bar{\nabla}_j K_{ki} - \bar{\nabla}_k K_{ji}), \tag{VIII.7.15}$$

$$R_{0i,0j} \equiv N(\bar{\partial}_0 K_{ij} + N K_{ik}K^k{}_j + \bar{\nabla}_i \partial_j N) \tag{VIII.7.16}$$

From these formulas, we obtain the following expressions for the Ricci curvature:

$$N R_{ij} \equiv N \bar{R}_{ij} - \bar{\partial}_0 K_{ij} + N K_{ij} K^h_h - 2N K_{ik}K^k{}_j - \bar{\nabla}_i \partial_j N, \tag{VIII.7.17}$$

$$N^{-1} R_{0j} \equiv \partial_j K^h_h - \bar{\nabla}_h K^h{}_j, \tag{VIII.7.18}$$

$$R_{00} \equiv N(\bar{\partial}_0 K^h_h - N K_{ij} K^{ij} + \bar{\Delta} N). \tag{VIII.7.19}$$

Also, with $\bar{R} := g^{ij} \bar{R}_{ij}$,

$$g^{ij} R_{ij} = \bar{R} - N^{-1}\bar{\partial}_0 K^h_h + (K^h_h)^2 - N^{-1}\bar{\Delta} N, \tag{VIII.7.20}$$

$$R \equiv -N^{-2} R_{00} + g^{ij} R_{ij} = \bar{R} + K_{ij}K^{ij} + (K^h_h)^2 - 2N^{-1}\bar{\partial}_0 K^h_h - 2N^{-1}\bar{\Delta} N, \tag{VIII.7.21}$$

and

$$S_{00} \equiv R_{00} - \frac{1}{2}g_{00}R \equiv \frac{1}{2}(R_{00} + g^{ij}R_{ij}).$$

It follows that

$$2N^{-2}S_{00} \equiv -2S^0_0 \equiv \bar{R} - K_{ij}K^{ij} + (K^h_h)^2. \tag{VIII.7.22}$$

VIII.7.2   Dynamical system with constraints
for $\bar{g}$ and K

We see in the above decomposition of the Ricci tensor that none of the components of the Einstein tensor contains the time derivatives of the lapse N or the shift β. One is thus led to consider the Einstein equations as a dynamical system with constraints for the two fundamental tensors $\bar{g}$ and K of the space slices M_t, while N and β are gauge variables.[27]

Constraints

The expressions for the Ricci tensor lead to the following **constraints for $\bar{g}$ and K** on space slices M_t for solutions of the Einstein equations:[28]

Hamiltonian constraint

$$2N^{-2}S_{00} \equiv -2S_0^0 \equiv \bar{R} - K_{ij}K^{ij} + (K_h^h)^2 = \rho, \quad \text{with} \quad \rho := -2T_0^0;$$

$$(\text{VIII.7.23})$$

for the usual classical sources, $\rho \geq 0$.

Momentum constraint

$$N^{-1}S_{0j} \equiv N^{-1}R_{0j} \equiv \partial_j K_h^h - \bar{\nabla}_h K^h{}_j = J_j, \quad \text{with} \quad J_j := -N^{-1}T_{0j}.$$

$$(\text{VIII.7.24})$$

These constraints coincide with those found previously by using the wave gauge and shown to be preserved under evolution for stress–energy tensor sources satisfying the conservation laws.

Evolution

The equations

$$R_{ij} \equiv \bar{R}_{ij} - \frac{\bar{\partial}_0 K_{ij}}{N} - 2K_{jh}K_i^h + K_{ij}K_h^h - \frac{\bar{\nabla}_j \partial_i N}{N} = \rho_{ij}, \quad (\text{VIII.7.25})$$

together with the definition

$$\bar{\partial}_0 g_{ij} = -2NK_{ij}, \quad (\text{VIII.7.26})$$

determine the derivatives transverse to M_t of $\bar{g}$ and K when these tensors are known on M_t, as well as the lapse N and shift β and the source ρ_{ij}. It seemed natural to look at these equations as evolution equations determining $\bar{g}$ and K, while N and β, which are projections of the tangent vector on the timeline respectively on e_0 and on the tangent space to M, are considered as gauge variables. These evolution equations have been used extensively in the past for numerical computations. The Cauchy problem for these equations is clearly well posed for analytic data—a physically unsatisfactory condition because analytic functions have non-localizable support. The data of lapse and shift does not seem to be a good gauge choice. However, it can be proved[29] that, given N and β, the operator R_{ij} given by (7.25), with K_{ij} given by (7.26), is a second-order differential system on the g_{ij} that is **Leray–Ohya**[30] hyperbolic and

[27] Darmois (1927) was the first to view the Einstein equations (in a Gaussian gauge $N = 1$, $\beta = 0$) as an evolution system for $\bar{g}$, K supplemented by initial constraints.

[28] These constraints correspond to the Gauss and Codazzi equations well known to geometers.

[29] See YCB-OUP2009, Chapter VIII, Section 3.

[30] See J. Leray and Y. Ohya (1968).

[31]Gevrey classes are C^∞ functions whose derivatives satisfy inequalities weaker than those satisfied by analytic functions and assure in this case the convergence of their expansion in Taylor series. In contrast to an analytic function, a Gevrey class is not determined by the values of all its derivatives at a given point, nor by its value in an open set; it can have compact support.

[32]Choquet-Bruhat and Ruggeri (1983) for zero shift; Choquet-Bruhat and York (1996) for arbitrary shift.

[33]Called an algebraic gauge in the first formulation of Choquet-Bruhat and Ruggeri, but now called densitizing the lapse; it is equivalent to a wave gauge choice for the time variable.

[34]Christodoulou and Klainerman (1989).

[35]Choquet-Bruhat and Noutcheguéme (1986) and, independently, Anderson and York (1999).

[36]A system of N first-order partial differential equations in N unknown scalar functions of $n+1$ variables with principal part

$$A_J^{I,\alpha} \frac{\partial u^J}{\partial x^\alpha}, \qquad I,J = 1,\ldots,N$$

is symmetric hyperbolic if the matrices A^α with (real) elements $A_J^{I,\alpha}$ are symmetric and the matrix A^0 is positive-definite. The Cauchy problem for such systems has one and only one solution, in relevant functional spaces (Friedrichs, 1954). See, for instance, YCB-OUP2009, Appendix IV.

causal. The Cauchy problem for such systems has solutions in **Gevrey classes**,[31] causally determined from initial data. However, the infinite number of derivatives required from a Gevrey class foreshadow the poor stability of results of numerical computation, as has been confirmed by numerical analysis.

More promising, appropriate combination of the evolution equation and the constraints have been shown to lead,[32] modulo a gauge choice[33] for N, to a quasidiagonal system of wave equations for K that together, with (7.1), give an hyperbolic causal system for $\bar{g}$ and K.

Another possibility, leading to a mixed elliptic–hyperbolic system, is to impose on N an elliptic equation that implies that the trace of K takes a given value.[34]

Preservation of constraints

As shown before for the wave gauge constraints, the geometrically formulated constraints (7.23) and (7.24) are preserved under evolution when the stress–energy tensor source satisfies the conservation laws. A simple proof follows.[35]

Theorem VIII.7.1 *If $R_{ij} - \rho_{ij} = 0$ holds, then the constraints satisfy a linear homogeneous first-order symmetric hyperbolic system.[36] If they are satisfied initially, then they are satisfied for all time.*

Proof. If $R_{ij} - \rho_{ij} = 0$, then we have, in the Cauchy adapted frame, with $\rho := g^{\alpha\beta}\rho_{\alpha\beta}$,

$$R - \rho = -N^2(R^{00} - \rho^{00}).$$

Hence

$$S^{00} - T^{00} = \frac{1}{2}(R^{00} - \rho^{00}) \quad \text{and} \quad R - \rho = -2N^2(S^{00} - T^{00}) = 2(S_0^0 - T_0^0)$$

and

$$S^{ij} - T^{ij} = -\frac{1}{2}\bar{g}^{ij}(R - \rho) = -\bar{g}^{ij}(S_0^0 - T_0^0).$$

With these identities, we may derive from the Bianchi identities and the conservation laws a linear homogeneous system for $\Sigma_0^i \equiv S_0^i - T_0^i$ and for $\Sigma_0^0 \equiv S_0^0 - T_0^0$ with principal parts

$$N^{-2}\partial_0\Sigma_0^i + \bar{g}^{ij}\partial_j\Sigma_0^0, \quad \text{and} \quad \partial_0\Sigma_0^0 + \partial_i\Sigma_0^i.$$

Since this system can be made symmetric hyperbolic, it has a unique solution, which is zero if the initial values are zero.   ∎

Exercise VIII.7.2 *Show that Σ_0^i and Σ_0^0 satisfy a quasidiagonal homogeneous second-order system with principal part the wave operator of the spacetime metric g.*

VIII.7.3   Geometric Cauchy problem. Regularity assumptions

To give a geometric formulation of the solution of the Cauchy problem for the Einstein equations with data on a manifold M, we introduce the following definitions (in the usual case, $n = 3$).

Definition VIII.7.1

1. An ***initial data set*** is a triple $(M, \bar{g}, K)$ *with* $(M, \bar{g})$ *a Riemannian n-dimensional manifold and K a symmetric 2-tensor on M.*

2. *An Einsteinian vacuum **development** of $(M, \bar{g}, K)$ is a Lorentzian $(n+1)$-manifold (V, g) solution of the vacuum Einstein equations such that M is an embedded submanifold of V and $\bar{g}$ is induced by g on M, while K is the extrinsic curvature of M in (V, g).*

The following existence theorem holds as a direct consequence of previously stated and known results on hyperbolic differential equations. The weakest hypotheses on initial data using only classical Sobolev spaces and energy estimates are when $n = 3$, $\bar{g} \in H_3^{\mathrm{loc}}(M)$, $K \in H_2^{\mathrm{loc}}(M)$. Recall that $H_s^{\mathrm{loc}}(M)$ is a space of tensor fields with components that, along with their derivatives of order less than or equal to s, are square-integrable in relatively compact domains of local coordinates covering M.[37] The local uniqueness up to isometries is a consequence of the geometric uniqueness given in Theorem VIII.5.1.

Theorem VIII.7.2   *(existence and local uniqueness up to isometries) An initial data set $(M, \bar{g}, K)$ for the Einstein vacuum equations satisfying the constraints admits a vacuum Einsteinian development (V, g). This development is locally unique, up to isometries.*

Remark VIII.7.3   *The hypothesis $g \in H_3^{\mathrm{loc}}$ when $n = 3$ is not sufficient to ensure the uniqueness of geodesics issuing from one point with a given tangent. This uniqueness would be ensured if the Christoffel symbols were Lipschitzian, which is not the case for H_3^{loc} metrics on three-dimensional space.*

Lowering the regularity required of data is conceptually important for the understanding of the mathematics and the physics of the theory, and essential in the study of global problems. Geometric hyperbolic evolution systems involving the Riemann curvature tensor instead of wave gauges have been considered and have allowed for the broadening of admissible functional spaces for the initial data. Klainerman, Rodnianski, and Szeftel[38] have conjectured that sufficient conditions for the existence of a Lorentzian metric solution of the vacuum Einstein equations in a four-dimensional neighbourhood of a 3-manifold M supporting initial data $(\bar{g}, K)$ are that $Ricci(\bar{g}) \in L^2$, that $\bar{\nabla} K \in L^2$ locally on M, and that the volume radius[39] of $(M, \bar{g})$ is strictly positive. In several long and difficult papers using refined functional analysis, they have proved their conjecture interpreted as a continuation argument for the Einstein equations; that is, the spacetime constructed by evolution from smooth

[37] Various, more intrinsic, equivalent definitions can be found, for instance, in YCB-OUP2009, Appendix III, Section 3.7.

[38] Klainerman, Rodnianski, and Szeftel (2012).

[39] The volume radius is the lower bound of the quotient by r^3 of the volume of geodesic balls of radius r in $(M, \bar{g})$.

data can be smoothly continued, together with a time foliation, as long as the curvature of the foliation and the first covariant derivatives of its second fundamental form remain L^2-bounded on the leaves of the foliation.

VIII.8   Solution of the constraints by the conformal method

The conformal method, initiated by Lichnerowicz and developed by Choquet-Bruhat and York, has been intensively used to prove existence of solutions of the constraints. Variants of it are now used in numerical computations of initial data for the two-body problem. It has the advantage of turning the momentum constraint into a linear system independent of the conformal factor for a weighted extrinsic curvature when the mean extrinsic curvature τ is zero or constant.[40] The conformal factor then satisfies a semilinear elliptic operator with principal part the Laplace operator of the conformal metric. Given a TT (transverse traceless) tensor, the momentum constraint can be written as a linear elliptic system for a vector field.[41] The method was used in early numerical computations, essentially by taking conformally flat metrics.

VIII.8.1   Conformally formulated (CF) constraints

When the sources ρ and J are known, the unknowns in the constraints (7.24) and (7.23) are the metric $\bar{g}$ and the tensor K. It is mathematically clear that these scalar and vector equations do not have a unique solution, even geometrically. Physically, this fact corresponds to the property of Einsteinian gravity that it is its own source. Roughly speaking, 'radiation data' should also be given, the constraints then becoming an elliptic system for a scalar function and a vector field. The geometric meaning of the lower-order terms permits discussion of the existence and uniqueness of solutions on general manifolds. We treat for simplicity of notation the usual physical case $n = 3$.[42]

The **Hamiltonian constraint** (7.23) reads, for general n,

$$\bar{R} - (|K|_{\bar{g}}^2 + \rho - \tau^2) = 0, \quad \bar{R} := R(\bar{g}), \quad \tau := \mathrm{tr}_{\bar{g}}\, K. \qquad \text{(VIII.8.1)}$$

This equation is turned into an elliptic equation for a scalar function φ by considering the metric $\bar{g}$ as given up to a conformal factor. In the case $n = 3$, one sets

$$\bar{g}_{ij} = \varphi^4 \gamma_{ij},$$

with γ a given metric on M and φ a function to be determined.

The scalar curvatures of the conformal metrics $\bar{g}$ and γ are found by straightforward computation to be linked when $n = 3$ by the formula

$$R(\bar{g}) \equiv \varphi^{-5}\left[\varphi R(\gamma) - 8\Delta_\gamma \varphi\right]. \qquad \text{(VIII.8.2)}$$

---

[40] Lichnerowicz (1944) for $\tau = 0$; York (1972) for $\tau = $ constant.

[41] Choquet-Bruhat (1971) for $\tau = 0$; York (1972) for $\tau = $ constant.

[42] The case of arbitrary n is treated in detail in YCB-OUP2009, Chapter VII, Section 3.

Exercise VIII.8.1   *Prove an analogous formula for general $n \geq 3$ with $\varphi^{4/(n-2)}$.*

Hint: The choice φ^4 when $n = 3$ is made to eliminate quadratic terms in first derivatives of φ.

The identity (8.2) implies the following theorem:

Theorem VIII.8.1   *When γ, K, and the source ρ are known, the Hamiltonian constraint is a semilinear second-order elliptic equation for φ, linear also in first derivatives, which, in the case $n = 3$, reads*

$$8\Delta_\gamma \varphi - R(\gamma)\varphi + (|K|_{\bar{g}}^2 - \tau^2 + \rho)\varphi^5 = 0. \qquad \text{(VIII.8.3)}$$

To solve the **momentum constraint**, we split the unknown K into a weighted traceless part $\tilde{K}$ and its trace τ. In the case $n = 3$, we set

$$K_{ij} = \varphi^{-2}\tilde{K}_{ij} + \frac{1}{3}\bar{g}_{ij}\tau. \qquad \text{(VIII.8.4)}$$

Equivalently, with indices in K and $\tilde{K}$ being respectively lifted with $\bar{g}^{ij}$ and γ^{ij},

$$K^{ij} = \varphi^{-10}\tilde{K}^{ij} + \frac{1}{3}\bar{g}^{ij}\tau, \qquad \bar{g}^{ij} = \varphi^{-4}\gamma^{ij} . \qquad \text{(VIII.8.5)}$$

The tensor $\tilde{K}$ is symmetric and traceless; indeed,

$$\mathrm{tr}_\gamma \tilde{K} \equiv \gamma^{ij}\tilde{K}_{ij} = \varphi^{-2}\bar{g}^{ij}\left(K_{ij} - \frac{1}{3}\bar{g}_{ij}\tau\right) = 0.$$

Straightforward computation shows that the momentum constraint becomes, with D the covariant derivative in the given metric γ,

$$D_i\tilde{K}^{ij} = \frac{2}{3}\varphi^6\gamma^{ij}\partial_i\tau + \varphi^{10}J. \qquad \text{(VIII.8.6)}$$

This equation has the interesting property that it does not contain φ when J is zero and τ is constant on M, that is, when M is be a submanifold with constant mean extrinsic curvature in the ambient spacetime. We have the following theorem:

Theorem VIII.8.2   *In the case $\tau = $ constant and $J = 0$, the symmetric 2-tensor $\tilde{K}^{ij} := \varphi^{-10}K^{ij}$ is a **TT tensor** (transverse, traceless) and the momentum constraint reduces to linear homogeneous sytem for $\tilde{K}$:*

$$D_j\tilde{K}^{ij} = 0, \qquad \gamma_{ij}\tilde{K}^{ij} = 0. \qquad \text{(VIII.8.7)}$$

Exercise VIII.8.2   *Show that the space of TT tensors is the same for two conformal metrics.*

Returning to the Hamiltonian constraint, we compute, using the definition of $\tilde{K}$,

$$|K|_{\bar{g}}^2 := \bar{g}^{ih}\bar{g}^{jk}K_{ij}K_{hk} = \varphi^{-12}|\tilde{K}|_\gamma^2 + \frac{1}{3}\tau^2. \qquad \text{(VIII.8.8)}$$

Exercise VIII.8.3   *Prove this formula.*

Hints: $\tilde{K}$ is traceless and $\gamma^{ih}\gamma^{jk}\gamma_{ij}\gamma_{hk} = \delta^h_j \delta^j_h = 3$.

Exercise VIII.8.4   *Write down the corresponding formula for arbitrary n.*

The Hamiltonian constraint now reads

$$8\Delta_\gamma \varphi - R(\gamma)\varphi + \varphi^{-7}|\tilde{K}|^2_\gamma + \left(-\frac{2}{3}\tau^2 + \rho\right)\varphi^5 = 0. \qquad (VIII.8.9)$$

It is a semilinear elliptic equation for φ when γ (a Riemannian metric), τ, ρ, and $\tilde{K}$ are known. It is called the **Lichnerowicz equation**.[43]

A scaling of the momentum of the sources by setting, when $n = 3$,[44]

$$J = \varphi^{-10}\tilde{J},$$

with $\tilde{J}$ considered as a known quantity, permits the extension of the decoupling property to the non-vacuum case. The momentum constraints now read

$$D_j \tilde{K}^{ij} = \tilde{J}^i.$$

The scalar part ρ of the sources can also be scaled. The scaling proposed by York, justified by physical considerations (at least in the case $n = 3$ for electromagnetic field sources), is

$$\rho = \varphi^{-8}\tilde{\rho},$$

with $\tilde{\rho}$ considered as a given function. The Hamiltonian constraint then reads

$$8\Delta_\gamma \varphi - R(\gamma)\varphi + |\tilde{K}|^2_\gamma \varphi^{-7} + \tilde{\rho}\varphi^{-3} - \frac{2}{3}\tau^2 \varphi^5 = 0. \qquad (VIII.8.10)$$

VIII.8.2   Elliptic system

We complete the writing of the CF constraints as an elliptic partial differential equation system on the initial manifold by a treatment of the momentum constraint[45] analogous to that indicated in Section VIII.6. We can split the tensor $\tilde{K}$ into an element of the kernel of the homogeneous part of the momentum constraint operator, i.e. a TT tensor Y for the given metric γ, and an element of the range of the dual operator, the conformal Lie derivative Z of a vector field X; that is, we set

$$\tilde{K}_{ij} = Y_{ij} + Z_{ij}, \quad \text{with} \quad D_i Y^{ij} = 0, \qquad (VIII.8.11)$$

and, when $n = 3$,

$$Z_{ij} := (\mathcal{L}_{\gamma,\text{conf}}X)_{ij} \equiv D_i X_j + D_j X_i - \frac{2}{3}\gamma_{ij} D_k X^k. \qquad (VIII.8.12)$$

A tensor of this form satisfies the momentum constraint if and only if the vector X satisfies the second-order system

$$\Delta_{\gamma,\text{conf}}X := D \cdot (\mathcal{L}_{\gamma,\text{conf}}X) = D \cdot \tilde{J}.$$

[43]It was obtained by Lichnerowicz (1944) for $n = 3$ and extended to general n by Choquet-Bruhat (1996). Scaling of the sources was introduced by York (1972). We will still refer to as Lichnerowicz equations all the equations deduced from the Hamiltonian constraint by the conformal method.

[44]York (1972).

[45]Choquet-Bruhat (1971a) and York (1972); Fischer and Marsden (1979), Choquet-Bruhat and York (1980), Choquet-Bruhat, Isenberg and York (2000) and see further references in Choquet-Bruhat and York (2002). For the coupling with a scalar field see Choquet-Bruhat, Isenberg and Pollack (2007).

Exercise VIII.8.5 *Show that this system is a linear elliptic system for X when γ and $\tilde{J}$ are known.*

The search for the arbitrary TT tensor Y_{ij} can itself be done through the data of an arbitrary traceless tensor U_{ij} by setting $Y := \mathcal{L}_{\gamma,\mathrm{conf}}V + U$ and imposing the requirement that the vector V satisfy the following elliptic system:

$$\Delta_{\gamma,\mathrm{conf}}V \equiv D \cdot (\mathcal{L}_{\gamma,\mathrm{conf}}V) = -D \cdot U.$$

The traceless tensor Y is then also transverse, i.e. satisfies $D \cdot Y = 0$. The arbitrary data in the extrinsic curvature K is in this scheme the symmetric traceless tensor U^{ij}. The physical extrinsic curvature is given by

$$K_{ij} = \varphi^{-2}\left[\mathcal{L}_{\gamma,\mathrm{conf}}(X+V)^{ij} + U^{ij}\right] + \frac{1}{3}\bar{g}^{ij}\tau.$$

The splitting of the solution $\tilde{K}$ into a given traceless tensor U and the conformal Lie derivative of an unknown vector X depends on the choice of γ, not only on its conformal class.[46]

[46] More refined splittings have been given by York under the name of conformal thin sandwich.

Exercise VIII.8.6 *Prove this statement.*

VIII.8.3   Physical comment

The conformal method is a mathematical, geometric convenience. The tilde ($\sim$) quantities are not directly observable. There is a large arbitrariness in the choice of their scaling, although some justifications are given a posteriori, at least for electromagnetic field sources. The tilde quantities play the role of parameters to construct initial data solutions of the constraints that can, hopefully, be used for evolution and lead to results that can be confronted with observations.

VIII.9   Motion of a system of compact bodies

There is now a wealth of new results from observations of motions of stars and galaxies obtained by powerful Earth-based and satellite telescopes. Also, a new generation of gravitational wave detectors (LIGO, GEO, TAMA, and VIRGO) are now operational, and it is hoped that within the next few years they will reach sensitivities that will allow for the first time observations of gravitational radiation. The prime targets of these observations are the motions of compact binaries: black holes and neutron stars. A fundamental problem for Einsteinian gravitation theory is the modelling of the motion and the gravitational wave emission of systems of compact bodies. For many years, considerable effort has been spent towards this goal, with little success until fairly recently, because of the nonlinearity of the equations, their hyperbolic–elliptic

[47]The problem becomes arduous when more than two bodies are involved.

character, and the insufficient stability of the solution of the chosen evolution equations, in particular the so-called ADM ones for $\bar{g}$ and K. In Newtonian theory, two bodies represented by pointlike masses describe conics with focus situated at the centre of gravity.[47] In Einsteinian gravitation, the dynamics of two-body systems is already too complex to be solved exactly, and prediction of the gravitational radiation they will emit in their inspiralling and merger can be obtained only by refined approximation methods (analytical or numerical) after a clever choice of unknowns and gauges, and splitting of equations between constraints and evolution. The mathematical well-posedness of the considered system is of course essential for the stability and reliability of the results. Refined analytical results have been obtained in the last decade both by the high-order post-Newtonian approach and by using the effective one-body (EOB) method. In parallel, thanks to the advent of extremely powerful computers, accurate numerical simulations of two-body systems have been obtained using various numerical codes. The results of the two types of methods (analytical and numerical) have been compared and found to be in remarkable agreement.[48]

[48]For recent comparisons, see, for instance, Damour, Nagar, and Bernuzzi (2013) and Hinder et al. (2014).

VIII.9.1   Effective one-body (EOB) method

[49]See Damour and Nagar (2011).

The EOB formalism[49] is an analytical approach that replaces the usual post-Newtonian expanded approximations to the motion and radiation of binary systems by *resummed expressions*. This formulation allows one to describe the dynamics of binary black hole systems up to the moment of the coalescence of the two black holes, and is further completed by the description of the final ringdown gravitational wave signal emitted by the distorted black hole formed during the coalescence process. This formalism thereby provides a quasi-analytical description of the entire waveform emitted by a binary black hole system from early inspiralling up to the final ringdown. It has also been extended to the description of the coalescence of binary neutron star systems.

VIII.9.2   Numerical Relativity

Although its origins lie in the 1960s, it is in the years since 2005 that Numerical Relativity has seen considerable expansion, owing to the enormous growth in the power of computers and to the development of stable codes describing the coalescence of compact binary systems. This has permitted the modelling of more realistic situations than before, although difficulties remain. Two main codes are now in use: one

[50]See Pretorius (2007) and references therein.

is based on damped generalized wave coordinates,[50] while the other,

[51]Baumgarte–Shapiro–Shibata–Nakamura.

called BSSN,[51] is a modification of the 3+1 decomposition using conformal weights. The code BSSNz^4 improves the stability of results by introducing four more unknowns linked with the harmonicity functions.

[52]See, for instance, Rezzolla and Zanotti (2013).

Numerical Relativity is a science in itself, outside of the scope of this book.[52]

VIII.10   Global properties

VIII.10.1   Global hyperbolicity and global uniqueness

A fundamental notion in the study of global properties of solutions of hyperbolic partial differential equations is that of **global hyperbolicity**, defined by Jean Leray[53] for general hyperbolic differential equations as compactness[54] (in a functional space) of paths joining two points whose tangent belong to the cone determining the domain of dependence of solutions of the Cauchy problem; that is, causal paths in the case of relativistic causal systems. Global hyperbolicity forbids, in particular, the existence of closed causal curves.[55] Later, Penrose[56] introduced what he called **strong causality**: a Lorentzian manifold (V, g) is said to be strongly causal if any neighbourhood of any point x contains a smaller neighbourhood such that no causal curve penetrates in it more than once. It can be proved[57] that global hyperbolicity is equivalent to strong causality together with compactness of the subsets of V that are intersections of the past of any point x with the future of any other point y, traditionally denoted by $I_-(x) \cap I_+(y)$.

A very useful definition due to R. Geroch is that of a **Cauchy hypersurface** in a Lorentzian manifold (V, g), that is, a submanifold S of codimension 1 intersected once and only once by each inextendible timelike curve. Geroch[58] proved that the existence of a Cauchy surface is equivalent to global hyperbolicity. He also proved the important property that the support of the spacetime is then a product $S \times R$.

With these definitions, one can complete the existence theorem by a global uniqueness theorem proved by Choquet-Bruhat and Geroch;[59] it uses in particular Zorn's lemma. Proofs without the use of this lemma have been published recently.[60]

Theorem VIII.10.1 *(existence and geometric global uniqueness) A vacuum Einsteinian development (V, g) of an initial data set $(M, \bar{g}, K)$, satisfying the vacuum Einstein constraints, exists and is unique (up to an isometry) in the class of maximal[61] globally hyperbolic spacetimes. The manifold M embedded in (V, g) is a Cauchy hypersurface.*

VIII.10.2   Global existence

A solution of the vacuum Einstein equations is called global if it is a complete Lorentzian manifold. It is generally called singular if it is incomplete. Incompleteness can result from the appearance of a curvature singularity (with the **Kretschmann scalar**[62] tending to infinity on a causal curve for a finite value of the canonical parameter) or another phenomenon.[63] Proving either global existence of solutions or, on the contrary, their incompleteness and the formation of singularities, are difficult problems and the subject of active research. Some remarkable achievements linked with the special properties of the Einstein equations, fundamental in physics but quite particular nonlinear geometric

[53] Leray (1953). The notion was applied and described in detail for Lorentzian manifolds by Choquet-Bruhat (1968).

[54] The empty set is considered as compact.

[55] A path is a mapping from an interval of R into the manifold; a curve is the image of a path in the manifold. A closed causal curve can be covered by an infinite sequence of causal paths joining two points, a non-compact set.

[56] See, for instance, Penrose (1968).

[57] See Choquet-Bruhat (1968).

[58] Geroch (1970). For a detailed proof, see, for instance, Ringström (2009).

[59] Choquet-Bruhat and Geroch (1969).

[60] See Ringström (2009), corrected by Ringström in <http://www.math.kth.se/~hansr/mghd.pdf> and in Sbierski (2013).

[61] Which cannot be embedded in a bigger one.

[62] The square of the Riemann tensor.

[63] See Problem IV.11.4 on Taub spacetime.

differential equations, have been obtained in the last twenty years with the use of advanced and intricate mathematics. The details of these achievements are outside of the scope of this book (in fact of any book of reasonable size!). We will only give some definitions and quote some results with references to original papers.

A future global existence theorem for data near Minkowskian data on a manifold tending to a null manifold at infinity was given by H. Friedrich[64] using an original conformal construction and the Newman–Penrose formalism.[65]

A breakthrough in the problem of global existence for initial data on a spacelike manifold was the proof in 1989 by Christodoulou and Klainerman[66] of the global nonlinear stability of Minkowski spacetime—that is, the construction of complete globally hyperbolic solutions of the vacuum Einstein equations with asymptotically Euclidean initial data near data for the Minkowski spacetime.[67] They used in particular a double null foliation and delicate estimates of the Riemann tensor.[68] The proof of the global nonlinear stability of Minkowski spacetime was obtained later by Lindblad and Rodnianski[69] through wave coordinates and the use of special properties (a kind of generalized **null condition**) of the Einstein equations. The global existence result has been extended to cases of Einstein equations with field sources, and even to the case or irrotational fluid sources.[70]

Proofs of global existence have been obtained for small initial data on several categories of compact manifolds with symmetries by: Moncrief, by Choquet-Bruhat and Moncrief, and by Choquet-Bruhat, and, in a case without symmetries, by Andersson and Moncrief.[71]

VIII.11   Singularities and cosmic censorship conjectures

The problem of the formation of singularities from generic initial data was attacked by Penrose and Hawking in the 1970s.[72] Inspired by the Schwarzschild solution, they discussed what they called the **strong** and the **weak cosmic censorship conjectures**. These conjectures concern generic Einsteinian spacetimes with physically reasonable sources. A generic spacetime can be understood as a spacetime with no isometry group, or as a spacetime that is stable (in some sense to be defined), under small perturbations. Reasonable sources are physical sources that have a hyperbolic, causal evolution and do not have their own singularities (shocks, shell crossings, etc.).

VIII.11.1   Strong cosmic censorship conjecture

The strong cosmic censorship conjecture[73] aims at proving the deterministic character of General Relativity at the classical (non-quantum) level. It can be formulated as follows:

---

[64]Friedrich (1986).

[65]This formalism uses on a four-dimensional manifolds Weyl spinors represented by objects with 2-index components.

For a tensorial formulation of the Friedrich conformal system, see Choquet-Bruhat and Novello (1987), CB-DM II, Chapter V, Section 7.

[66]See Christodoulou and Klainerman (1993). See also Klainerman and Nicolo (2003) and Bieri (2007).

[67]This theorem showed that Einstein's intuition that the only global asymptotically Euclidean vacuum solution of his equations is Minkowski spacetime required one more hypothesis to be true, namely a faster fall-off of the metric at spatial infinity, which implies the vanishing of the ADM mass.

[68]Their proof has been notably simplified by Lydia Bieri. For the use of equations satisfied by the Riemann tensor, see Problem I.14.5 in Chapter I.

[69]Lindblad and Rodnianski (2005).

[70]Rodnianski and Speck (2009).

[71]See articles and references in Chruściel and Friedrich (2004).

[72]See Hawking and Ellis (1973).

[73]Suggested in private discussions by Geroch and Penrose in 1969, and formalized by Eardley and Moncrief (1981).

Strong cosmic censorship conjecture *The maximal globally hyperbolic vacuum Einsteinian development of generic[74] initial data is inextendible as a vacuum, even non-globally hyperbolic, Einsteinian spacetime.[75]*

The inextendibility is sometimes conjectured to hold for all developments, not necessarily solutions of the Einstein equations—this is certainly the case when incompleteness is due to the appearance of infinite curvature.

Examples of non-globally hyperbolic extensions of a globally hyperbolic spacetime with initial data on S^3 are provided by the Taub–NUT spacetimes.[76] However, these extensions do not provide a counterexample to the strong cosmic censorship conjecture, because the Taub spacetime, because of its symmetries, is not generic. V. Moncrief and J. Isenberg have shown that some qualitative features of Taub–NUT spacetime imply in fact the existence of an isometry group.

[74]Loosely speaking 'generic' means 'without special properties'. In a more precise mathematical sense, it can be interpreted that 'generic initial data' form a dense subset of the set of possible initial data in some relevant topology, see for instance discussion in Isenberg and Moncrief (2002).

[75]The formulation leaves open the smoothness required of extendibility and also the hypotheses on initial data. It was pointed out by P. Chruściel that the requirement that the initial manifold $(M, \bar{g})$ be complete is not sufficient to make the conjecture plausible—initial data for Minkowski spacetime on a hyperboloid are then a trivial counterexample.

[76]See Problem IV.11.4 in Chapter IV.

VIII.11.2 Weak cosmic censorship conjecture

The original idea of Penrose came from the study of spherical gravitational collapse where a black hole forms, hiding the singularity to timelike observers. The **weak cosmic censorship conjecture**[77] is that in generic Einsteinian spacetimes with physically reasonable sources, it is not possible for there to form any **naked singularity**, that is, a singularity visible by an observer; in other words, the *past of no point x contains a future causal curve that is inextendible.*

[77]The weak and strong cosmic conjectures are independent.

Note that the big bang is not a counterexample to this conjecture—it has no past, and hence does not correspond to any future inextendible causal curve. Nor is the Schwarzschild metric with $m < 0$ on the manifold $(R^3 - \{0\}) \times R$ a counterexample to the 'non-naked singularity' conjecture, because it is not considered to be a physically meaningful metric.

A more elaborate version of the weak cosmic censorship conjecture uses the Penrose definition of conformal null infinity—it raises technical difficulties regarding the existence and smoothness of the boundary of a Penrose diagram for general spacetimes.

Christodoulou has obtained in a series of papers complete results supporting the weak cosmic censorship conjecture for the Einstein–scalar equations with spherical symmetry.

An important theorem due to Penrose links the existence of singularities with trapped surfaces.[78] In a four-dimensional spacetime, a **trapped surface** S is a two-dimensional compact spacelike surface without boundary (typically a sphere) such that the two families of future-directed null geodesics orthogonal to S have a positive convergence whether they are directed inwards or outwards. Penrose's theorem says that in a non-compact spacetime, a trapped surface always hides an incompleteness of the spacetime.

[78]Penrose (1965).

[79] Christodoulou (2009).

[80] Klainerman, Luk, and Rodnianski (2013).

In a 500-page book[79] Christodoulou proves the formation of trapped surfaces due to 'short-pulse' data on a characteristic cone. Christodoulou's[80] construction has recently been extended by Klainerman, Luk, and Rodnianski to data on a null geodesic segment and his proof has been simplified.

VIII.12   Problems

VIII.12.1   Symmetric hyperbolic systems

The system of N first-order linear partial differential equations on $R^n \times R$,

$$\mathcal{M}^\alpha \frac{\partial}{\partial x^\alpha} u + \mathcal{A} u + f = 0, \quad \text{i.e.} \quad \mathcal{M}^{\alpha, IJ} \frac{\partial}{\partial x^\alpha} u_I + \mathcal{A}^{IJ} u_I + f^J = 0, \tag{VIII.12.1}$$

with $u = (u_I, I = 1, \dots, N)$ a set of N unknown functions, $f = (f^I, I = 1, \dots, N)$ a set of given functions, and $\mathcal{M}^\alpha$ and $\mathcal{A}$ given $N \times N$ matrices, is called **symmetric hyperbolic (SH)** if the matrices $\mathcal{M}^\alpha$ are symmetric ($\mathcal{M}^{\alpha, IJ} = \mathcal{M}^{\alpha, JI}$) and the quadratic form defined by the matrix $\mathcal{M}^0$, the coefficient of $\partial/\partial x^0$, is positive-definite; that is,

$$\mathcal{M}^0(u, u) \equiv \mathcal{M}^{0, IJ} u_I u_J > 0 \quad \text{for all } u \neq 0.$$

1. Define a function called the **energy of** u at time $x^0 = t$ by the integral

$$E_t(u) := \frac{1}{2} \int_{x^0 = t} \mathcal{M}^0(u, u) \, d^n x, \quad d^n x := dx^1 \dots dx^n. \tag{VIII.12.2}$$

Show that under appropriate smoothness and boundary conditions, $E_t(u)$ satisfies the **energy equality**

$$E_T(u) = E_{t_0}(u) + \int_{t_0}^T \int_{x^0 = t} \left[\frac{1}{2} \left(\frac{\partial}{\partial x^\alpha} \mathcal{M}^{\alpha IJ} \right) u_J u_I \right.$$
$$\left. + \mathcal{A}^{IJ} u_J u_I + f^J u_J \right] d^n x \, dt. \tag{VIII.12.3}$$

2. Show that the energy equality implies an **integral energy inequality** of the form

$$E_T(u) \leq E_{t_0}(u) + \int_{t_0}^T \left[C_1(t) E_t(u) + C_2(t) E_t^{\frac{1}{2}}(u) \right] dt. \tag{VIII.12.4}$$

Deduce from it a bound of the energy.

Solution

1. It holds that

$$u_J \mathcal{M}^{\alpha, IJ} \frac{\partial}{\partial x^\alpha} u_I \equiv u_I \mathcal{M}^{\alpha, JI} \frac{\partial}{\partial x^\alpha} u_J,$$

and hence, if the matrices $\mathcal{M}^\alpha$ are symmetric (i.e. $\mathcal{M}^{\alpha, IJ} = \mathcal{M}^{\alpha, JI}$), then

$$u_J \mathcal{M}^{\alpha, IJ} \frac{\partial}{\partial x^\alpha} u_I = \frac{1}{2} \mathcal{M}^{\alpha, IJ} \frac{\partial}{\partial x^\alpha} (u_I u_J)$$

$$\equiv \frac{1}{2} \frac{\partial}{\partial x^\alpha} (u_I u_J \mathcal{M}^{\alpha, IJ}) - \frac{1}{2} u_I u_J \frac{\partial}{\partial x^\alpha} \mathcal{M}^{\alpha, IJ}.$$

Under appropriate smoothness and fall-off conditions at infinity, the integral on R^n of $(\partial/\partial x^i)(u_I u_J \mathcal{M}^{i, IJ})$ vanishes owing to Stokes's formula. There remains the equality

$$\int_{t_0}^T \int_{x^0 = t} \left\{ \frac{1}{2} \left[\frac{\partial}{\partial x^0} (u_I u_J \mathcal{M}^{\alpha, IJ}) - u_I u_J \frac{\partial}{\partial x^\alpha} \mathcal{M}^{\alpha, IJ} \right] \right.$$

$$\left. + \mathcal{A}^{IJ} u_I u_J + f^J u_J \right\} dx^n \, dx^0 = 0,$$

which gives the energy equality after performing in the first term the integrations first with respect to x^n and then over x^0.

2. Assume that $\mathcal{M}^0$ is uniformly positive-definite, i.e.

$$\mathcal{M}^{0, IJ} u_I u_J \geq C_0 |u|^2, \quad C_0 > 0, \text{ a constant,}$$

while $(\partial/\partial x^\alpha)\mathcal{M}^{\alpha, IJ}$ and $\mathcal{A}^{IJ}$ are uniformly bounded and f^J is square-integrable for each t. It is straightforward to deduce an integral energy inequality of the given form from the energy equality using elementary properties of integrals and the Cauchy–Schwarz inequality

$$\left| \int_{R^n} u f \, d^n x \right| \leq \left(\int_{R^n} |u|^2 \, d^n x \right)^{\frac{1}{2}} \left(\int_{R^n} |f|^2 \, d^n x \right)^{\frac{1}{2}}.$$

To prove the energy inequality, one uses a general theorem[81] for differential equations that says that if $f(t, y)$ is a function continuous in t and Lipschitzian in y, then the differential equation

$$z' = f(t, z)$$

[81] See the proof for instance in Choquet (2006).

has one and only one maximal solution taking a given initial value. Any C^1 function y satisfying the inequalities

$$y' \leq f(t, y), \quad y(t_0) \leq z(t_0)$$

satisfies $y(t_0) \leq z(t_0)$.

This theorem implies that if $y(t)$ satisfies the integral inequality

$$y(T) \leq y(t_0) + \int_{t_0}^T \left[C_1 y(t) + C_2 y(t)^{\frac{1}{2}} \right] dt, \qquad \text{(VIII.12.5)}$$

with C_1 and C_2 non-negative, then $y(t)$ is less than or equal to the function z satisfying the corresponding equality[82] and such that $z(t_0) \geq y(t_0)$.

In the considered case, $y(t) := E_t(u)$, one finds that z is the solution of an equation of the form

$$z' = C_1 z + C_2 z^{\frac{1}{2}},$$

i.e. setting $z = \zeta^2$, the linear equation

$$2\zeta' = C_1 \zeta + C_2,$$

which has a solution with initial data $\zeta_0 := y_0^{\frac{1}{2}}$ continuous and bounded for all finite t. Writing down this solution is left as an exercise for the reader.

VIII.12.2   The wave equation as a symmetric hyperbolic system

1. The wave equation with mass a constant m for a scalar function u on a Lorentzian manifold (V, g) is

$$g^{\alpha\beta}\nabla_\alpha\partial_\beta u - mu + f = 0.$$

For $g = \eta$, the Minkowski metric, write this as a symmetric hyperbolic system.

For m a constant, compare the 'mathematical energy' of u defined by (12.2) with the physical energy as it would be defined from the wave equation by multiplying it with $\partial u/\partial t$.

2. Extend the study to the case when (V, g) is a sliced Lorentzian manifold.

Solution

1. Recall our signature $(-, + + \ldots +)$. The equation reads

$$-\eta^{\alpha\beta}\nabla_\alpha\partial_\beta u + mu \equiv \partial^2_{00}u - \sum_i \partial^2_{ii}u + mu = f, \quad \partial_\alpha := \frac{\partial}{\partial x^\alpha}.$$

Set $u_\alpha = \partial_\alpha u$ and consider the system $\mathcal{M}^\alpha \partial_\alpha U$ with unknowns $U_I = (u, u_\alpha)$:

$$\partial_0 u = u_0, \quad \partial_0 u_i - \partial_i u_0 = 0, \quad \partial_0 u_0 - \sum_i \partial_i u_i + mu = f$$

The principal operator $\mathcal{M}^\alpha \partial_\alpha$ is the matrix

$$\begin{pmatrix} \partial_0 & & & & \\ & \partial_0 & & & -\partial_1 \\ & & \partial_0 & & -\partial_1 \\ & & & \partial_0 & -\partial_1 \\ & -\partial_1 & -\partial_2 & -\partial_3 & \partial_0 \end{pmatrix}.$$

The matrices $\mathcal{M}^\alpha$ are obviously symmetric and $\mathcal{M}^0$, the unit matrix, is positive-definite.

Multiplication of the wave equation by $\partial u/\partial t$ gives by simple computation the identity, if m is a constant:

$$\frac{\partial u}{\partial t}\left[\frac{\partial^2 u}{\partial t^2} - \sum_i \frac{\partial^2 u}{(\partial x^i)^2} + mu - f\right]$$

$$\equiv \frac{1}{2}\frac{\partial}{\partial t}\left[\left(\frac{\partial u}{\partial t}\right)^2 + \sum_i \left(\frac{\partial u}{\partial x^i}\right)^2 + mu^2\right] - \sum_i \frac{\partial}{\partial x^i}\left(\frac{\partial u}{\partial t}\frac{\partial u}{\partial x^i}\right) = \frac{\partial u}{\partial t}f.$$

Hence, by integration when $f \equiv 0$, using Stokes's formula,

$$\frac{1}{2}\int_{M_t}\left[\left(\frac{\partial u}{\partial t}\right)^2 + \sum_i \left(\frac{\partial u}{\partial x^i}\right)^2 + mu^2\right]dx^1\ldots dx^n = \text{constant if } f \equiv 0.$$

The integrand is interpreted as the physical energy density at time t, positive if $m \geq 0$ and zero for $m = 0$ only if $u = $ constant.

2. For a non-flat Lorentzian metric on a sliced manifold, we take an orthonormal frame adapted to the slicing,

$$g = \eta_{\alpha\beta}\theta^\alpha\theta^\beta,$$

with

$$\theta^0 = N\,dt, \quad \theta^i = a^i_j\,dx^j + \beta^i\,dt, \quad \text{hence} \quad \partial_0 = \frac{\partial}{\partial t} - \beta^i\frac{\partial}{\partial x^i}, \quad \partial_i = A^j_i\frac{\partial}{\partial x^j}.$$

The wave operator reads

$$g^{\alpha\beta}\nabla_\alpha\partial_\beta u \equiv \eta^{\alpha\beta}(\partial_\alpha\partial_\beta u + \gamma^\lambda_{\alpha\beta}\partial_\lambda u).$$

Introducing the new unknowns $u_\alpha = \partial_\alpha u$ and the identities

$$\partial_0 u_i - \partial_i u_0 = C^\alpha_{i0}u_\alpha$$

gives for the unknowns $U = (u, u_\alpha)$ a linear system with operator in matrix form

$$\mathcal{M}^\alpha \partial_\alpha U + AU.$$

The matrices $\mathcal{M}^\alpha$ are the same as in the Minkowski case, and hence symmetric, and the matrix coefficient of $\partial/\partial t$ in $\mathcal{M}^0$ is the Minkowski $\mathcal{M}^0$ metric. The same reasoning applies as previously.

VIII.12.3   The evolution set of Maxwell equations as a first-order symmetric hyperbolic system

The evolution set of Maxwell equations on Minkowski spacetime M^4, with unknowns (E, H), comprises (see Chapter II)

$$\frac{\partial E^1}{\partial t} = \frac{\partial H^3}{\partial x^2} - \frac{\partial H^2}{\partial x^3} - j, \quad \frac{\partial H^1}{\partial t} = \frac{\partial E^2}{\partial x^3} - \frac{\partial E^3}{\partial x^2}, \quad \text{(VIII.12.6)}$$

together with the equations obtained by circular permutation of the indices 1, 2, 3. The characteristic matrix $\mathcal{M}^\alpha \partial_\alpha$ is

$$\begin{pmatrix} \partial_0 & & & & -\partial_2 & \partial_3 \\ & \partial_0 & & -\partial_2 & \partial_1 & 0 \\ & & \partial_0 & \partial_3 & & -\partial_1 \\ & -\partial_2 & -\partial_3 & \partial_0 & & \\ -\partial_2 & \partial_1 & & & \partial_0 & \\ \partial_3 & 0 & -\partial_1 & & & \partial_0 \end{pmatrix}.$$

It is symmetric, and $\mathcal{M}^0$, the unit matrix, is positive-definite. Note that in this case the energy densities corresponding to symmetric hyperbolicity and to physics coincide, both being equal to $\frac{1}{2}(|E|^2 + |H|^2)$.

VIII.12.4   Conformal transformation of the CF constraints

Show that two different choices of the given metrics γ and γ' lead to equivalent conformally formulated constraints if the sources are appropriately chosen.

VIII.12.5   Einstein equations in dimension $2 + 1$

The vacuum Einstein equations are trivial in the case $n = 2$ in the sense that $Ricc(g) = 0$ implies that the spacetime metric g is locally flat when $n + 1 = 3$. However, the $(2 + 1)$-dimensional Einstein theory has a topological content: (V, g) is not necessarily the Minkowski spacetime M_3. In particular, $V = S \times R$ with S a two-dimensional compact surface can be a Lorentzian flat manifold with S the torus T^2 or a surface of genus greater than 1. On the other hand, $(2 + 1)$-dimensional, non-flat, Einstein equations with sources appear for spacetimes that admit a one-parameter spacelike isometry group.[83]

Write a conformal formulation of the constraints in the case $n = 2$ by setting[84]

$$\bar{g} = \exp(2\lambda)\gamma \quad \text{and} \quad \bar{K}^{ij} = \exp(4\lambda)\tilde{K}^{ij} + \frac{1}{2}\bar{g}^{ij}\tau.$$

[83]See Choquet-Bruhat and Moncrief (2001) and Choquet-Bruhat (2004).

[84]Moncrief (1986).

Solution

Elementary computation gives

$$\bar{\nabla}_i K^{ij} - \bar{g}^{ij}\partial_i \tau \equiv \exp(4\lambda)D_i \tilde{K}^{ij} - \frac{1}{2}\bar{\gamma}^{ij}\exp(-2\lambda)\partial_i\tau.$$

The constraints split again into a linear system for $\tilde{K}$ and a semilinear equation for λ when $\tilde{K}$ is known, if τ is a given constant and the momentum of the sources is zero, or properly weighted.

VIII.12.6 Electrovac Einsteinian spacetimes, constraints

The Einstein equations with electromagnetic source are (see Chapter IV)

$$S_{\alpha\beta} = \tau_{\alpha\beta} \equiv F_\alpha{}^\lambda F_{\beta\lambda} - \frac{1}{4}g_{\alpha\beta}F^{\lambda\mu}F_{\lambda\mu}, \qquad \text{(VIII.12.7)}$$

where the electromagnetic 2-form satisfies the Maxwell equations

$$dF \equiv 0, \quad \text{i.e} \quad \nabla_\alpha F_{\beta\gamma} + \nabla_\gamma F_{\alpha\beta} + \nabla_\beta F_{\gamma\alpha} = 0, \qquad \text{(VIII.12.8)}$$

and, in vacuo,

$$\delta F = 0, \quad \text{i.e.} \quad \nabla_\alpha F^{\alpha\beta} = 0. \qquad \text{(VIII.12.9)}$$

The electromagnetic initial data on a hypersurface M are a 2-form $\bar{F}$ and a vector field $\bar{E}$. The 2-form $\bar{F}$ is the form induced on M by the electromagnetic field F, while $\bar{E}$ is the electric vector field on M relative to the unit normal n to M in the spacetime metric.

1. Show that in a Cauchy adapted frame, the components of $\bar{E}$ are

$$\bar{E}^i := \bar{N}\bar{F}^{0i}, \quad \text{with } N \text{ the lapse of the metric } g. \qquad \text{(VIII.12.10)}$$

2. Show that a solution of the Maxwell equations must satisfy on M the constraints

$$d\bar{F} = 0, \quad \text{i.e} \quad \bar{\nabla}_h \bar{F}_{ij} + \bar{\nabla}_j \bar{F}_{hi} + \bar{\nabla}_i \bar{F}_{jh} = 0, \qquad \text{(VIII.12.11)}$$

and

$$\bar{\nabla}_i \bar{E}^i = 0. \qquad \text{(VIII.12.12)}$$

Solution

1. $E^\beta = n_\alpha F^{\alpha\beta}$, $n_\alpha = N\delta^0_\alpha$, i.e $E^0 = 0$, $E^i = NF^{0i}$.
2. $dF = 0$ and $\delta F = 0$ on M imply

$$\partial_h \bar{F}_{ij} + \partial_j \bar{F}_{hi} + \partial_i \bar{F}_{jh} \equiv \bar{\nabla}_h \bar{F}_{ij} + \bar{\nabla}_j \bar{F}_{hi} + \bar{\nabla}_i \bar{F}_{jh} = 0$$

and

$$\overline{\nabla_\alpha F^{\alpha 0}} \equiv \overline{\partial_i F^{i0}} + \Gamma^\alpha_{\alpha i}F^{i0} + \Gamma^0_{\alpha\lambda}F^{\alpha\lambda} \equiv \frac{1}{\sqrt{\text{Det }g}}\overline{\sqrt{g}\partial_i F^{i0}} + F^{i0}\partial_i\sqrt{g} = 0,$$

that is, using $\operatorname{Det} g \equiv N^2 \det \bar{g}$,

$$\partial_i(\bar{N}\sqrt{\operatorname{Det}\bar{g}}\,\bar{F}^{0i}) \equiv \bar{\nabla}\bar{E}^i = 0. \qquad \text{(VIII.12.13)}$$

VIII.12.7   Electrovac Einsteinian spacetimes, Lorenz gauge

The vector potential A is a locally defined 1-form such that

$$F = dA, \quad \text{i.e.} \quad F_{\alpha\beta} = \partial_\alpha A_\beta - \partial_\beta A_\alpha. \qquad \text{(VIII.12.14)}$$

The first Maxwell equation $dF = 0$ is automatically satisfied. The potential A is said to be in Lorenz[85] gauge if it has zero divergence:

$$\nabla_\alpha A^\alpha \equiv 0. \qquad \text{(VIII.12.15)}$$

[85] It seems that this gauge, although used by Lorentz, was first introduced by another mathematician named Lorenz (without a 't').

Show that, in vacuo, the second Maxwell equation reduces in wave coordinates and Lorentz gauge to a quasidiagonal semilinear system of wave equations for the components of A of the form

$$g^{\alpha\lambda}\partial_\lambda\partial_\alpha A_\beta = f_\beta(g, \partial g, \partial A). \qquad \text{(VIII.12.16)}$$

State a local-in-time existence and uniqueness theorem for a solution of the Cauchy problem for the Einstein–Maxwell system

Solution

$$\nabla^\alpha F_{\alpha\beta} \equiv g^{\alpha\lambda}(\partial_\lambda F_{\alpha\beta} - \Gamma^\mu_{\lambda\beta}F_{\alpha\mu} - \Gamma^\mu_{\lambda\alpha}F_{\mu\beta}) = 0.$$

In wave coordinates, it holds that $g^{\alpha\lambda}\Gamma^\mu_{\lambda\alpha} = 0$, and the above system reads, in terms of A,

$$g^{\alpha\lambda}\left[\partial_\lambda\partial_\alpha A_\beta - \partial_\lambda\partial_\beta A_\alpha - \Gamma^\mu_{\lambda\beta}(\partial_\alpha A_\mu - \partial_\mu A_\alpha)\right] = 0.$$

By elementary manipulations, these equations become

$$g^{\alpha\lambda}\partial_\lambda\partial_\alpha A_\beta - \partial_\beta(g^{\alpha\lambda}\partial_\lambda A_\alpha) + (\partial_\beta g^{\alpha\lambda})\partial_\lambda A_\alpha - \Gamma^\mu_{\lambda\beta}(\partial_\alpha A_\mu - \partial_\mu A_\alpha) = 0. \qquad \text{(VIII.12.17)}$$

If A satisfies the gauge condition $g^{\alpha\lambda}\partial_\lambda A_\alpha = 0$, they reduce to a quasidiagonal semilinear system of wave equations for the components of A, of the form

$$g^{\alpha\lambda}\partial_\lambda\partial_\alpha A_\beta = f_\beta(g, \partial g, \partial A). \qquad \text{(VIII.12.18)}$$

The equation $g^{\alpha\lambda}\partial_\lambda A_\alpha = 0$ is equivalent in wave coordinates, where $g^{\alpha\lambda}\Gamma^\mu_{\alpha\lambda} \equiv 0$, to the Lorenz gauge condition, because

$$\nabla_\lambda A^\lambda \equiv g^{\alpha\lambda}\nabla_\lambda A_\alpha \equiv g^{\alpha\lambda}(\partial_\lambda A_\alpha - \Gamma^\mu_{\lambda\alpha}A_\mu).$$

Equations (12.18) together with the Einstein equations in wave gauge, with source the Maxwell tensor of F, constitute a quasidiagonal quasilinear system of wave equations for the pair g, A. Local existence and uniqueness for a solution of the Cauchy problem for the pair (g, A) in wave and Lorenz gauge results from the general theorem quoted before. The proof that the constructed (g, A) satisfies the gauge conditions if the initial data satisfy the constraints follows lines analogous to those given in Section VIII.4, using, in addition, the identity $\nabla_\alpha \nabla_\beta F^{\alpha\beta} \equiv 0$.

Various other gauge conditions can be used on A to solve the Maxwell equations: for example the temporal gauge $A_0 = 0$ or the Coulomb gauge $\nabla_i A^i = 0$. They have proved useful in different domains.

All these gauges, like the Lorenz gauge, generalize to Yang–Mills fields. They lead to equations with the same principal parts as in the case of electromagnetism, but in this case only semilinear.[86]

[86]That is to say, only linear in the principal terms—in the case of the Yang–Mills system, the first-order derivatives.

VIII.12.8   Wave equation for F

1. Show that the Maxwell equations satisfied by the electromagnetic 2-form F on a Lorentzian manifold (V, g) imply that F satisfies on V a quasidiagonal, quasilinear system of wave equations with coefficients depending on the curvature of g.

2. Extend the result to the Yang–Mills case.

Solution

1. The Maxwell equations in vacuuo, $dF = 0$, $\delta F = 0$, imply

$$(\delta d + d\delta) F = 0; \qquad (\text{VIII.12.19})$$

that is,[87] by straightforward computation using the Ricci identity,

$$g^{\alpha\beta} \nabla_\alpha \nabla_\beta F_{\lambda\mu} + R^\alpha_\mu F_{\alpha\lambda} - R^\alpha_\lambda F_{\alpha\mu} + 2R^\alpha{}_\lambda{}^{,\beta}{}_\mu F_{\alpha\beta} = 0. \quad (\text{VIII.12.20})$$

[87]See, for instance, the expression for this operator for an arbitrary p-form in CB-DMI, V B 4.

2. The Yang–Mills gauge- and metric-covariant derivative of F is defined by

$$\hat{\nabla} F := \nabla F + [A, F], \qquad (\text{VIII.12.21})$$

with ∇ the Riemannian covariant derivative and $[.,.]$ the bracket in the Lie algebra corresponding to the considered Yang–Mills model. The Yang–Mills equations in vacuo are

$$\hat{\nabla}_\alpha F_{\beta\gamma} + \hat{\nabla}_\gamma F_{\alpha\beta} + \hat{\nabla}_\beta F_{\gamma\alpha} = 0 \qquad (\text{VIII.12.22})$$

and

$$\hat{\nabla}_\alpha F^{\alpha\beta} = 0. \qquad (\text{VIII.12.23})$$

These equations imply the following second-order semilinear equation for F, depending on A:

$$\hat{\nabla}^\lambda \hat{\nabla}_\lambda F_{\alpha\beta} - 2[F^\gamma_\alpha, F_{\gamma\beta}] = 0, \qquad (\text{VIII.12.24})$$

where

$$\hat{\nabla}^\lambda \hat{\nabla}_\lambda F_{\alpha\beta} \equiv \nabla^\lambda \nabla_\lambda F_{\alpha\beta} + 2\nabla_\gamma [A^\gamma, F_{\alpha\beta}] - [\nabla_\gamma A^\gamma, F_{\alpha\beta}]$$
$$+ [A_\gamma, [A^\gamma, F_{\alpha\beta}]]. \tag{VIII.12.25}$$

This equation for F plays an essential role in the proof of the global existence of a solution of the Cauchy problem for the Yang–Mills equations by Eardley and Moncrief.[88]

[88]Eardley and Moncrief (1982). See also the survey article by Choquet-Bruhat (1983).

VIII.12.9   Wave equation for the Riemann tensor

Show that the Riemann tensor of an Einsteinian spacetime of arbitrary dimension satisfies a quasidiagonal semilinear system of wave equations[89] in the spacetime metric.

[89]Bel (1958).

Solution

The Riemann tensor satisfies the Bianchi identities

$$\nabla_\alpha R_{\beta\gamma,\lambda\mu} + \nabla_\gamma R_{\alpha\beta,\lambda\mu} + \nabla_\beta R_{\gamma\alpha,\lambda\mu} \equiv 0. \tag{VIII.12.26}$$

One deduces from this identity and the Ricci identity an identity of the form

$$\nabla^\alpha \nabla_\alpha R_{\beta\gamma,\lambda\mu} + \nabla_\gamma \nabla^\alpha R_{\alpha\beta,\lambda\mu} + \nabla_\beta \nabla^\alpha R_{\gamma\alpha,\lambda\mu} + S_{\beta\gamma,\lambda\mu} \equiv 0, \tag{VIII.12.27}$$

where $S_{\beta\gamma,\lambda\mu}$ is a homogeneous quadratic form in the Riemann tensor:

$$S_{\beta\gamma,\lambda\mu} \equiv \{R_\gamma{}^\rho R_{\rho\beta,\lambda\mu} + R^\alpha{}_{\gamma,\beta}{}^\rho R_{\alpha\rho,\lambda\mu} + [(R^\alpha{}_{\gamma,\lambda}{}^\rho R_{\alpha\beta,\rho\mu}) - (\lambda \to \mu)]\}$$
$$- \{\beta \to \gamma\}. \tag{VIII.12.28}$$

Returning to the Bianchi identities gives, by contraction,

$$\nabla_\alpha R_{\beta\gamma,}{}^\alpha{}_\mu + \nabla_\gamma R_{\alpha\beta,}{}^\alpha{}_\mu + \nabla_\beta R_{\gamma\alpha,}{}^\alpha{}_\mu \equiv 0. \tag{VIII.12.29}$$

Therefore, using the symmetry $R_{\alpha\beta,\lambda\mu} \equiv R_{\lambda\mu,\alpha\beta}$,

$$\nabla_\alpha R^\alpha{}_{\beta,\lambda\mu} + \nabla_\mu R_{\lambda\beta} - \nabla_\lambda R_{\mu\beta} \equiv 0. \tag{VIII.12.30}$$

If the Ricci tensor $R_{\alpha\beta}$ satisfies the Einstein equations $R_{\alpha\beta} = \rho_{\alpha\beta}$, then the previous identities imply equations of the form

$$\nabla^\alpha \nabla_\alpha R_{\beta\gamma,\lambda\mu} + S_{\beta\gamma;\lambda\mu} = J_{\beta\gamma,\lambda\mu}, \tag{VIII.12.31}$$

where $J_{\beta\gamma,\lambda\mu}$ depends on the sources $\rho_{\alpha\beta}$ and is zero in vacuum:

$$J_{\beta\gamma,\lambda\mu} \equiv \nabla_\gamma (\nabla_\mu \rho_{\lambda\beta} - \nabla_\lambda \rho_{\mu\beta}) - (\beta \to \gamma). \tag{VIII.12.32}$$

Note that (12.26) and (12.29) modulo the Einstein equations are analogous to the Maxwell equations for the electromagnetic 2-form F:

$$dF = 0, \quad \delta F = J, \tag{VIII.12.33}$$

where J is the electric current.

VIII.12.10   First-order symmetric hyperbolic system for the Riemann tensor, Bel–Robinson energy

Write in a Cauchy adapted frame a first-order symmetric differential system with constraints satisfied by the Riemann tensor of a $(3 + 1)$-dimensional Einsteinian spacetime.

Solution

In a coframe θ^0, θ^i where $g_{0i} = 0$ the Bianchi identities (12.26) with $\{\alpha\beta\gamma\} = \{ijk\}$ and the equations (12.30) with $\beta = 0$ do not contain derivatives ∂_0 of the Riemann tensor. We call them 'Bianchi constraints'. The remaining equations, called from here on 'Bianchi equations', read as follows:

$$\nabla_0 R_{hk,\lambda\mu} + \nabla_k R_{0h,\lambda\mu} + \nabla_h R_{k0,\lambda\mu} = 0, \qquad \text{(VIII.12.34)}$$

$$\nabla_0 R^0{}_{i,\lambda\mu} + \nabla_h R^h{}_{i,\lambda\mu} = \nabla_\lambda \rho_{\mu i} - \nabla_\mu \rho_{\lambda i} \equiv J_{\lambda\mu i}, \qquad \text{(VIII.12.35)}$$

where the pair $(\lambda\mu)$ is either $(0j)$ or (jl), with $j < l$. There are three of one or the other of these pairs if the space dimension $n = 3$.

Equations (12.34) and (12.35) are, for each given pair $(\lambda\mu, \lambda < \mu)$, a first-order system for the components $R_{hk,\lambda\mu}$ and $R_{0h,\lambda\mu}$. If we choose at a point of the spacetime an orthonormal frame, then the principal operator is diagonal by blocks; each block corresponding to a choice of a pair $(\lambda\mu, \lambda < \mu)$ is a symmetric 6×6 matrix that reads

$$\mathcal{M} = \begin{pmatrix} \partial_0 & 0 & 0 & \partial_2 & -\partial_1 & 0 \\ 0 & \partial_0 & 0 & 0 & \partial_3 & -\partial_2 \\ 0 & 0 & \partial_0 & -\partial_3 & 0 & \partial_1 \\ \partial_2 & 0 & -\partial_3 & \partial_0 & 0 & 0 \\ -\partial_1 & \partial_3 & 0 & 0 & \partial_0 & 0 \\ 0 & -\partial_2 & \partial_1 & 0 & 0 & \partial_0 \end{pmatrix}.$$

The numerically valued matrix $\mathcal{M}^t$ of coefficients of the operator $\partial/\partial t$ corresponding to the Bianchi equations relative to the Cauchy adapted frame is proportional to the unit matrix, with coefficient N^{-2}, and hence is positive-definite; thus the proof is complete. The energy corresponding to the first-order symmetric hyperbolic system (12.34), (12.35) is the **Bel–Robinson energy**.

VIII.12.11   Schwarzschild trapped surface

Show that a 2-surface $t =$ constant, $r = r_0$ of the Schwarzschild spacetime is not trapped if $r_0 > 2m$. Show that it is trapped if $r_0 \leq m$.

Relativistic fluids

[1]As said before (Chapter IV), this property motivated Einstein in the choice of his equations.

IX.1 Introduction

A **fluid matter source** in a domain of a spacetime (V, g) is such that there exists in this domain a timelike vector field u, called the unit velocity, satisfying $g(u, u) \equiv g_{\alpha\beta} u^\alpha u^\beta = -1$, whose trajectories are the flow lines of matter. A moving Lorentzian orthonormal frame is called a **proper frame** if its timelike vector is u. In a proper frame, the unit velocity has components $u^0 = 1$, $u^i = 0$.

One may also consider null fluids, with flow lines trajectories of a null vector field u, i.e. such that $u^\alpha u_\alpha = 0$.

The Einstein equations with fluid source are

$$S_{\alpha\beta} \equiv R_{\alpha\beta} - \frac{1}{2} g_{\alpha\beta} R = T_{\alpha\beta}, \qquad (IX.1.1)$$

where the stress–energy (momentum) tensor T is deduced from the equivalence principle and its expression in Special Relativity (see Chapters II and III) for the considered type of fluid. The conservation equations

$$\nabla_\alpha T^{\alpha\beta} = 0, \qquad (IX.1.2)$$

which in Special Relativity resulted from the physical laws of conservation of energy and momentum, are in General Relativity[1] a consequence of the Bianchi identities. The conservation equations (1.2) must sometimes be completed by equations satisfied by other physical quantities appearing in T (e.g. the electromagnetic field).

In this chapter, we will describe general properties of perfect fluids. When appropriate definitions are given, a number of these properties generalize well-known properties of non-relativistic perfect fluids. However, the equivalence of mass and energy in Relativity introduces a number of fundamental differences. The relativistic causality principle, limitation by the speed of light of the speed of any macroscopically transmitted signal, leads also to new considerations.

We will only briefly touch on the case of dissipative fluids, which are still subject to controversies.

IX.2 Case of dust

A fluid source is called **pure matter** or **dust** if in a proper frame, it has neither momentum nor stresses; hence, in a proper frame, the only non-vanishing component of the stress–energy tensor is $T_{00} = r$, the proper mass density. In an arbitrary frame,

$$T_{\alpha\beta} = ru_\alpha u_\beta, \quad u^\alpha u_\alpha = -1, \qquad (IX.2.1)$$

where u is the unit flow velocity.

At usual laboratory, and often at astronomical, scales, this dust stress–energy tensor is a good approximation of a general fluid stress–energy tensor, since the mass energy is of order c^2 (c being the velocity of light), with respect to the other forms of energy.

The conservation laws for the dust stress–energy tensor read

$$\nabla_\alpha T^{\alpha\beta} \equiv u^\beta \nabla_\alpha(ru^\alpha) + ru^\alpha \nabla_\alpha u^\beta = 0. \qquad (IX.2.2)$$

They give, using the property $u^\beta u_\beta = -1$, and hence $u_\beta \nabla_\alpha u^\beta = 0$, the continuity equation (conservation of matter)

$$\nabla_\alpha(ru^\alpha) = 0 \qquad (IX.2.3)$$

and the geodesic motion of the particles

$$u^\alpha \nabla_\alpha u^\beta = 0. \qquad (IX.2.4)$$

Similar equations are obtained for a null-dust model where $u^\alpha u_\alpha = 0$.

The geometric initial data for the spacetime metric g on an initial manifold M are a Riemannian metric $\bar{g}$ and a symmetric 2-tensor K. The initial data for a dust source are a scalar function $\bar{r}$ on M and a tangent vector field $\bar{v}$ to M. A solution (V, g, r, u) of the coupled Einstein–dust equations is an Einsteinian development of the initial data set $(M, \bar{g}, K, \bar{r}, \bar{v})$ if $\bar{g}$ and K are respectively the induced metric and the second fundamental form of M as an embedded submanifold in (V, g), while $\bar{r}$ is the function induced by r on M and $\bar{v}$ is the value on M of the dust velocity with respect to the proper frame of an observer with timelike vector orthogonal to M in (V, g). In local coordinates such that the values on M of the shift and the lapse of the development are respectively $\bar{\beta} = 0$ and $\bar{N} = 1$, it holds that $\bar{v}^i = (\bar{u}^0)^{-1} \bar{u}^i$, where $\bar{u}^\alpha$ are the components of u in the considered coordinate system at points of M.

It can be proved[2] that an initial data set $(M, \bar{g}, K, \bar{r}, \bar{v})$ satisfying the constraints for the Einstein equations with dust source admits a globally hyperbolic and maximal Einsteinian development[3] (V, g), unique up to isometries, with dust source (r, u).

The problem of how to use the general equations of fluid evolution to determine the motion of **isolated bodies** is a long standing one, which has received only very partial answers. In the pure matter case, it can be proved[4] that a solution can be found for isolated bodies, i.e. when the support ω of $\bar{r}$ is the disconnected union of compact sets, by giving to $\bar{v}$ arbitrary values outside ω (see Problem IX.20.2).

[2] Fourès (Choquet)-Bruhat (1958). For more refined statements on the required regularity, see YCB-OUP2009, Chapter IX.

[3] Non-complete in general.

[4] Choquet-Bruhat and Friedrich (2006).

IX.3 Charged dust

[5]This model is seldom valid in the real world, where opposite charges interact.

The stress–energy tensor of charged pure matter[5] (dust) is the sum of the stress–energy tensor of the matter and the Maxwell tensor of the electromagnetic field F:

$$T_{\alpha\beta} = ru_\alpha u_\beta + \tau_{\alpha\beta}, \qquad \text{(IX.3.1)}$$

with

$$\tau_{\alpha\beta} = F_\alpha{}^\lambda F_{\beta\lambda} - \frac{1}{4}g_{\alpha\beta}F^{\lambda\mu}F_{\lambda\mu}. \qquad \text{(IX.3.2)}$$

The Maxwell equations are,

$$dF = 0 \quad \text{and} \quad \nabla \cdot F = J, \quad \text{i.e.} \quad \nabla_\alpha F^{\alpha\beta} = J^\beta := qu^\beta, \qquad \text{(IX.3.3)}$$

where J is the convection electric current of the charge density q. They imply the conservation of electric charge:

$$\nabla_\alpha(qu^\alpha) = 0. \qquad \text{(IX.3.4)}$$

We introduce the electromagnetic potential A, a 1-form such that (for simplicity, we suppose that A exists globally on the considered domain)

$$F = dA. \qquad \text{(IX.3.5)}$$

We take A in Lorenz gauge, i.e. such that

$$\delta A = 0. \qquad \text{(IX.3.6)}$$

The Maxwell equations then read as a wave equation for A, namely

$$\nabla_\alpha \partial^\alpha A^\beta - R^\beta{}_\lambda A^\lambda = J^\beta = qu^\beta. \qquad \text{(IX.3.7)}$$

Modulo the Maxwell equations (and $u^\alpha u_\alpha = -1$), the stress–energy conservation equations are equivalent to

$$\nabla_\alpha(ru^\alpha) = 0 \qquad \text{(IX.3.8)}$$

and

$$ru^\alpha \nabla_\alpha u^\beta + qu_\lambda F^{\beta\lambda} = 0. \qquad \text{(IX.3.9)}$$

Equations (3.4) and (3.8) imply

$$u^\alpha \partial_\alpha \left(\frac{q}{r}\right) = 0;$$

that is, the specific charge q/r is constant along the flow lines; it is constant throughout the spacetime that we construct if it is constant initially. We will make this simplifying (though not necessary) hypothesis, and set

$$q = kr,$$

with k some given constant. Equation (3.9) can then be replaced by

$$u^\alpha \nabla_\alpha u^\beta + ku_\lambda F^{\beta\lambda} = 0. \qquad \text{(IX.3.10)}$$

The proof of the existence of a solution (local in time) of the Cauchy problem for the full Einstein–Maxwell–dust system when the initial data satisfy the Einstein and Maxwell constraints with dust source follows the same lines as in the vacuum Einstein–Maxwell case. Geometric global uniqueness can also be proved.

IX.4   Perfect fluid

IX.4.1   Stress–energy tensor

We have written in Chapter IV the stress–energy tensor T of a perfect fluid in General Relativity:

$$T_{\alpha\beta} = \mu u_\alpha u_\beta + p(g_{\alpha\beta} + u_\alpha u_\beta), \qquad \text{(IX.4.1)}$$

where u is the unit flow vector, while μ and p are respectively the energy and pressure densities. In a **proper** frame, it holds that $T_{00} = \mu$, $T_{0i} = 0$, and $T_{ij} = p\delta_{ij}$, the isotropic-in-space pressure tensor. For classical fluids, μ and p are non-negative.[6]

Exercise IX.4.1 *Show that in a spacetime of dimension $n + 1$, the Einstein equations with a perfect fluid source can be written as*

$$R_{\alpha\beta} = \rho_{\alpha\beta}, \quad \text{with} \quad \rho_{\alpha\beta} \equiv (\mu + p)u_\alpha u_\beta + \frac{1}{n-1}g_{\alpha\beta}(\mu - p).$$

Hint: We have shown in Chapter IV (equation IV.2.5) that, with n the dimension of space (in the classical case, $n = 3$),

$$\rho_{\alpha\beta} \equiv T_{\alpha\beta} - \frac{1}{n-1}g_{\alpha\beta}T_\lambda^\lambda.$$

We have here

$$T \equiv g^{\alpha\beta}T_{\alpha\beta} = np - \mu. \qquad \text{(IX.4.2)}$$

The **energy–momentum vector** relative to a timelike vector X is

$$P_X^\alpha := T^{\alpha\beta}X_\beta \equiv (\mu + p)u^\alpha u^\beta X_\beta + pX^\alpha. \qquad \text{(IX.4.3)}$$

For $X = -u$, the components of P are $P^0 = \mu$, the energy density, and $P^i = 0$, zero momentum density.

We mention the following properties, which use a terminology frequently referred to, although misleading since weak and strong conditions are unrelated in the following propositions.

Proposition IX.4.1 *If μ and p are non-negative, while X and Y are causal with the same time orientation, then*

1. *The scalar $T^{\alpha\beta}X_\alpha Y_\beta$ is non-negative; one says that a perfect fluid satisfies the **weak energy condition**.*

2. *The scalar[7] $\rho_{\alpha\beta}X^\alpha X^\beta$ is non-negative if X is causal; such a perfect fluid satisfies the **strong energy condition**.*

[6] Quantum or cosmological phenomena can lead to the appearance of negative pressures.

[7] Recall that the Einstein equations can be written as $R_{\alpha\beta} = \rho_{\alpha\beta}$, $\rho_{\alpha\beta} := T_{\alpha\beta} - \frac{1}{n-1}Tg_{\alpha\beta}$.

3. If $\mu > p$ and if X is past timelike, then the energy–momentum vector P_X is future timelike like u; such a perfect fluid satisfies the **dominant energy condition**.

Exercise IX.4.2 *Prove these properties.*

Hint: Compute these quantities in a proper frame and for Part 2 use the property $(X^0)^2 \geq \sum (X^i)^2$.

IX.4.2   Euler equations

The **conservation laws** of a perfect fluid read

$$\nabla_\alpha T^{\alpha\beta} \equiv (\mu + p)u^\alpha \nabla_\alpha u^\beta + g^{\alpha\beta}\partial_\alpha p + u^\beta[\nabla_\alpha(\mu + p)u^\alpha] = 0. \quad \text{(IX.4.4)}$$

By taking the contracted product with u_β, one deduces from these equations the **energy equation**

$$(\mu + p)\nabla_\alpha u^\alpha + u^\alpha \partial_\alpha \mu = 0, \quad \text{(IX.4.5)}$$

and, using this equation, one obtains the **equations of motion**

$$(\mu + p)u^\alpha \nabla_\alpha u^\beta + (g^{\alpha\beta} + u^\alpha u^\beta)\partial_\alpha p = 0. \quad \text{(IX.4.6)}$$

The set of equations (4.4), (4.5) is called the Euler equations.

Exercise IX.4.3 *Show that the condition* $u^\alpha u_\alpha = -1$ *is conserved along the flow lines.*

Hint: The equations of motion imply $u^\alpha \partial_\alpha(u^\beta u_\beta) = 0$.

IX.5   Thermodynamics

IX.5.1   Conservation of rest mass

In relativity, mass and energy are the same entity. However, there exists[8] another scalar function of physical interest, namely the particle number density r (better called the rest mass density if there are different kinds of particles with non-zero rest mass). In the absence of chemical reactions or quantum phenomena, conservation of particle number implies the equation

$$\nabla_\alpha(r u^\alpha) = 0. \quad \text{(IX.5.1)}$$

The quantity r^{-1} plays the role of a **specific volume**.

IX.5.2   Definitions. Conservation of entropy

The difference between the total energy density μ and the rest mass density r is called the internal energy density. One denotes by ε the specific internal energy density; that is, one sets

$$\mu = r(1 + \varepsilon). \quad \text{(IX.5.2)}$$

[8]See Taub (1959).

In the case of local thermodynamic equilibrium (reversible thermo-dynamics), one defines a specific entropy density S and an absolute temperature T by extending to relativistic perfect fluids the identity of the first law of thermodynamics, namely

$$TdS := d\varepsilon + pd(r^{-1}). \qquad (IX.5.3)$$

The thermodynamic quantities $\mu, p, S,$ and T are spacetime scalar functions.

Theorem IX.5.1   *In a perfect fluid, the thermodynamic identity (5.3) and the matter conservation equation (5.1) imply the conservation of the entropy density along the flow lines:*[9]

[9]See Pichon (1965).

$$u^\alpha \partial_\alpha S = 0. \qquad (IX.5.4)$$

Proof. The identity (5.3) and the definition (5.2) of ε give that

$$Tu^\alpha \partial_\alpha S \equiv u^\alpha \partial_\alpha \varepsilon - r^{-2} p u^\alpha \partial_\alpha r \equiv u^\alpha r^{-1} \partial_\alpha \mu - (\mu + p) r^{-2} u^\alpha \partial_\alpha r,$$

and hence, using the energy equation (5.1),

$$Tu^\alpha \partial_\alpha S \equiv -r^{-2}(\mu + p)\nabla_\alpha(ru^\alpha). \qquad \blacksquare$$

IX.5.3   Equations of state $(n = 3)$

For a perfect fluid, only two thermodynamic scalars are independent; the others are linked to them by relations that are assumed to depend only on the nature of the given fluid. Usually, the general formula called the **equation of state** is the data of p as some function of μ and S:

$$p = p(S, \mu),$$

invertible as a function $\mu = \mu(p, S)$.

Two circumstances are of particular physical interest for General Relativity, astrophysics, and cosmology.

In **astrophysics**, one is inspired by what is known from classical flu-ids, with additional relativistic considerations. Particularly interesting cases are those of barotropic and polytropic fluids.

Barotropic fluids

A fluids is called **barotropic** when the equation of state reduces to

$$p = p(\mu).$$

The fluid dynamics is then governed by the energy and momentum equa-tions. The particle number conservation equation decouples from the others and can be solved after the fluid motion has been determined. Some physical situations that correspond to this model are the following:

(1) Very cold matter, including models of nuclear matter, such as in neutron stars.

(2) Ultrarelativistic fluids, i.e. fluids where the energy μ is largely dominated by the radiation energy. Then, by the Stefan–Boltzmann law, it holds that $\mu = KT^4$ and $p = \frac{1}{3}KT^4$, and hence

$$p = \frac{1}{3}\mu. \tag{IX.5.5}$$

The same equations for p and μ (with different constants K) and the equation of state (5.5) are valid for a fluid of neutrinos or electron–positron pairs.

Remark IX.5.1　*The stress–energy tensor of an ultrarelativistic fluid is traceless.*

Polytropic fluids

A fluids is called **polytropic** if it obeys an equation of state of the form

$$p = f(S)r^\gamma.$$

Several physical situations correspond to this case.

Frequently in stellar situations, only the internal energy ε and pressure p are dominated by radiation; then $\varepsilon = KrT^4$ and

$$p = \frac{1}{3}KT^4, \quad \text{hence} \quad p = \frac{1}{3}r\varepsilon. \tag{IX.5.6}$$

On the other hand, the thermodynamic identity together with the expressions ε and p imply

$$dS = \frac{4K}{3}d(r^{-1}T^3), \quad \text{hence} \quad S = \frac{4KT^3}{3r}. \tag{IX.5.7}$$

Eliminating T between (5.6) and (5.7) gives the polytropic equation of state of index $\gamma = \frac{4}{3}$:

$$p = \frac{K}{3}\left(\frac{3S}{4K}\right)^{\frac{4}{3}} r^{\frac{4}{3}}, \quad \text{with} \quad \mu = 3p + r. \tag{IX.5.8}$$

More refined equations of state adapted to various physical situations have also been considered.

In **cosmology**, new information obtained from modern ground-based and satellite telescopes has brought puzzling questions. It has been known for a long time that radiation energy is presently only a very small fraction of the energy content of the cosmos, but it has recently been found[10] that baryonic matter itself represents at present only about 4% of this energy content. Another 24% is constituted by dark matter, whose nature is conjectural, and the remaining 72% by what is called 'dark energy', whose nature is a mystery.

In the early universe of the big-bang models, at very high temperature, the energy content of the sources can be roughly modelled as an ultrarelativistic fluid. For later times, after the formation of galaxies,

[10]In particular by analysing data from the Planck satellite (see Chapter VII).

cosmologists in general have assumed, for simplicity, a linear equation of state independent of entropy:

$$p = (\gamma - 1)\mu. \tag{IX.5.9}$$

In order that the speed of sound waves not be greater than the speed of light (see Section IX.6), it has been supposed that $1 \leq \gamma \leq 2$. The case $\gamma = 1$ corresponds to dust. In the case $\gamma = 2$, the fluid is called stiff, or incompressible. In a stiff fluid, the speeds of sound and light are equal.

IX.6   Wave fronts and propagation speeds

In the spacetime of Galilean–Newtonian mechanics a wave front is a 2-surface in space that propagates with time. Its propagation speed at a point of space and an instant of time is the quotient by an infinitesimal absolute time interval δt of the infinitesimal distance between two wave fronts measured in the direction orthogonal to them, at times t and $t + \delta t$. In an Einsteinian spacetime, a wave front is a three-dimensional timelike or null submanifold. The definition of its propagation speed depends on the choice of an observer, and requires some thought.

Let us give first some definitions.

IX.6.1   Characteristic determinant

The **characteristic polynomial** of a linear partial differential equation at a point $x \in R^{n+1}$ is the polynomial obtained by replacing in its higher-order terms the partial derivatives $\partial/\partial x^\alpha$ by the components of a covariant vector.

In the case of a system of N first-order partial differential equations[11] with N unknowns U^I and principal part

$$\sum_{I=1}^{N} A_I^{J,\alpha} \frac{\partial U^I}{\partial x^\alpha}, \quad J = 1, \ldots, N,$$

the **characteristic polynomial** $\Phi(x, p)$ is the determinant of the $N \times N$ matrix with elements $A_J^{J,\alpha} p_\alpha$:

$$\Phi(x, p) := \mathrm{Det} \left[A_I^{J,\alpha}(x) p_\alpha \right].$$

At a given point x, the equation $\Phi(x, p) = 0$ defines a cone in the cotangent space called the **characteristic cone**.

The **wave fronts** associated to a system of first-order partial differential equations on a spacetime are the submanifolds of this spacetime whose normals are roots of the characteristic determinant; that is, wave fronts satisfy in local coordinates an equation $f = 0$, where f is a solution of the first-order nonlinear partial differential equation, called the **eikonal** equation,

[11]For the definition for more general systems, see, for instance, YCB-OUP2009, Appendix IV, Section 2.

$$\Phi\left(x, \frac{\partial f}{\partial x}\right) = 0.$$

Generically, discontinuities of the derivatives of the unknowns of the order appearing in these principal parts can occur only across such submanifolds.

By general results on first-order partial differential equations,[12] a wave front is generated by the bicharacteristics of the eikonal equation, also called rays, which are solutions of the ordinary differential system

[12]See for instance CB-DMI, IV Chapter 7.

$$\frac{dx^\alpha}{\partial\Phi/\partial p_\alpha} = -\frac{dp_\alpha}{\partial_\alpha\Phi/\partial x^\alpha} = d\lambda, \qquad (IX.6.1)$$

where λ is a parameter, called the canonical parameter. The tangents to the rays issuing from a given point x generate a cone in the tangent space to the spacetime at x, called the wave cone, which is dual to the characteristic cone. The wave cone is the envelope of the hyperplanes whose normals (in the metric g) belong to the characteristic cone. A wave front at x is tangent to the wave cone at x along the direction of a ray.

Exercise IX.6.1   *Show that $\Phi(x,p)$ is constant along a ray.*

Exercise IX.6.2   *The eikonal equation associated to the wave operator $\Box_g$ on a spacetime (V,g) is*

$$\frac{1}{2}g^{\alpha\beta}\frac{\partial f}{\partial x^\alpha}\frac{\partial f}{\partial x^\beta} = 0. \qquad (IX.6.2)$$

Show that the light cone and the characteristic cone of a spacetime (V,g) can be identified by the usual identification of the tangent and cotangent spaces to V through the metric g.

IX.6.2   Wave front propagation speed

The propagation speed of a smooth wave front with respect to an observer is the propagation speed of the tangent plane to the wave front with respect to the proper Lorentz frame of this observer, with the following definition (one could give analogous definitions for higher dimensions, but to be clearer we prefer to stick to $3 + 1$).

Definition IX.6.1   *The **propagation speed** at a point x of a three-dimensional hyperplane P with respect to an orthonormal Lorentzian frame at x is the velocity with respect to this frame of the vector of greatest slope in P, orthogonal to the 2-plane[13] intersection of P with the space hyperplane $X^0 = 0$.*

[13]Two 3-hyperplanes in four-dimensional spacetimes intersect generically along a 2-plane whose normal in a 3-plane is a vector.

Lemma IX.6.1   *The propagation speed V of a hyperplane with normal ν with respect to a Lorentzian frame at x with time vector u is given by the formula*

$$|V|^2 = \frac{(u^\alpha \nu_\alpha)^2}{(u^\alpha u^\beta + g^{\alpha\beta})\nu_\alpha \nu_\beta}. \qquad \text{(IX.6.3)}$$

Proof. The vector of greatest slope in P with respect to the Lorentz frame with time vector u is the vector in P orthogonal to the 2-plane I that is the intersection of P and the hyperplane orthogonal to u. This intersection I satisfies the equations

$$u^\alpha X_\alpha - 0, \quad \nu^\alpha X_\alpha = 0. \qquad \text{(IX.6.4)}$$

Choose for the first space vector in the Lorentzian frame the projection of ν on the hyperplane orthogonal to u; then (6.4) read

$$X_0 = 0, \quad \nu^0 X_0 + \nu^1 X_1 = 0. \qquad \text{(IX.6.5)}$$

A vector in I therefore has components $X_0 = 0$, $X_1 = 0$, with X_2, X_3 arbitrary. The vector of greatest slope Y in P orthogonal to I is uniquely determined by the conditions

$$Y \in P, \quad \text{i.e.} \quad \nu^0 Y_0 + \nu^1 Y_1 = 0, \quad \text{hence} \quad Y_1 = -\frac{\nu^0}{\nu^1}Y_0, \qquad \text{(IX.6.6)}$$

and the orthogonality with I, that is, $Y_2 = Y_3 = 0$.
   The propagation speed of P with respect to the considered Lorentz frame is

$$|V| = \left|\frac{Y_1}{Y_0}\right| = \left|\frac{\nu^0}{\nu^1}\right|. \qquad \text{(IX.6.7)}$$

The given formula (6.3) takes the form (6.7) in a Lorentz frame where the time axis is the unit vector u and the normal ν to P has components $\nu_2 = \nu_3 = 0$. ∎

IX.6.3   Case of perfect fluids

The Euler and entropy conservation equations of a perfect fluid are of first order. Using the entropy equation (5.4) in the energy equation (4.5), we see that the Euler–entropy equations written for the unknowns S, p, u have a characteristic matrix composed of two blocks around the diagonal. One of these blocks corresponds to the entropy S and its conservation law (5.4); it reduces to the linear form[14]

$$a \equiv u^\alpha X_\alpha.$$

The other block corresponds to the unknowns p and u and the equations (4.4) and (4.5). It is the following 5×5 matrix[15] with $\rho \equiv \mu + p$, $\mu'_p = \partial\mu/\partial p$:

[14]To avoid confusion with pressure, we denote by X the vector previously denoted by p.

[15]The four components of u are considered as independent unknowns. The identity $g(u, u) = -1$ is preserved by the flow.

$$\mathcal{M} \equiv \begin{pmatrix} a & \rho\mu'_p X_0 & \rho\mu'_p X_1 & \rho\mu'_p X_2 & \rho\mu'_p X_3 \\ X^0 + au^0 & \rho a & 0 & 0 & 0 \\ X^1 + au^1 & 0 & \rho a & 0 & 0 \\ X^2 + au^2 & 0 & 0 & \rho a & 0 \\ X^3 + au^3 & 0 & 0 & 0 & \rho a \end{pmatrix}.$$

The determinant of this matrix is computed to be

$$- \rho^4 a^3 D, \tag{IX.6.8}$$

with D quadratic in X and given by

$$D \equiv (\mu'_p - 1)(u^\alpha X_\alpha)^2 + \mu'_p X^\alpha X_\alpha. \tag{IX.6.9}$$

We see that a perfect fluid has two types of wave fronts:

- Matter wave fronts, $f = $ constant, such that

$$u^\alpha \partial_\alpha f = 0. \tag{IX.6.10}$$

They are submanifolds generated by the flow lines. Their propagation speed for a comoving observer (i.e. in a proper rest frame of the fluid) is zero.

- Sound wave fronts, whose normals satisfy $D = 0$. In a proper rest frame $g_{\alpha\beta} = \eta_{\alpha\beta}$ (the Minkowski metric), $u^0 = 1, u^i = 0$. The corresponding eikonal equation reads

$$- \mu'_p (\partial_0 f)^2 + \sum_i (\partial_i f)^2 = 0. \tag{IX.6.11}$$

The propagation speed of these wave fronts is, assuming $\mu'_p > 0$,

$$|V| = (\mu'_p)^{-\frac{1}{2}};$$

it is less than the speed of light,[16] as expected from a relativistic theory, if and only if

$$\mu'_p \geq 1. \tag{IX.6.12}$$

A fluid such that $\mu'_p = 1$ is called an **incompressible** or **stiff** fluid. In such a fluid, sound waves propagate with the speed of light.

[16]Recall that we use geometric units; i.e. the speed of light is equal to 1.

IX.7   Cauchy problem for the Euler and entropy system

The **Cauchy problem** is the determination of the solution of a partial differential system from its initial data. Consider first one linear[17] partial differential equation of order $m+1$ with unknown a function U on R^{n+1} where coordinates are denoted by x^0, x^i, $i = 1, \ldots, n$. The initial data for U on the submanifold $x^0 = 0$ are the values for $x^0 = 0$ of U and its partial derivatives with respect to x^0 of order less than or equal to m. A Cauchy problem is said to be **well posed**, for initial data in some

[17]That is, linear in the highest-order, $m + 1$, terms.

functional space, if it admits in a neighbourhood of $x^0 = 0$ one and only one solution depending continuously on the initial data. The equation is called **hyperbolic** if it admits a well-posed Cauchy problem with data in spaces where only a finite number of derivatives are involved[18] and if the solution of this Cauchy problem manifests a finite propagation speed. Hyperbolic well-posedness depends only on the coefficients of the principal terms, i.e. terms of order $m + 1$. For a **quasilinear** differential equation (i.e. one that is linear only in the highest $(m+1)$-order terms), the hyperbolicity depends on the values of the coefficients of these terms for the given initial data. By contrast, if the equation is **semilinear** (i.e. the coefficients of the highest-order terms do not depend on the unknowns), hyperbolicity does not depend on the initial data.[19]

In General Relativity, **causality** is an essential feature, saying that the properties of physical phenomena at one point depend only on the past of this point, determined by the causal paths[20] of the Lorentzian spacetime metric g. The definitions of Cauchy problem and hyperbolicity extend to systems of partial differential equations, in particular to quasidiagonal systems. The wave gauge reduced vacuum Einstein equations are a quasidiagonal and quasilinear second-order partial differential systems as studied in Chapter VIII; they are a hyperbolic system for data on a Lorentzian metric on a spacelike submanifold. They are a special case of a quasidiagonal **Leray hyperbolic systems**.[21]

IX.7.1   The Euler and entropy equations as a Leray hyperbolic system

It can be shown[22] that the Euler equations are equivalent to a quasidiagonal system with principal terms in the diagonal either $u^\alpha \partial_\alpha$ or the third-order operator

$$u^\alpha \left[(\mu'_p - 1) u^\lambda u^\beta + \mu'_p g^{\lambda\beta} \right] \partial_\lambda \partial_\beta \partial_\alpha. \qquad \text{(IX.7.1)}$$

The characteristic cone at a point is

$$u^\alpha X_\alpha \left[(\mu'_p - 1) u^\lambda u^\beta + \mu'_p g^{\lambda\beta} \right] X_\lambda X_\beta = 0. \qquad \text{(IX.7.2)}$$

It is the union of a hyperplane $u^\alpha X_\alpha = 0$, spacelike if u is timelike, reading (in a proper frame for u)

$$(P) := \{X_0 = 0\},$$

and a second-order cone (C), reading (in a proper frame for u)

$$(C) := \left\{ -(X_0)^2 + \mu'_p \sum (X_i)^2 = 0 \right\}.$$

It is easy to see by drawing a figure that if $\mu'_p > 0$, then any straight line passing through a point y in the interior of (C), which is

$$-(y_0)^2 + \mu'_p \sum (y_i)^2 \leq 0,$$

[18] Mathematicians often use Sobolev spaces H_s.

[19] One calls hyperquasilinear equations of order $m+1$ in which the coefficients of the highest-order terms do not depend on derivatives of the unknowns of order $> m - 1$. The Einstein equations are hyperquasilinear.

[20] In other words, physicists say that the speed of light is a maximum for the propagation of all observables.

[21] Leray (1953). See also Dionne (1962): this article by a student of Leray refines his results in an important way. It treats only the case of one equation, but the results extend trivially to quasidiagonal systems.

[22] Choquet-Bruhat (1966).

[24]For details, see YCB-OUP2009, Appendix IV, Section 2 or Leray (1953).

[25]Friedrichs (1954). The Leray hyperbolic systems are much more general, containing equations of arbitrary orders, but some symmetric hyperbolic systems do not satisfy the criteria for Leray hyperbolicity.

[26]Anile (1989).

[27]A FOSH system, for barotropic fluids, was found in Special Relativity by K. O. Friedrichs. Its construction in General Relativity was sketched by Rendall (1992).

but not passing through the vertex, cuts the union $(P) \cup (C)$ at three distinct points.[23] This is the criterion of Leray hyperbolicity for a third-order operator, or a quasidiagonal system of such operators.[24]

IX.7.2   First-order symmetric hyperbolic systems

An alternative to the Leray approach to proving the well-posedness of a Cauchy problem for evolution equations is the Friedrichs theory[25] of **first-order symmetric hyperbolic (FOSH) systems**. The first-order linear system of partial differential equations on R^{n+1},

$$M_{IJ}^t \frac{\partial U^J}{\partial t} + M_{IJ}^i \frac{\partial U^J}{\partial x^i} = f_I,$$

is called **symmetric** if the $n+1$, $N \times N$ matrices $\mathcal{M}^t$, $\mathcal{M}^i$ with elements M_{IJ}^t, M_{IJ}^i are symmetric (i.e. $M_{IJ}^\alpha = M_{JI}^\alpha$). It is hyperbolic with respect to the $x^0 \equiv t$ coordinate, and is then described as simply **symmetric hyperbolic**, if, in addition, the matrix $\mathcal{M}^t$ is a positive-definite quadratic form. The Cauchy problem is then well posed if its initial data are given on a submanifold $S_0 := \{x^0 = \text{constant}\}$. In a quasilinear system, the matrices $\mathcal{M}^\alpha$ are functions of the unknowns U, but not of their derivatives. The symmetric hyperbolicity depends then on the values of these unknowns. The Cauchy problem with data on S_0 is then well posed if the system is symmetric hyperbolic when in the coefficients the unknowns are replaced by their initial data. Note that in general, however, the solution exists only in a neighbourhood of S_0.

It is straightforward to write for wave equations equivalent FOSH systems (see Problem VIII.12.10 in Chapter VIII). In the case of the Euler equations, the procedure is more subtle but may be preferred for some numerical computations. To obtain FOSH systems, one can apply general methods inaugurated by Lax, developed by Boillat and Ruggeri, and explained in the book by Anile.[26] These authors use a convex functional and auxiliary unknowns. In the case of relativistic perfect fluids, one can perform a direct computation to show the equivalence of the Euler–entropy system to a FOSH system.[27] The idea is to take as unknowns the pressure p, denoted by U^0, and the space components u^i, denoted by U^I, $I = 1, 2, 3$. The component u^0 is determined through the identity

$$u^\alpha u_\alpha = -1. \tag{IX.7.3}$$

From this identity, one deduces

$$\nabla_\alpha u^0 = -\frac{u_i \nabla_\alpha u^i}{u_0}. \tag{IX.7.4}$$

The energy–continuity equation yields the following first-order evolution equation for p and u^i:

$$\nabla_i u^i - \frac{u_i \nabla_0 u^i}{u_0} + \mu_p' u^\alpha \frac{\partial_\alpha p}{\mu + p} = 0. \tag{IX.7.5}$$

The combination of the equations of motion of indices 0 and i,

$$u^\alpha \nabla_\alpha u^i + (g^{\alpha i} + u^\alpha u^i)\frac{\partial_\alpha p}{\mu + p} - \frac{u^i}{u^0}\left[u^\alpha \nabla_\alpha u^0 + (g^{\alpha 0} + u^\alpha u^0)\frac{\partial_\alpha p}{\mu + p}\right] = 0,$$

reduces to another first-order evolution equation for p and u^i:

$$u^\alpha\left(\nabla_\alpha u^i + \frac{u^i u_j}{u^0 u_0}\nabla_\alpha u^j\right) + \left(g^{\alpha i} - \frac{u^i}{u^0}g^{\alpha 0}\right)\frac{\partial_\alpha p}{\mu + p} = 0. \qquad \text{(IX.7.6)}$$

Setting $A := (\mu + p)^{-1}$, the principal matrix of the system (7.5), (7.6) reads $\mathcal{M} \equiv \mathcal{M}^\alpha \partial_\alpha$,

$$:= \begin{pmatrix} A^2 \mu'_p u^\alpha \partial_\alpha & A\left(\partial_1 - \dfrac{u_1}{u_0}\partial_0\right) & A\left(\partial_2 - \dfrac{u_2}{u_0}\partial_0\right) & A\left(\partial_3 - \dfrac{u_3}{u_0}\partial_0\right) \\[2ex] A\left(\partial^1 - \dfrac{u^1}{u^0}\partial^0\right) & u^\alpha\left(1 + \dfrac{u^1 u_1}{u^0 u_0}\right)\partial_\alpha & \dfrac{u^2 u_1}{u^0 u_0}u^\alpha \partial_\alpha & \dfrac{u^1 u_3}{u^0 u_0}u^\alpha \partial_\alpha \\[2ex] A\left(\partial^2 - \dfrac{u^2}{u^0}\partial^0\right) & \dfrac{u^1 u_2}{u^0 u_0}u^\alpha \partial_\alpha & u^\alpha\left(1 + \dfrac{u^2 u_2}{u^0 u_0}\right)\partial_\alpha & \dfrac{u^2 u_3}{u^0 u_0}u^\alpha \partial_\alpha \\[2ex] A\left(\partial^3 - \dfrac{u^3}{u^0}\partial^0\right) & \dfrac{u^1 u_3}{u^0 u_0}u^\alpha \partial_\alpha & \dfrac{u^1 u_3}{u^0 u_0}u^\alpha \partial_\alpha & u^\alpha\left(1 + \dfrac{u^3 u_3}{u^0 u_0}\right)\partial_\alpha \end{pmatrix}.$$

$$\text{(IX.7.7)}$$

It can be proved that the system is equivalent to a symmetric system with $\partial_0 = \partial/\partial t$ and with $\mathcal{M}^t$ positive-definite[28] if u is timelike and $\mu'_p \geq 1$; that is, to a FOSH system.

[28]See YCB-OUP2009, Chapter IX, Section 13.

Exercise IX.7.1   *Show that (7.5), (7.6) are equivalent to the original Euler–entropy equations.*

IX.8   Coupled Einstein–Euler–entropy system

IX.8.1   Initial data

An initial data set for the Einstein–Euler system with a given equation of state will be the usual data for the Einstein equations together with data for the fluid source.

We have seen that an initial data set for the Einstein equations is a triple $(M, \bar{g}, K)$ with M a 3-manifold, $\bar{g}$ a properly Riemannian metric, and K a symmetric 2-tensor on M. A spacetime (V, g) is said to admit these initial data if M can be embedded in V as a submanifold M_0 with induced metric $\bar{g}$ and extrinsic curvature K. If (V, g) is a solution of the Einstein equations with source the stress–energy tensor T, then $(\bar{g}, K)$ must satisfy the following constraints:

$$R(\bar{g}) - K \cdot K + (\operatorname{tr} K)^2 = 2\rho, \qquad \text{(IX.8.1)}$$

$$\bar{\nabla} \cdot K - \bar{\nabla}\operatorname{tr} K = J. \qquad \text{(IX.8.2)}$$

In a Cauchy adapted frame, where the equation of $M \equiv M_0$ in V is $x^0 \equiv t = 0$ and the time axis is orthogonal to M_0, one has, in the case of a perfect fluid,

$$\rho \equiv N^2 T^{00} \equiv N^2 (\mu + p)(u^0)^2 - p, \qquad (\text{IX.8.3})$$

$$J^i = N T^{i0} \equiv N(\mu + p) u^i u^0,$$

$$N u^0 = (1 + g_{ij} u^i u^j)^{\frac{1}{2}}. \qquad (\text{IX.8.4})$$

Using $u^\alpha u_\alpha = -1$, that is,

$$N u^0 = (1 + g_{ij} u^i u^j)^{\frac{1}{2}}, \qquad (\text{IX.8.5})$$

we see that on M_0 the quantities ρ and J depend only on the values $\bar{\mu}$ and $\bar{p}$ of μ and p on M_0, and on the projection v of u, with components $v^i = u^i$, on M_0; they do not depend on the choice of lapse and shift. The initial data on M_0 for a perfect fluid with equation of state $\mu = \mu(p, S)$ are two scalars $\bar{p}$ and $\bar{S}$ and a tangent vector $\bar{v}$. Data for the third-order system are obtained by using the restriction to M_0 of the Euler equations and its first-order derivative in the direction of u. This computation now requires a choice for initial lapse and shift, since this is also required for the solution of the reduced Einstein equations.

IX.8.2   Evolution

The Einstein equations reduced in harmonic gauge, with perfect fluid source, read as a second-order quasidiagonal system for the metric whose principal part is the wave operator of the spacetime metric while the fluid variables appear at order zero. It is straightforward to show that, together with the Euler–entropy equations, they form a Leray hyperbolic system provided that $\mu'_p > 0$. However, the system is causal (i.e. the domain of dependence is determined by the light cone) only if $\mu'_p \geq 1$. It is also straightforward to write in that case the coupled system as a FOSH system. We have already described in Chapter VIII the proof that a solution of the reduced system is a solution of the full system if the initial data satisfy the constraints.

IX.9   Dynamical velocity

IX.9.1   Fluid index and Euler equations

Important properties of non-relativistic fluids generalize to relativistic fluids if one introduces a spacetime vector called the **dynamical velocity**[29] linked with both the kinematic unit vector u and the thermodynamic quantities.

[29]See Lichnerowicz (1955) and references therein. The dynamical velocity was used by Fourès (Choquet)-Bruhat (1958) for the first proof of the well-posedness of the Cauchy problem for relativistic perfect fluids, coupled or not with the Einstein equations.

In the case of barotropic fluids, i.e. with equation of state $\mu = \mu(p)$, the simplest way is to define a function $f(p)$, called the index of the fluid, by

$$f(p) := \exp\left[\int_{p_0}^{p} \frac{dp}{\mu(p) + p}\right] \qquad \text{(IX.9.1)}$$

and the dynamical 4-velocity by

$$C_\alpha = f u_\alpha, \quad \text{hence} \quad C^\alpha C_\alpha = -f^2, \; C^\beta \nabla_\alpha C_\beta = -f \partial_\alpha f. \qquad \text{(IX.9.2)}$$

Theorem IX.9.1   *For a barotropic fluid, the Euler equations (4.4) and (4.5) are equivalent to the equation*

$$C^\alpha (\nabla_\alpha C_\beta - \nabla_\beta C_\alpha) = 0 \qquad \text{(IX.9.3)}$$

and

$$\nabla_\alpha C^\alpha + (\mu'_p - 1)\frac{C^\alpha C^\beta}{C^\lambda C_\lambda} \nabla_\alpha C_\beta = 0, \quad \text{with } \mu'_p \equiv \frac{\partial \mu}{\partial p}. \qquad \text{(IX.9.4)}$$

Exercise IX.9.1   *Prove this theorem.*

Exercise IX.9.2   *Show that the flow lines of a perfect fluids are geodesics of a metric confomal to the spacetime metric.*

Hint: Show that $\tilde{g} = F^2 g$ implies

$$C^\alpha \tilde{\nabla}_\alpha C_\beta = C^\alpha \nabla_\alpha C_\beta + C^\alpha C_\alpha \partial_\beta(\log F).$$

Corollary IX.9.1   *A barotropic relativistic fluid is incompressible* $(\mu'_p = 1)$ *if and only if*[30]

$$\nabla_\alpha C^\alpha = 0. \qquad \text{(IX.9.5)}$$

[30]This property of incompressible fluids generalizes the classical one for Newtonian fluids, $\partial_i v^i = 0$, implied in this case by the constancy of density.

Assuming an equation of state and expressing p as a function of f, and hence of C, the equations (9.4), (9.5) read as a first-order differential system for C.

Exercise IX.9.3   *Write these equations for the equation of state* $p = (\gamma - 1)\mu$.

IX.9.2   Vorticity tensor and Helmholtz equations

The vorticity tensor is defined through the dynamical velocity as the antisymmetric 2-tensor

$$\Omega_{\alpha\beta} \equiv \nabla_\alpha C_\beta - \nabla_\beta C_\alpha.$$

It results from the equations of motion (9.3) that this tensor is orthogonal to the velocity; indeed, these equations read

$$C^\alpha \Omega_{\alpha\beta} = 0. \qquad \text{(IX.9.6)}$$

Theorem IX.9.2   *(relativistic Helmholtz equations) The Lie derivative of the vorticity tensor Ω with respect to the dynamic velocity C vanishes:*

$$\mathcal{L}_C \Omega = 0.$$

Proof. Since Ω is an exterior 2-form, the differential of the 1-form C, it is a closed form, satisfying in local coordinates the identity

$$\nabla_\alpha \Omega_{\beta\gamma} + \nabla_\gamma \Omega_{\alpha\beta} + \nabla_\beta \Omega_{\gamma\alpha} \equiv 0,$$

and hence

$$C^\alpha \nabla_\alpha \Omega_{\beta\gamma} + C^\alpha \nabla_\gamma \Omega_{\alpha\beta} + C^\alpha \nabla_\beta \Omega_{\gamma\alpha} \equiv 0. \qquad \text{(IX.9.7)}$$

The equations of motion (9.3) imply, by the derivation ∇_γ,

$$C^\alpha \nabla_\gamma \Omega_{\alpha\beta} + \Omega_{\alpha\beta} \nabla_\gamma C^\alpha = 0.$$

Using this equation, the identity (9.7) gives the equation

$$C^\alpha \nabla_\alpha \Omega_{\beta\gamma} - \Omega_{\alpha\beta} \nabla_\gamma C^\alpha - \Omega_{\gamma\alpha} \nabla_\beta C^\alpha \equiv 0;$$

that is, using the expression for the Lie derivative of a covariant tensor and the antisymmetry of Ω,

$$(\mathcal{L}_C \Omega)_{\beta\gamma} \equiv C^\alpha \nabla_\alpha \Omega_{\beta\gamma} + \nabla_\gamma C^\alpha \Omega_{\beta\alpha} + \nabla_\beta C^\alpha \Omega_{\alpha\gamma} = 0. \qquad \text{(IX.9.8)}$$

∎

We have shown that the vorticity tensor Ω satisfies a linear differential homogeneous system along the flow lines, and hence we have the following corollary:

Corollary IX.9.2 *If a smooth (barotropic) flow has vanishing vorticity on a 3-submanifold transversal to the flow lines, it has a vanishing vorticity on the domain of spacetime spanned by these flow lines.*

IX.9.3   General perfect fluid enthalpy h

For non-barotropic fluids with equation of state depending on the entropy S, one introduces the **enthalpy** h, defined through the thermodynamic identity

$$dh \equiv r^{-1} dp + T dS, \qquad \text{(IX.9.9)}$$

or, equivalently, because of the thermodynamic identity (5.3),

$$dh \equiv d[r^{-1}(\mu + p)], \quad \text{we set} \quad h := r^{-1}(\mu + p), \qquad \text{(IX.9.10)}$$

and now define the dynamical velocity by

$$C^\alpha := h u^\alpha, \quad \text{setting} \quad \Omega_{\alpha\beta} := \nabla_\alpha C_\beta - \nabla_\beta C_\alpha. \qquad \text{(IX.9.11)}$$

Exercise IX.9.4 *Show that*

$$C^\alpha \nabla_\beta C_\alpha = -h \partial_\beta h. \qquad \text{(IX.9.12)}$$

Exercise IX.9.5 *Show that the equations of motion are equivalent to*

$$(hT)^{-1} C^\alpha \Omega_{\alpha\beta} = \partial_\beta S. \qquad \text{(IX.9.13)}$$

IX.10   Irrotational flows

IX.10.1   Definition and properties

A flow with zero vorticity is called **irrotational**, its trajectories are orthogonal hypersurfaces, since the equations $\nabla_\alpha C_\beta - \nabla_\beta C_\alpha = 0$ imply that there exists on spacetime (at least locally) a function Φ such that

$$C_\alpha = \partial_\alpha \Phi.$$

Theorem IX.10.1   *The irrotational flow of a perfect fluid is governed by a quasilinear wave-type equation.*[31]

[31] Fourès (Choquet)-Bruhat (1958).

Proof.   When the flow is irrotational, (9.3) is satisfied identically while (9.4) reads

$$\left[g^{\alpha\beta} + (1 - \mu'_p)u^\alpha u^\beta \right] \nabla_\alpha \partial_\beta \Phi = 0. \qquad (IX.10.1)$$

In this equation, the quantities u and p are given functions of Φ and $\partial\Phi$. Indeed, by definition,

$$g^{\alpha\beta}\partial_\alpha\Phi\partial_\beta\Phi \equiv -f^2 \quad \text{and} \quad u^\alpha \equiv \frac{\partial^\alpha\Phi}{f},$$

while p can be expressed in terms of f, and hence of $\partial\Phi$, by inverting the relation (9.1).

The characteristic cone of this quasilinear second-order differential equation for Φ is given by the quadratic form

$$\left[g^{\alpha\beta} + (1 - \mu'_p)u^\alpha u^\beta \right] X_\alpha X_\beta = 0, \qquad (IX.10.2)$$

whose dual is the sound cone. This quadratic form is of Lorentzian signature if and only if $\mu'_p > 0$; the equation satisfied by Φ is then hyperbolic. It is causal (the sound cone is interior to the light cone) if $\mu'_p \geq 1$.   ∎

Remark IX.10.1   *In the case of a stiff fluid, $\mu'_p = 1$, the equation for Φ reduces to the usual linear wave equation.*

Exercise IX.10.1   *Check the quadratic form (10.2) using (6.9).*

IX.10.2   Coupling with the Einstein equations

The Einstein equations in harmonic gauge with source given by an irrotational flow, together with the equation for Φ, form a second-order quasidiagonal system whose principal parts are the wave operator of the spacetime metric or the sound wave operator. It is a causal hyperbolic system if $\mu'_p \geq 1$.

Remark IX.10.2   *The solution can be interpreted as an Einsteinian spacetime with source an irrotational flow as long as $\partial\Phi$ is timelike, i.e. $g^{\alpha\beta}\partial_\alpha\Phi\partial_\beta\Phi < 0$.*

Exercise IX.10.2   *Formulate the Cauchy problem and a local existence theorem for the Einstein equations coupled with a perfect fluid in irrotational flow.*

IX.11   Equations in a flow-adapted frame

[32]Cattaneo (1959), Ferrarese (1963), and Ferrarese and Bini (2007).

It was stressed long ago by Cattaneo and Ferrarese[32] that the natural physical quantities are timelines and not spacelike submanifolds. Using the same kind of formalism as given in Chapter VII for Cauchy adapted frames, but for Lorentzian frames with time axis tangent to the timelines (time-adapted frames) Friedrich[33] has written the perfect fluid equations as a symmetric system for the pressure and the connection coefficients. This system is hyperbolic with respect to a time function, but only under additional appropriate conditions.[34]

[33]Friedrich (1998).

[34]See YCB-OUP2009, Chapter IX, Section 14, or Choquet-Bruhat and York (1998), which uses the Riemann tensor, instead of the Weyl tensor used by Friedrich.

IX.12   Shocks

A shock is a discontinuity in the fluid variables across a timelike n-manifold ($n = 3$ in the classical case). The stress–energy tensor is then discontinuous; its derivative is meaningful only in a generalized sense.

The relativistic **Rankine–Hugoniot equations** express the vanishing of the divergence of the stress–energy tensor in the space of generalized functions (distributions); they are

$$n_\alpha[T^{\alpha\beta}] = 0,$$

where n is the spacelike normal to the timelike shock front Σ and $[T^{\alpha\beta}]$ is a measure with support Σ. A deep study of the formation of shocks for fluids in Special Relativity has been made by Christodoulou.[35]

[35]Christodoulou (2007).

IX.13   Charged fluids

IX.13.1   Equations

The stress–energy tensor of a charged perfect fluid is the sum of the stress–energy tensor of the fluid variables and the Maxwell tensor of the electromagnetic field F:

$$T_{\alpha\beta} = T^{(\text{fluid})}_{\alpha\beta} + \tau_{\alpha\beta}, \qquad (\text{IX.13.1})$$

with

$$T^{(\text{fluid})}_{\alpha\beta} := (\mu + p)u_\alpha u_\beta + pg_{\alpha\beta} \qquad (\text{IX.13.2})$$

and

$$\tau_{\alpha\beta} = F_\alpha{}^\lambda F_{\beta\lambda} - \frac{1}{4}g_{\alpha\beta}F^{\lambda\mu}F_{\lambda\mu}. \qquad (\text{IX.13.3})$$

The state of the charged fluid is governed by the Einstein equations

$$S_{\alpha\beta} = T_{\alpha\beta}, \qquad (\text{IX.13.4})$$

the Maxwell equations, with J the electric current,

$$dF = 0 \quad \text{and} \quad \nabla \cdot F = J, \quad \text{i.e.} \quad \nabla_\alpha F^{\alpha\beta} = J^B, \qquad (\text{IX.13.5})$$

and the conservation equations, which are

$$\nabla_\alpha T^{\alpha\beta} \equiv \nabla_\alpha(T^{\alpha\beta}_{(\text{fluid})} + \tau^{\alpha\beta}) = 0. \qquad \text{(IX.13.6)}$$

Using previous formulas, we see that these conservation equations read

$$(\mu + p)u^\alpha\nabla_\alpha u^\beta + g^{\alpha\beta}\partial_\alpha p + u^\beta[\nabla_\alpha(\mu + p)u^\alpha] + J_\lambda F^{\beta\lambda} = 0. \quad \text{(IX.13.7)}$$

Classically, the electric current J is the sum of a convection and a conduction current:

$$J_\lambda = qu_\lambda + \sigma E_\lambda, \quad E_\lambda = u^\mu F_{\mu\lambda}, \qquad \text{(IX.13.8)}$$

where q is a scalar function (the electric charge density), σ is the electric conductivity (in general assumed to be constant), and E is the electric field.

The current J satisfies the conservation equation

$$\nabla_\alpha J^\alpha = 0. \qquad \text{(IX.13.9)}$$

As for uncharged fluids, taking the contracted product of (13.7) with u gives a continuity equation

$$(\mu + p)\nabla_\alpha u^\alpha + u^\alpha\partial_\alpha\mu - \sigma E^\alpha E_\alpha = 0 \qquad \text{(IX.13.10)}$$

and equations of motion

$$(\mu+p)u^\alpha\nabla_\alpha u^\beta + g^{\alpha\beta}\partial_\alpha p + u^\beta(u^\alpha\partial_\alpha p + \sigma E^\alpha E_\alpha) + J_\lambda F^{\beta\lambda} = 0. \quad \text{(IX.13.11)}$$

Modulo initial conditions, as in the vacuum case (see Chapter VIII), the Maxwell equations are equivalent to wave-type equations

$$(d\delta + \delta d)F = dJ. \qquad \text{(IX.13.12)}$$

One can add to these equations the law of particle number conservation

$$\nabla_\alpha(ru^\alpha) = 0 \qquad \text{(IX.13.13)}$$

and consider that μ is a function of p and r. If the equation of state is given by μ as a function of the pressure p and entropy S, then (13.13) must be replaced by a thermodynamic law.

IX.13.2   Fluids with zero conductivity

When the conductivity is zero, the Lorentz force reduces to qE, which is orthogonal to the flow vector u and hence does not furnish any work. One says that the fluid is non-dissipative. The first law of thermodynamics and particle number conservation then imply, as in the uncharged case, the law of conservation of entropy:

$$u^\alpha\partial_\alpha S = 0.$$

The Euler–entropy equations are then expected to be a hyperbolic system. This is confirmed by the following theorem:

Theorem IX.13.1   *The Einstein equations in wave gauge with sources an electromagnetic field of potential in Lorentz gauge and a charged perfect fluid with equation of state $\mu = \mu(p, S)$ and with zero conductivity ($\sigma \equiv 0$) are, if $\mu'_p > 0$, a hyperbolic system for g, q, u, p, and the electromagnetic potential A. This system is causal if $\mu'_p \geq 1$.*

A solution of the Cauchy problem for the full Einstein–Maxwell–charged fluid system, for initial data satisfying the constraints, can be deduced by methods analogous to those used for uncharged fluids. Apart from the addition of electromagnetic wave fronts, the wave fronts are the same as for uncharged fluids.

IX.13.3   Fluids with finite conductivity

When the conductivity σ is finite, the Lorentz force is no longer orthogonal to u: it works under the fluid flow. The fluid is called dissipative and its properties are expected to be different from those of non-dissipative fluids. Indeed, experiments show, for example, that no shock wave propagates in charged mercury, a liquid with non-zero conductivity. This fact suggests that for such a fluid, the Cauchy problem is not well posed in spaces of functions with a finite number of derivatives. Indeed the first law of thermodynamics now gives the entropy equation (S increases along the flow lines if $\sigma > 0$, as physically predicted)

$$Tu^\alpha \partial_\alpha S = \sigma E^\alpha E_\alpha.$$

The equation expressing conservation of the electric current J,

$$\nabla_\alpha(qu^\alpha + \sigma u_\lambda F^{\alpha\lambda}) = 0, \qquad (IX.13.14)$$

contains derivatives of the electromagnetic field F, and hence second derivatives of the potential A. The characteristic determinant of the Euler–entropy–electromagnetic system is the same as in the case of zero conductivity, but the characteristic matrix is no longer diagonalizable; the system is *not hyperbolic in the sense defined before*.

Fluids with finite, non-zero, conductivity are a physical example[36] of **Leray–Ohya hyperbolic systems** (which the authors called 'non-strictly hyperbolic'). Such systems have solutions in Gevrey classes, spaces of C^∞ functions whose derivatives satisfy inequalities weaker than the inequalities satisfies by the derivatives of analytic functions that ensure the convergence of their Taylor series. These Gevrey classes enjoy the property fundamental in Relativity of being able to possess a compact support without being identically zero. This permits the limitation by the speed of light of signal solutions of Leray–Ohya hyperbolic systems.

It can be proved that if $\mu'_p \geq 1$, then the reduced Einstein–Maxwell–Euler–entropy system of a charged perfect fluid with finite, non-zero,

[36]Choquet-Bruhat (1965).

conductivity is Leray–Ohya hyperbolic (and causal) but not hyperbolic (in the strict sense).[37]

[37] K. O. Friedrichs has obtained for a charged perfect fluid with non-zero finite conductivity a symmetric hyperbolic system, but by modifying the equations in a way that is not physically justified.

IX.14   Magnetohydrodynamics

In the case where the electric conductivity is so large that it can be considered as infinite, the proper-frame electric field $E_\alpha = u^\beta F_{\alpha\beta}$ becomes negligible. The case of a zero electric field is called magnetohydrodynamics.[38] It plays a fundamental role in plasma physics.

[38] Equations for magnetohydrodynamics in Special Relativity, keeping terms neglected in classical magnetohydrodynamics, were obtained by Hoffman and Teller (1950). In General Relativity, they were first written down and studied by Bruhat (Choquet-Bruhat) (1960).

IX.14.1   Equations

In three space dimensions, for a fluid with **infinite conductivity** $\sigma = \infty$, the electromagnetic field reduces, in the local frame defined by the fluid velocity, to the magnetic vector H. The second Maxwell equation $\delta F = J$ is replaced by

$$E^\alpha := u_\beta F^{\beta\alpha} = 0. \tag{IX.14.1}$$

From the definition of the vector H (orthogonal to u),

$$H^\alpha = u_\beta (F^*)^{\beta\alpha}, \tag{IX.14.2}$$

one deduces

$$(F^*)^{\beta\alpha} = H^\alpha u^\beta - H^\beta u^\alpha. \tag{IX.14.3}$$

A straightforward calculation gives the following expression for the Maxwell tensor:

$$\tau_{\alpha\beta} \equiv \left(u_\alpha u_\beta + \frac{1}{2} g_{\alpha\beta} \right) |H|^2 - H_\alpha H_\beta. \tag{IX.14.4}$$

The first Maxwell equation $dF = 0$ becomes

$$\nabla_\alpha (F^*)^{\beta\alpha} \equiv \nabla_\alpha (H^\alpha u^\beta - H^\beta u^\alpha) = 0. \tag{IX.14.5}$$

Modulo these equations, the divergence of τ is found to be

$$\nabla_\alpha \tau^{\alpha\beta} = \left(u^\alpha u^\beta + \frac{1}{2} g^{\alpha\beta} \right) \partial_\alpha |H|^2 + |H|^2 \nabla_\alpha (u^\alpha u^\beta) - \nabla_\alpha (H^\alpha H^\beta). \tag{IX.14.6}$$

Straightforward computation shows that it is a vector orthogonal both to u and to H.

The Lorentz force $\nabla_\alpha \tau^{\alpha\beta}$ being orthogonal to u, the continuity equation for a fluid of infinite conductivity is the same as for an uncharged fluid, namely

$$\nabla_\alpha [(\mu + p) u^\alpha] - u^\alpha \partial_\alpha p = 0. \tag{IX.14.7}$$

The equations of motion read

$$(\mu + p) u^\alpha \nabla_\alpha u^\beta + (g^{\alpha\beta} + u^\beta u^\alpha) \partial_\alpha p + \nabla_\alpha \tau^{\alpha\beta} = 0. \tag{IX.14.8}$$

IX.14.2   Wave fronts

A straightforward computation shows that the characteristic polynomial of the first-order system (14.5), (14.7), (14.8) factorizes as

$$[(\mu + p)a^2 - b^2]^2 D, \quad \text{with} \quad a := u^\alpha X_\alpha, \ \ b := H^\alpha X_\alpha, \qquad \text{(IX.14.9)}$$

and with D a fourth-order polynomial in X:

$$D \equiv (\mu'_p - 1)(\mu + p)a^4 + [(\mu + p + |H|^2 \mu'_p)a^2 - b^2] X^\alpha X_\alpha. \qquad \text{(IX.14.10)}$$

The fourth-order cone $D = 0$ is called the **magneto-acoustic cone**.

The second-order cone, which enters as a double factor in (14.9),

$$(\mu + p)a^2 - b^2 = 0, \qquad \text{(IX.14.11)}$$

is called the **Alfvén cone**. It is composed of two hyperplanes

$$[(\mu + p)^{\frac{1}{2}} u^\alpha \pm H^\alpha] X_\alpha = 0. \qquad \text{(IX.14.12)}$$

The corresponding wave fronts, tangent to the dual of a Alfvén plane, are called **Alfvén waves**.

In the proper frame of the fluid, the normals ν to the Alfvén waves satisfy the equations

$$(\mu + p)^{\frac{1}{2}} \nu_0 \pm H^i \nu_i = 0, \qquad \text{(IX.14.13)}$$

The propagation speed of the Alfvén waves is calculated using the general formula

$$|V|^2 = \frac{(u^\alpha \nu_\alpha)^2}{(u^\alpha u^\beta + g^{\alpha\beta}) \nu_\alpha \nu_\beta}. \qquad \text{(IX.14.14)}$$

One finds that

$$|V_{\text{Alf}}| = (\mu + p)^{-\frac{1}{2}} |H_{\tilde{\nu}}|, \qquad \text{(IX.14.15)}$$

where H_ν is the scalar product of H (a space vector) with the projection on space (normed to 1) of the normal ν to the wave front.

The wave fronts whose normals lie on the magnetoacoustic cone $D = 0$ are called **magnetoacoustic** wave fronts. The equation of the magnetoacoustic cone is a polynomial of order 4 in ν, the normal to a wave front, which reads, in a proper rest frame of the fluid,

$$\mu'_p (\mu + p + H^2)(\tilde{\nu}_0)^4 - \tilde{\nu}_0^2 (\mu + p + |H|^2 \mu'_p + H_{\tilde{\nu}}^2) + H_{\tilde{\nu}}^2 = 0, \ \ \text{(IX.14.16)}$$

where we have denoted

$$\tilde{\nu} = \frac{\nu}{\sum_i [\nu_i^2]^{1/2}} \quad \text{and} \quad H_{\tilde{\nu}} = H_i \tilde{\nu}^i. \qquad \text{(IX.14.17)}$$

We deduce from this formula, solved for the unknown $\tilde{\nu}_0^2$, the two magnetoacoustic wave speeds. Both are well defined, because $|H_{\tilde{\nu}}| \leq |H|$, and less than 1, the speed of light, if $\mu'_p \geq 1$. They are

$$|V|^2_{\text{MA}} = \frac{\mu + p + |H|^2 \mu'_p + H_{\tilde{\nu}}^2 \pm [(\mu + p + |H|^2 \mu'_p + H_{\tilde{\nu}}^2)^2 - 4 H_{\tilde{\nu}}^2 \mu'_p (\mu + p + H^2)]^{\frac{1}{2}}}{2 \mu'_p (\mu + p + H^2)}. \qquad \text{(IX.14.18)}$$

The rapid wave speed corresponds to the plus sign and the slow wave speed to the minus sign.

Study of the characteristic determinant shows[39] that the system is a Leray–Ohya-type hyperbolic system—the fact that Alfvén wave fronts are tangent to the magnetosonic wave fronts does not permit one to conclude its (strict) Leray hyperbolicity. However, the system has been proved to be symmetrizable hyperbolic[40], both by the general method of Lax–Boillat and Ruggeri, and directly (at least in special Relativity) by K.O. Friedrichs. This hyperbolicity could be foreseen physically by the existence of shock waves[41] and of high-frequency waves[42] propagating according to first-order differential equations.

IX.15   Yang–Mills fluids (quark–gluon plasmas)

Although Yang–Mills charges are not manifest at ordinary scales, plasmas of quarks and gluons exist under extreme circumstances.[43] Their properties have analogies, but also differences, with those of electromagnetic plasmas. The equations look formally the same. The Maxwell stress–energy tensor is replaced by the Yang–Mills stress–energy tensor and the electric current by the Lie-algebra-valued Yang–Mills current

$$J^\alpha = \gamma u^\alpha + \sigma F^{\beta\alpha} u_\beta,$$

where γ is a function on spacetime taking its values in the same Lie algebra as the Yang–Mills field F and σ is a number, the fluid conductivity.

The mathematical properties of **Yang–Mills fluids** are[44] quite similar to those of electrically charged fluids, at least locally, in the case of finite conductivity.

In the case of **infinite conductivity**, however, a remarkable property occurs: the wave fronts do not split into analogies of Alfvén waves and acoustic waves, but are at each point tangent to an undecomposable sixth-order cone.[45]

IX.16   Viscous fluids

There have been many discussions of equations for relativistic viscous fluids and a relativistic heat equation. Most proposals for such equations were motivated by the desire to obtain a causal theory, that is, one described by partial differential equations of hyperbolic type, with a causal domain of dependence, whereas the usual Fourier law of heat transfer as well as the viscous fluid Navier–Stokes equations of Newtonian mechanics are of parabolic type, corresponding to an infinite propagation speed. We will see in Chapter X some results for the approximation beyond perfect fluids obtained in the framework of Relativity.

---

[39]See Choquet-Bruhat (1966).

[40]See Anile (1989).

[41]General relativistic equations were given by Bruhat (Choquet-Bruhat) (1960) and were studied in depth by Lichnerowicz (1967).

[42]Anile and Greco (1978).

[43]Extremely briefly in Earth-based laboratories. Astrophysicists have conjectured that they are present in the cores of neutron stars.

[44]See Choquet-Bruhat (1992).

[45]Choquet-Bruhat (1992b, 1993).

IX.16.1   Generalized Navier–Stokes equations

The Einstein equivalence principle would give us an expression for the stress–energy tensor of a viscous fluid in General Relativity if we knew of such as expression in Special Relativity. However, this is not the case—there is no general consensus about such a tensor in Special Relativity, everyone being influenced in his choice by his own background. The difference between various authors begins already at the physical definition of the flow vector[46] u. We will stick to the simplest definition, considering u as the flux density vector for the particle number r. This quantity satisfies the conservation law

$$\nabla_\alpha(ru^\alpha) = 0. \tag{IX.16.1}$$

It was remarked previously that the components T^{00}, T^{0i}, T^{ij} of the stress–energy tensor T in a proper frame represented respectively the energy, momentum, and stresses of the fluid with respect to this frame. For perfect fluids, the expressions for these quantities were taken as their classical analogues, which themselves result from first principles.

In non-relativistic mechanics, the stress tensor of a viscous fluid is obtained by considering it as a linearized perturbation σ of the perfect fluid stress tensor, a perturbation due to the deformation tensor of the flow lines. In inertial coordinates of absolute space, this deformation tensor is

$$D_{ij} := \frac{\partial v_i}{\partial x^j} + \frac{\partial v_j}{\partial x^i}. \tag{IX.16.2}$$

One writes the **classical Navier–Stokes** equations for viscous fluids by introducing the expansion and the shear of the flow lines (in Euclidean space E^3). The contribution of rotational flow is discarded, on the basis that it corresponds to rigid motions, which do not generate stresses. The **Newtonian stress tensor of a viscous fluid** is then written as the sum of the stress tensor $p\delta_{ij}$ of a perfect fluid and a symmetric 2-tensor given by

$$\sigma_{ij} := \lambda\delta_{ij}\frac{\partial v^k}{\partial x^k} + \nu\left(D_{ij} - \frac{2}{3}\delta_{ij}\frac{\partial v^k}{\partial x^k}\right). \tag{IX.16.3}$$

The scalar coefficients, in general considered as constants, are called the bulk viscosity and the shear viscosity.

The extension of the tensor σ to Relativity is ambiguous, owing on the one hand to the absence of absolute space and on the other hand to the equivalence of mass with energy, and hence with the work of the friction forces due to viscosity. The following tensor has been proposed as a *relativistic generalization* of the classical one:

$$T_{\alpha\beta} := (\mu + p)u_\alpha u_\beta + pg_{\alpha\beta} + \sigma_{\alpha\beta}, \tag{IX.16.4}$$

with σ the **viscosity stress tensor**, linear in the first derivatives of u and orthogonal to u,

$$\sigma_{\alpha\beta} = \lambda\pi_{\alpha\beta}\nabla_\rho u^\rho + \nu\pi_\alpha^\rho\pi_\beta^\sigma(\nabla_\rho u_\sigma + \nabla_\rho u_\sigma), \tag{IX.16.5}$$

[46]Landau and Lifshitz (1987) link u with some density of energy, choosing u such that the stress tensor is orthogonal to u.

where π is the projection tensor[47]

$$\pi_{\alpha\beta} := g_{\alpha\beta} + u_\alpha u_\beta. \qquad \text{(IX.16.6)}$$

Some authors add to T a 'momentum term'

$$u_\alpha q_\beta + u_\beta q_\alpha, \qquad \text{(IX.16.7)}$$

supposed to represent a heat flow resulting from friction. The problem is then to write an equation to determine q and justify the symmetrization in (16.7).

The equations for $\nabla_\alpha T^{\alpha\beta}$ deduced from (16.4) and (16.5) are rather complicated. They have been studied in detail by G. Pichon[48] in the case when $\nabla_\alpha u^\alpha = 0$, called an incompressible fluid. He has shown that for large enough $\mu + p$, they are of parabolic type, corresponding therefore to infinite propagation speed.

IX.16.2   A Leray–Ohya hyperbolic system for viscous fluids

We have seen that the vector that leads to the generalization of classical properties of irrotational motion, for perfect fluids, is the dynamical velocity $C := h\, u$, with h the enthalpy (see Section IX.9.3). We define for this vector C the shear, the vorticity tensor, and the expansion of the congruence of its trajectories (the flow lines of the fluid) by the usual decomposition. In the case $n + 1 = 4$,

$$\nabla_\alpha C_\beta = \frac{1}{2}\Omega_{\alpha\beta} + \Sigma_{\alpha\beta} + \frac{1}{4}\Theta g_{\alpha\beta},$$

with Ω the vorticity tensor,

$$\Omega_{\alpha\beta} := \nabla_\rho C_\sigma - \nabla_\rho C_\sigma.$$

In the case of a perfect fluid, Ω is orthogonal to C (i.e. to u). The scalar Θ is the expansion of the congruence defined by the vector field C:

$$\Theta := \nabla_\alpha C^\alpha.$$

We have seen that Θ is zero if and only if the fluid is incompressible (in the relativistic sense). The symmetric tensor Σ is the shear, which has zero trace:

$$\Sigma_{\alpha\beta} := \frac{1}{2}(\nabla_\alpha C_\beta + \nabla_\beta C_\alpha) - \frac{1}{4}g_{\alpha\beta}\nabla_\lambda C^\lambda.$$

We propose to take as stress–energy tensor of a viscous fluid the tensor

$$T_{\alpha\beta} = (\mu + p)u_\alpha u_\beta + g_{\alpha\beta}p + \tilde{T}_{\alpha\beta}, \qquad \text{(IX.16.8)}$$

where $\tilde{T}_{\alpha\beta}$ is the part of T due to viscosity, a perturbation of the perfect fluid stress–energy tensor, linear in ∇C, of the form

$$\tilde{T}_{\alpha\beta} := \lambda\Theta g_{\alpha\beta} + \nu\Sigma_{\alpha\beta}, \qquad \text{(IX.16.9)}$$

[47] Note that the coefficients λ and ν do not have quite the same interpretation as in Newtonian mechanics.

[48] Pichon (1965).

with λ and ν viscosity coefficients depending on the particular fluid under consideration. The conservation laws $\nabla_\alpha T^{\alpha\beta} = 0$ are then second-order equations for the dynamical velocity C with principal part

$$\lambda g^{\alpha\beta}\nabla_\alpha\nabla_\lambda C^\lambda + \nu\nabla_\alpha \left[\frac{1}{2}(\nabla^\alpha C^\beta + \nabla^\beta C^\alpha) - \frac{1}{4}g^{\alpha\beta}\nabla_\lambda C^\lambda\right]$$

$$= \frac{\nu}{2}(\nabla_\alpha\nabla^\alpha C^\beta + \nabla_\alpha\nabla^\beta C^\alpha) + \left(\lambda - \frac{1}{4}\nu\right)\nabla^\beta\nabla_\alpha C^\alpha.$$

The 4×4 characteristic matrix has elements

$$\frac{\nu}{2}X_\alpha X^\alpha C^\beta + aX_\alpha X^\beta C^\alpha, \quad a := \frac{\nu}{4} + \lambda.$$

To compute the characteristic determinant, we choose a frame where $X^2 = X^3 = 0$. The matrix then reads

$$\begin{pmatrix} \frac{\nu}{2}X^\alpha X_\alpha + aX_0 X^0 & aX_1 X^0 & 0 & 0 \\ aX_0 X^1 & \frac{\nu}{2}X^\alpha X_\alpha + aX_1 X^1 & 0 & 0 \\ 0 & 0 & \frac{\nu}{2}X^\alpha X_\alpha & 0 \\ 0 & 0 & 0 & \frac{\nu}{2}X^\alpha X_\alpha \end{pmatrix}.$$

$$(\text{IX.16.10})$$

The characteristic determinant is then

$$\Delta(X) \equiv P(X)\left(\frac{\nu}{2} + a\right)^2 (X^\alpha X_\alpha)^2,$$

with

$$P(X) := \left(\frac{\nu}{2}X^\alpha X_\alpha + aX_0 X^0\right)\left(\frac{\nu}{2}X^\alpha X_\alpha + aX_1 X^1\right) - a^2 X^1 X_1 X^0 X_0,$$

i.e.

$$P(X) := \left(\frac{\nu}{2}X^\alpha X_\alpha\right)^2 + \frac{\nu}{2}X^\alpha X_\alpha(aX_\alpha X^\alpha) \equiv \frac{\nu}{2}\left(\frac{\nu}{2} + a\right)(X^\alpha X_\alpha)^2.$$

The proposed system is Leray–Ohya hyperbolic (Gevrey class of index 2) and causal, a satisfactory property for a relativistic theory.

It can be conjectured that a solution of the proposed relativistic Navier–Stokes equations converges to a solution of the perfect fluid equations when λ and ν tend to zero.

IX.17   The heat equation

We personally think that heat transfer is a collective effect; the Fourier equation corresponds in fact to some asymptotic steady state.[49] It is not surprising that it does not translate readily into relativistic causal equations. The classical heat equation appears after the study of the Brownian motion, unfortunately the extension to Relativity of the study of Brownian motion is still in its infancy, owing to several difficulties, in particular the lack of absolute time.

[49] A singular perturbation with higher derivative has been proposed by Vernotte to give an account of the few moments occurring before the steady state is attained. Vernotte's idea has been extended to General Relativity by Cattaneo.

The Müller–Ruggeri extended thermodynamics theory offers a way to treat dissipative phenomena in a relativistic context (see Chapter X).

IX.18   Conclusion

The relativistic theory of perfect fluids is mathematically and physically[50] satisfactory. With proper definitions, the theorems of classical (i.c. non-relativistic) perfect fluid theory are valid. Global mathematical problems remain mostly open, but this is also the case for classical fluids in spite of recent progress. A book by Christodoulou[51] treats in depth three-dimensional perfect fluids in special Relativity up to shock formation.

[50]In spite of problems with the choice of an equation of state for the fluid, particularly when the whole cosmos is concerned, although it is unlikely that its content can be modelled by a perfect fluid.

[51]Christodoulou (2007).

IX.19   Solution of Exercise IX.6.2

Light cone: The light rays issuing from a given point satisfy the equations

$$\frac{dx^\alpha}{d\lambda} = g^{\alpha\beta} p_\beta, \qquad g^{\alpha\beta} p_\alpha p_\beta = 0. \qquad \text{(IX.19.1)}$$

Hence, with the usual identification of the tangent and cotangent space to V through the metric g,

$$g_{\alpha\beta} v^\alpha v^\beta = 0, \qquad v^\alpha := g^{\alpha\beta} p_\beta.$$

IX.20   Problems

IX.20.1   Specific volume

Show by integrating the equation

$$\nabla_\alpha(r u^\alpha) = 0$$

and using the Stokes formula that the t-dependent space scalar

$$\bar{r} \equiv r u^0 N \qquad \text{(IX.20.1)}$$

represents the density of flow lines crossing a spacelike submanifold, where N denotes the spacetime lapse.

IX.20.2   Motion of isolated bodies

Denote by (V, g, r, u) the manifold, spacetime metric, and perfect fluid flow solution of the Einstein equation with perfect fluid source with initial data $(\bar{g}, K, \bar{v}, \bar{r})$ on a manifold M satisfying the hypothesis of the existence and uniqueness theorem. Assume that the support of the initial density $\bar{r}$ is the closure $\bar{\omega}$ of some open subset $\omega \subset M$, which has several disconnected components, spaces occupied by material bodies at the initial time.

Show that the solution does not depend on the 'unphysical' data of $\bar{v}$ where $\bar{r}$ vanishes.

Extend the property to charged dust.

Solution

Equations (2.4) and (2.3) show that the support of r is contained in the geodesic tube Ω generated by geodesics of g tangent to $\bar{u}$ issuing from points in $\bar{\omega}$. Let $\bar{v}_1$ be another initial value for u, such that $\bar{v}_1 = \bar{v}$ on $\bar{\omega}$. The geodesic tube Ω_1 of g corresponding to $\bar{v}_1$ coincides with Ω; therefore, $u_1 = u$ in Ω. Since $r = 0$ outside Ω, the dust stress–energy tensors $ru \otimes u$ and $ru_1 \otimes u_1$ with $u_1 = u$ in Ω coincide on the whole of V; therefore, the triple (g, r, u_1) satisfies the Einstein dust system on V. By the uniqueness theorem, it coincides, up to isometry, with another solution (g_1, r_1, u_1) taking the same initial values, i.e. $g_1 = g$, $r_1 = r$ on V and $u_1 = u$ in the domain $\Omega_1 = \Omega$ of the presence of a fluid flow.

The theorem on the motion of isolated bodies extends to the case of charged dust if we reasonably make $\bar{q}/\bar{r} = 0$ outside the support of $\bar{r}$ because the flow lines of u depend only on g and F, while the right-hand sides of (3.10) vanish outside the support of $\bar{r}$.

IX.20.3   Euler equations for the dynamic velocity

Recall that

$$C^\alpha C_\alpha = -f^2, \quad \text{hence} \quad C^\alpha \nabla_\beta C_\alpha = -f \partial_\beta f.$$

Write the equations of motion and the energy equation of a perfect fluid in terms of the dynamic velocity.

Solution

We deduce from the definition of the index f that

$$\frac{\partial \log f}{\partial p} = \frac{1}{\mu + p}, \quad \text{hence} \quad \frac{\partial p}{\partial \log f} = \mu + p, \quad \frac{\partial p}{\partial f^2} = \frac{\mu + p}{2 f^2},$$

$$\frac{\partial p}{\partial f^2} = \frac{\partial p}{\partial \log f} \frac{\partial \log f}{\partial f} \left(\frac{\partial f^2}{\partial f} \right)^{-1} = (\mu + p) \frac{1}{f} \frac{1}{2f}.$$

Hence

$$\partial_\alpha p = \frac{\partial p}{\partial f^2} 2 f \partial_\alpha f, \quad \text{hence} \quad \frac{\partial_\alpha p}{\mu + p} = f^{-1} \partial_\alpha f = -f^{-2} C^\beta \nabla_\alpha C_\beta.$$

Recall that the Euler equations of motion are

$$u^\alpha \nabla_\alpha u_\beta + (g^\alpha_\beta + u^\alpha u_\beta)\frac{\partial_\alpha p}{\mu + p} = 0.$$

We have

$$u^\alpha \nabla_\alpha u_\beta \equiv f^{-1} C^\alpha \nabla_\alpha (f^{-1} C_\beta) \equiv f^{-2}(C^\alpha \nabla_\alpha C_\beta - f^{-2} C^\alpha C_\beta f \partial_\alpha f)$$
$$\text{(IX.20.2)}$$

and

$$(g^\alpha_\beta + u^\alpha u_\beta)\frac{\partial_\alpha p}{\mu + p} = (g^\alpha_\beta + u^\alpha u_\beta) f^{-2} f \partial_\alpha f \equiv f^{-2}(f \partial_\beta f + f^{-2} C^\alpha C_\beta f \partial_\alpha f),$$
$$\text{(IX.20.3)}$$

which reads

$$u^\alpha \nabla_\alpha u_\beta + (g^\alpha_\beta + u^\alpha u_\beta)\frac{\partial_\alpha p}{\mu + p} \equiv f^{-2}(C^\alpha \nabla_\alpha C_\beta - C^\alpha \nabla_\beta C_\alpha),$$

and hence, for the equations of motion,

$$C^\alpha \Omega_{\alpha\beta} = 0. \qquad \text{(IX.20.4)}$$

The energy (also called continuity) equation for a barotropic fluid reads

$$\nabla_\alpha u^\alpha + u^\alpha \mu'_p \frac{\partial_\alpha p}{\mu + p} = 0, \qquad \mu'_p := \frac{d\mu}{dp}. \qquad \text{(IX.20.5)}$$

We have

$$\nabla_\alpha u^\alpha + u^\alpha \mu'_p \frac{\partial_\alpha p}{\mu + p} \equiv \nabla_\alpha (f^{-1} C^\alpha) + f^{-1} C^\alpha \mu'_p f^{-2} f \partial_\alpha f$$
$$\equiv f^{-1} \left[\nabla_\alpha C^\alpha + C^\alpha f^{-2}(-1 + \mu'_p) \right].$$

Hence, for the energy equation,

$$\nabla_\alpha C^\alpha + f^{-2}(-1 + \mu'_p) C^\alpha \partial_\alpha f = 0.$$

This reduces to $\nabla_\alpha C^\alpha = 0$ if $\mu'_p = 1$ (incompressible fluids).

IX.20.4   Hyperbolic Leray system for the dynamical velocity

1. Prove that the equations satisfied by the dynamical velocity C,

$$C^\alpha \Omega_{\alpha\beta} = 0 \quad \text{with} \quad \Omega_{\alpha\beta} := \nabla_\alpha C_\beta - \nabla_\beta C_\alpha$$

and

$$\nabla_\alpha C^\alpha = (\mu'_p - 1)\frac{C^\alpha}{C^\lambda C_\lambda} C^\beta \nabla_\alpha C_\beta,$$

imply, when $\delta\Omega$ is known, that C satisfies a quasidiagonal system of second-order differential equations that is hyperbolic if $d\mu/dp > 0$.

2. Show that C satisfies a quasilinear quasidiagonal third-order system, which is hyperbolic and causal if $d\mu/dp > 1$.

Solution

1. For C considered as a 1-form, we have the identity (see Chapter I)

$$\{(d\delta + \delta d)C\}_\gamma \equiv -g^{\alpha\beta}\nabla_\alpha\nabla_\beta C_\gamma + R^\alpha_\gamma C_\alpha. \qquad \text{(IX.20.6)}$$

The definitions of the rotational and the energy equations give

$$dC = \Omega \quad \text{and} \quad \delta C := -\nabla_\alpha C^\alpha = -(\mu'_p - 1)\frac{C^\alpha}{C^\lambda C_\lambda}C^\beta\partial_\alpha C_\beta. \qquad \text{(IX.20.7)}$$

It results from (20.6) and (20.7) that

$$g^{\alpha\beta}\nabla_\alpha\nabla_\beta C_\lambda - R^\alpha_\lambda C_\alpha = -\nabla_\gamma\left[(\mu'_p - 1)\frac{C^\alpha}{C^\lambda C_\lambda}C^\beta\partial_\alpha C_\beta\right] + \nabla^\alpha\Omega_{\alpha\gamma}. \qquad \text{(IX.20.8)}$$

We compute

$$\nabla_\gamma\left[(\mu'_p - 1)\frac{C^\alpha}{C^\lambda C_\lambda}C^\beta\nabla_\alpha C_\beta\right] \equiv A(\mu'_p - 1) + B\partial_\gamma(\mu'_p - 1),$$

where we have set

$$A := \nabla_\gamma\left(\frac{C^\alpha}{C^\lambda C_\lambda}C^\beta\nabla_\alpha C_\beta\right), \quad \text{and} \quad B := \frac{C^\alpha C^\beta}{C^\lambda C_\lambda}\nabla_\alpha C_\beta.$$

It holds that

$$A \equiv \frac{C^\alpha C^\beta}{C^\lambda C_\lambda}\nabla_\gamma\nabla_\alpha C_\beta + \left(\nabla_\gamma\frac{C^\alpha C^\beta}{C^\lambda C_\lambda}\right)(\nabla_\alpha C_\beta).$$

By the antisymmetry of the Riemann tensor, we have

$$C^\alpha C^\beta\nabla_\gamma\nabla_\alpha C_\beta \equiv C^\alpha C^\beta(\nabla_\alpha\nabla_\gamma C_\beta + R_{\gamma\alpha,\beta\rho}C^\rho) \equiv C^\alpha C^\beta\nabla_\alpha\nabla_\gamma C_\beta.$$

By the definition of the vorticity Ω,

$$C^\alpha C^\beta\nabla_\alpha\nabla_\gamma C_\beta \equiv C^\alpha C^\beta\nabla_\alpha(\Omega_{\gamma\beta} + \nabla_\beta C_\gamma)$$
$$\equiv C^\alpha C^\beta\nabla_\alpha\nabla_\beta C_\gamma + C^\beta C^\alpha\nabla_\alpha\Omega_{\gamma\beta},$$

with, by the Helmholtz equation,

$$C^\alpha\nabla_\alpha\Omega_{\gamma\beta} = \Omega_{\beta\alpha}\nabla_\gamma C^\alpha + \Omega_{\alpha\gamma}\nabla_\beta C^\alpha. \qquad \text{(IX.20.9)}$$

Therefore,

$$A \equiv \frac{1}{C^\lambda C_\lambda}\left[C^\alpha C^\beta\nabla_\alpha\nabla_\beta C_\gamma + C^\beta(\Omega_{\beta\alpha}\nabla_\gamma C^\alpha + \Omega_{\alpha\gamma}\nabla_\beta C^\alpha)\right]$$
$$+ \left(\nabla_\gamma\frac{C^\alpha C^\beta}{C^\lambda C_\lambda}\right)(\nabla_\alpha C_\beta).$$

We deduce from these computations, since Ω is of first order in C, that when $\nabla^\alpha\Omega_{\alpha\gamma}$ is known, the dynamical velocity C satisfies a system of quasidiagonal second-order equations with principal operator

$$(\Box_{g,C}C)_\gamma := \left[g^{\alpha\beta} + (\mu'_p - 1)\left(\frac{C^\alpha C^\beta}{C^\lambda C_\lambda}\right)\right]\nabla_\alpha\nabla_\beta C_\gamma. \qquad \text{(IX.20.10)}$$

The first-order terms in C are

$$D_\gamma := (\mu_p' - 1) \left[-\frac{\Omega_{\gamma\beta}}{C^\lambda C_\lambda} C^\alpha \nabla_\alpha C^\beta + \left(\nabla_\gamma \frac{C^\alpha C^\beta}{C^\lambda C_\lambda} \right) (\nabla_\alpha C_\beta) \right].$$

In a proper frame, the corresponding characteristic polynomial is

$$- \mu_p' X_0^2 + \sum (X_i)^2. \qquad\qquad \text{(IX.20.11)}$$

It is hyperbolic if $\mu_p' > 0$, causal if $\mu_p' \geq 1$. It reduces to the Minkowski polynomial for $\mu_p' = 1$.

2. Since $\nabla^\alpha \Omega_{\alpha\beta}$ is second-order in C, the principal part of the full operator does not reduce to the operator (20.10). To obtain a hyperbolic operator for C, we take the derivative of (20.4) in the direction of C. We have

$$C^\lambda \nabla_\lambda \nabla_\alpha \Omega^\alpha{}_\beta \equiv \nabla^\alpha (C^\lambda \nabla_\lambda \Omega_{\alpha\beta}) - (\nabla^\alpha C^\lambda) \nabla_\lambda \Omega_{\alpha\beta}.$$

Hence, modulo the Helmholtz equations, $C^\lambda \nabla_\lambda \nabla_\alpha \Omega^\alpha{}_\beta$ is a quadratic polynomial in ∇C and $\nabla^2 C$, namely

$$\begin{aligned} E_\beta &:= C^\lambda \nabla_\lambda \nabla_\alpha \Omega^\alpha{}_\beta \\ &= \nabla^\alpha \left(\Omega_{\beta\lambda} \nabla_\alpha C^\lambda + \Omega_{\lambda\alpha} \nabla_\beta C^\lambda \right) - \left(\nabla^\alpha C^\lambda \right) \nabla_\lambda \Omega_{\alpha\beta}. \end{aligned} \qquad \text{(IX.20.12)}$$

We have proved that $C^\alpha (\nabla_\alpha \Box_{C,g} C + \delta\Omega)$ is a quasidiagonal third-order system with characteristic polynomial in a proper frame

$$f X_0 \left[-\mu_p' X_0^2 + \sum (X_i)^2 \right].$$

The lower second-order terms in C are

$$C^\alpha \nabla_\alpha D_\gamma + E_\gamma.$$

IX.20.5   Geodesics of conformal metric

Show that the trajectories of the dynamical velocity $C := fu$ are geodesics of the metric $\tilde{g} := f^2 g$ conformal to the original spacetime metric g.

Solution

If $\tilde{g} = f^2 g$, it holds that

$$\begin{aligned} \tilde{\nabla}_\beta C_\alpha - \nabla_\beta C_\alpha &= (\Gamma^\lambda_{\beta\alpha} - \tilde{\Gamma}^\lambda_{\beta\alpha}) C_\lambda \\ &= -f^{-3} \left(\partial_\alpha f C_\beta + \partial_\beta f C_\alpha - g_{\alpha\beta} g^{\lambda\mu} \partial_\mu f C_\lambda \right), \end{aligned} \qquad \text{(IX.20.13)}$$

and hence, using (9.5) and (9.4),

$$C^\beta \tilde{\nabla}_\beta C_\alpha = C^\beta \nabla_\alpha C_\beta - f^{-2} \partial_\alpha f C^\beta C_\beta = 0. \qquad \text{(IX.20.14)}$$

IX.20.6   Cosmological equation of state $p = (\gamma - 1)\mu$

Take the cosmological equation of state $p = (\gamma - 1)\mu$.

1. Compute the index f.
2. Compute the energy equation.

Solution

1. If $p = (\gamma - 1)\mu$, then the index f is given by

$$f \equiv \exp\left[\int \frac{(\gamma - 1)\, dp}{\gamma p}\right],$$

and hence, up to an irrelevant multiplicative constant,

$$f \equiv p^{(\gamma - 1)/\gamma}, \qquad p = f^{\gamma/(\gamma - 1)}.$$

2. If $p = (\gamma - 1)\mu$, then

$$\mu'_p - 1 = \frac{2 - \gamma}{\gamma - 1},$$

and the energy equation reads

$$\nabla_\alpha C^\alpha + \frac{2 - \gamma}{\gamma - 1} \frac{C^\alpha C^\beta}{C^\lambda C_\lambda} \nabla_\alpha C_\beta = 0. \tag{IX.20.15}$$

Relativistic kinetic theory

X.1   Introduction

In kinetic theory, the matter, usually called a **plasma**, is composed of a collection of particles whose size is negligible at the considered scale: rarefied gases in the laboratory, stars in galaxies, or even clusters of galaxies at the cosmological scale. The number of particles is so great and their motion so chaotic that it is impossible to observe their individual motion.

We have introduced in Chapter IX equations for perfect fluids in General Relativity as a direct consequence of the perfect fluid equations in Special Relativity and the equivalence principle. We have seen that they are quite satisfactory mathematically as well as physically. We have also seen that there are no compelling macroscopic considerations leading to satisfactory relativistic equations for dissipative fluids, even in Special Relativity. The classical Navier–Stokes equations lead, like the Fourier law of heat transfer, to equations of parabolic type, and hence to an infinite propagation speed of signals, which is incompatible with relativistic causality. The formal generalizations of these equations with the aim of obtaining hyperbolic equations lack general justification and there is no consensus about them.

An a posteriori justification of the non-relativistic Euler and Navier–Stokes macroscopic equations is to deduce them from the motion of the fluid particles at the microscopic scale and the statistical hypothesis of classical kinetic theory. It presents no conceptual difficulty to extend the setting of kinetic theory to Special as well as to General Relativity. It is straightforward to construct an energy–momentum vector and a stress–energy tensor for kinetic matter and to couple the latter with the Einstein equations. The collective motion of collisionless particles is naturally modelled and leads to conservation equations for the stress–energy tensor. A case of particular interest in General Relativity is when the 'particles' are stars in galaxies, or even galaxies in clusters of galaxies. It is then appropriate to take the charges to be zero and the masses to have values[1] between two positive numbers.

When the particles undergo collisions, one can write a Boltzmann equation, and a coherent Einstein–Boltzmann system, for an appropriate choice of the collision cross-section. The problem is the choice of a physically reasonable cross-section; this problem arises already in the non-relativistic case, but is especially delicate when it is stars or galaxies that collide. The Boltzmann equation arises in Special Relativity for

[1]And not discrete values, while this is the case for electromagnetic plasmas.

plasmas of elementary particles with a finite number of distinct proper masses and charges; the gravitational field is then negligible under usual circumstances. However, the general relativistic Boltzmann equation is possibly important for obtaining equations for dissipative relativistic fluids as approximations.

Throughout this chapter, as in previous chapters, we assume that the spacetime (V, g) is an oriented and time-oriented Lorentzian manifold, the tangent bundle T_V is oriented by the orientation of V, and all fibres are positively oriented. We treat the $(n+1)$-dimensional case, specifying $n = 3$ when it enjoys special properties.

X.2   Distribution function

X.2.1   Definition

It is assumed that the state of the matter in a spacetime (V, g) is represented[2] by a 'one-particle distribution function'. This distribution function f is interpreted as the density of particles at a point $x \in V$ that have a momentum $p \in T_x V$, the tangent space to V at x. In a relativistic theory, the momentum p is a future timelike or null vector whose components define the energy and momentum of the particle with respect to a Lorentzian frame. With our signature conventions, it holds that[3] $g(p, p) = -m^2$, with m the rest mass of the particle. In view of applications to astrophysics, we do not assume a priori that it is the same for all particles. We state a definition:

Definition X.2.1   *A **distribution function** f is a non-negative scalar function on the so-called **phase space** $\mathcal{P}_V$, a subbundle of the tangent bundle T_V to the spacetime V:*

$$f : \mathcal{P}_V \to R \quad by \quad (x, p) \mapsto f(x, p), \quad with \quad x \in V, \quad p \in \mathcal{P}_x \subset T_x V. \tag{X.2.1}$$

If (V, g) is a Lorentzian manifold, then the fibre $\mathcal{P}_x$ at x is such that $g_x(p, p) \leq 0$ and, in a time-oriented frame, $p^0 \geq 0$.

Remark X.2.1   *If a particle has positive rest mass, then it holds that $p^0 > 0$ in a time-oriented frame.*

If a particle has zero rest mass, then p^0 can vanish only if p itself vanishes. The value $p = 0$ is, from the mathematical point of view, a singular point of the vector field X. On the other hand, relativistic physics says that particles with zero mass move with the speed of light, the component p^0 in an orthonormal frame is the energy with respect to an observer at rest in this frame, and quantum theory tells us that it does not vanish, i.e. $p^0 > 0$.

X.2.2   Interpretation

The physical meaning of the distribution function is that it gives a mean[4] 'presence number' density of particles in phase space. More precisely,

[2]Representation by a many-particle distribution function is possible, but difficult to handle. Representation by a one-particle distribution function is linked with the onset of chaos. The mathematical justification of the onset of chaos in relativistic dynamics is a largely open problem.

[3]In local coordinates, $g(p, p) \equiv g_{\alpha\beta} p^\alpha p^\beta$.

[4]In the sense of the Gibbs ensemble.

we denote by θ the $2(n+1)$-volume form on TV, i.e. with θ_x and θ_p respectively the volume forms on V and T_xV,

$$\theta := \theta_x \wedge \theta_p. \tag{X.2.2}$$

In local coordinates, θ_x and θ_p are given by

$$\theta_x = |\text{Det } g|^{\frac{1}{2}} dx^0 \wedge dx^1 \ldots \wedge dx^n, \quad \theta_p := |\text{Det } g|^{\frac{1}{2}} dp^0 \wedge dp^1 \ldots \wedge dp^n. \tag{X.2.3}$$

X.2.3   Moments of the distribution function

The moments of f are functions or tensors on V obtained by integration on the fibres of the phase space $\mathcal{P}_V$ of products of f by tensor products of p with itself.

Moment of order zero

This is by definition the integral on the fibre $\mathcal{P}_x$ of the distribution function:[5]

$$r_0(x) := \int_{\mathcal{P}_x} f\theta_p. \tag{X.2.4}$$

It is a 'density of presence' in spacetime.

First and second moments

The **first moment** of f is a vector field on V defined by

$$P^\alpha(x) := \int_{\mathcal{P}_x} p^\alpha f(x,p)\theta_p. \tag{X.2.5}$$

If the spacetime is time-oriented and the particles have non-negative rest mass and travel towards the future, i.e. if p^α is future causal (timelike or null), then the fibre $\mathcal{P}_x$ is included in the subset $p^0 > 0$. The vector fp^α is then also causal and future-directed, since $f \geq 0$, and the same is true of the first moment P^α.

Out of the first moment, one extracts a scalar $r \geq 0$ interpreted as the square of a specific rest mass density given by

$$r^2 := -P^\alpha P_\alpha. \tag{X.2.6}$$

If P is timelike, then $r > 0$ and one deduces from the first moment a unit vector U interpreted as the macroscopic flow velocity and given by

$$U^\alpha := r^{-1}P^\alpha. \tag{X.2.7}$$

A sufficient condition for P to be timelike is that the particles have positive mass, since then all p, and hence also P, are timelike.

The **second moment** of the distribution function f is the symmetric 2-tensor on spacetime given by

$$T^{\alpha\beta}(x) := \int_{\mathcal{P}_x} f(x,p)p^\alpha p^\beta \theta_p. \tag{X.2.8}$$

It is interpreted as the **stress–energy tensor of the distribution** f.

[5] Recall that the integral on an oriented $(n+1)$-manifold $\mathcal{P}_x$, with orientation defined by the order of coordinates $x^0, \ldots, x^n$, of the exterior differential form $f(p)\theta_p$ is equal to the calculus integral

$$\int_{\mathcal{P}_x} f(p)\theta_p = \int_{\mathcal{P}_x} f(p)\, dp^0 \ldots dp^n.$$

Theorem X.2.1   *If a distribution function f depends on p only through a scalar product $U_\alpha p^\alpha$, with U^α a given timelike vector on spacetime, then the first and second moments of f read respectively as the momentum vector and the stress–energy tensor of a perfect fluid. The unit flow velocity is collinear with U.*

Proof.   Assume that

$$f(x,p) \equiv F(x, U_\alpha(x)p^\alpha). \tag{X.2.9}$$

The first moment of f is then

$$P^\alpha(x) \equiv \int_{\mathcal{P}_x} p^\alpha F(x, U_\alpha(x)p^\alpha)\theta_p. \tag{X.2.10}$$

Take at x an orthonormal Lorentzian frame with time axis collinear with U. In this frame, $U_\alpha(x) = -\lambda(x)\delta_\alpha^0$, with $\lambda := (-U_\alpha U^\alpha)^{\frac{1}{2}}$, and hence we have the following:

1. The first moment of f has components at x such that

$$P^i(x) = \int_{\mathcal{P}_x} p^i F(x, -\lambda(x)p^0)\theta_p = 0, \tag{X.2.11}$$

   because the integrand is antisymmetric in the p^i while F is positive. The vector P is therefore collinear with the time axis. The component P^0 in the orthonormal frame,

$$P^0(x) = \int_{\mathcal{P}_x} p^0 F(x, -\lambda(x)p^0)\theta_p, \tag{X.2.12}$$

   is the macroscopic rest mass density (the presence number density if all particles have the same mass normalized to 1).

2. In the same frame, the same antisymmetry considerations imply that $T^{0i} = 0$ and $T^{ij} = 0$ for $i \neq j$. The equality of the components $T^{ii}, i = 1, \ldots, n$,

$$T^{ii} = \int_{\mathcal{P}_x} (p^i)^2 F(x, -\lambda(x)p^0)\theta_p, \tag{X.2.13}$$

   results from the invariance under rotations in momentum space of the function F and the volume element θ_p. The component T^{00} is the positive function

$$T^{00} = \int_{\mathcal{P}_x} (p^0)^2 F(x, -\lambda(x)p^0)\theta_p. \tag{X.2.14}$$

   The tensor T is therefore the stress–energy tensor of a perfect fluid with flow vector $u = \lambda^{-1}U$, specific energy given by (2.17), and pressure computed from 2.13.

∎

Remark X.2.2   *The physical interpretation of the first moment that we have given coincides with the interpretation chosen by Marle[6] and with Eckart's original definition.[7] For a discussion and interpretation*

[6]Marle (1969).
[7]Eckart (1940).

of the second moment in the general case (dissipative fluids), see the discussion in Marle's paper on pp. 137–143.

Higher moments

Higher moments are defined as totally symmetric tensors on V given by

$$M^{\alpha_1 \dots \alpha_p} := \int_{\mathcal{P}_x} f(x,p) p^{\alpha_1} \dots p^{\alpha_p} \theta_p. \qquad (\text{X.2.15})$$

They play an important role in the Müller–Ruggeri extended thermodynamics.

X.2.4 Particles of a given rest mass

Stars or galaxies do not have the same rest mass, but the original kinetic theory was constructed for the case of gases in laboratories, composed of molecules with the same rest mass m. In that case, the phase space over (V, g), denoted by $\mathcal{P}_{m,V}$, has for fibre $\mathcal{P}_{m,x}$ the **mass hyperboloid** (also called the **mass shell**)

$$\mathcal{P}_{\mathbf{m,x}} \equiv \mathcal{P}_x \cap \{g(p,p) = -m^2\}. \qquad (\text{X.2.16})$$

In the case of particles of a given mass m, the volume form $\theta_{m,p}$ on the mass shell $\mathcal{P}_{m,x}$ is, taking the p^i as local coordinates on $\mathcal{P}_{m,x}$ (then p^0 is a function of x and p^i),

$$\theta_{m,p} = \frac{|\mathrm{Det}\, g|^{\frac{1}{2}}}{p_0} dp^1 \wedge \dots \wedge dp^n. \qquad (\text{X.2.17})$$

Exercise X.2.1 *Prove this formula using the Leray definition*

$$d\left[\frac{1}{2}(g(p,p) - m^2)\right] \wedge \theta_{m,p} = \theta_p. \qquad (\text{X.2.18})$$

The moments can be defined by the same formula as in Section X.2.3, by integration on the mass shell and replacing θ_p by $\theta_{m,p}$.

One can also consider the case of N *different particles with different masses* m_I, $I = 1, 2, \dots, N$; there are then N different phase spaces $\mathcal{P}_{m_I,V}$, and different distribution functions f_I. The moments to consider are the sums of the corresponding moments; for example, the second moment is

$$T^{\alpha\beta}(x) := \sum_I \int_{\mathcal{P}_{m_a,x}} f_I(x,p) p^\alpha p^\beta \theta_{m_I,p}. \qquad (\text{X.2.19})$$

For details see the article by Marle.[8]

---

[8] Marle (1969).

X.3   Vlasov equations

In the **Vlasov models**, one assumes that the gas is so rarefied that the particle trajectories do not scatter (i.e. one neglects collisions between particles). Their motion is determined by the average fields they generate on spacetime.

X.3.1   General relativistic (GR)–Vlasov equation

In a curved spacetime of General Relativity, in the absence of non-gravitational forces, each particle follows a geodesic of the spacetime metric g. We have seen (Chapter I, Section I.8) that the differential system satisfied by a geodesic in the tangent space TV of a pseudo-Riemannian manifold (V, g) reads in local coordinates, with λ called a canonical affine parameter,

$$ p^\alpha := \frac{dx^\alpha}{d\lambda}, \quad \frac{dp^\alpha}{d\lambda} = G^\alpha, \quad \text{with } G^\alpha := -\Gamma^\alpha_{\lambda\mu} p^\lambda p^\mu, \tag{X.3.1}$$

where $\Gamma^\alpha_{\lambda\mu}$ are the Christoffel symbols of the metric g. In other words, the trajectory of a particle in TV is an element of the **geodesic flow** generated by the vector field $X = (p, G)$ whose components p^α, G^α in a local trivialization of TV over the domain of a chart of V are given by (3.1).

Exercise X.3.1   *Show by direct calculation that $X = (p, G)$ is indeed a vector field on the tangent bundle TV.*

In a collisionless model, the physical law of conservation of particles, together with the invariance of the volume form in TV under the geodesic flow (Liouville's theorem; see Problem X.14.1) imposes that the distribution function f be constant under this flow, that is, that it satisfy the following first-order linear differential equation, which we call the **GR–Vlasov** equation:

$$ \mathcal{L}_X f \equiv p^\alpha \frac{\partial f}{\partial x^\alpha} - \Gamma^\alpha_{\lambda\mu} p^\lambda p^\mu \frac{\partial f}{\partial p^\alpha} = 0. \tag{X.3.2}$$

We have interpreted the *scalar* $-g(p, p)$ as the square of the rest mass of the particle. The following lemma, the formulation of a property already seen in other contexts, makes this interpretation physically consistent:

Lemma X.3.1   *In a GR–Vlasov plasma, the scalar $g(p, p)$ is constant under the geodesic flow; that is,*

$$ \mathcal{L}_X \{g(p, p)\} := p^\alpha \frac{\partial g(p, p)}{\partial x^\alpha} + G^\alpha \frac{\partial g(p, p)}{\partial p^\alpha} \equiv 0, \quad G^\alpha := -\Gamma^\alpha_{\lambda\mu} p^\lambda p^\mu.$$

Exercise X.3.2   *Prove this identity.*

Hint: Use the values of the Christoffel symbols.

For particles with a given rest mass, it is convenient to take as coordinates on the phase space $\mathcal{P}_m$ the $2n+1$ numbers x^α, p^i. The distribution function f_m in such coordinates is

$$f_m(x^\alpha, p^i) = f(x^\alpha, p^i, p^0(p^i)). \qquad \text{(X.3.3)}$$

Exercise X.3.3   *Prove that $g_{\alpha\beta}p^\alpha p^\beta = -m^2$ implies*

$$\frac{\partial p^0}{\partial p^i} = -\frac{p_i}{p_0}, \quad \frac{\partial p^0}{\partial x^\alpha} = -\frac{p^\lambda p^\mu}{2p_0}\frac{\partial g_{\lambda\mu}}{\partial x^\alpha}. \qquad \text{(X.3.4)}$$

Deduce from this definition and identities that the Vlasov equation takes the form, with bounded coefficients,[9]

$$\frac{\partial f_m}{\partial x^0} + \frac{p^i}{p^0}\frac{\partial f_m}{\partial x^i} + \frac{G^i}{p^0}\frac{\partial f_m}{\partial p^i} = 0.$$

[9] Choquet-Bruhat (1971b).

X.3.2   EM–GR–Vlasov equation

When, in addition to gravitation, the *particles are subjected to some given force*, represented by a vector Φ tangent to the spacetime V, the vector Y tangent to the particle trajectories in the phase space $\mathcal{P}_V$ over a spacetime (V, g) is

$$Y = (p, G + \Phi), \quad \text{i.e.} \quad (Y^A) = (p^\alpha, G^\beta + \Phi^\beta). \qquad \text{(X.3.5)}$$

If the volume form θ is invariant under the flow of Y, that is, if $\mathcal{L}_Y\theta = 0$, then particle number conservation gives for the distribution function f a generalized GR–Vlasov equation

$$\mathcal{L}_Y f \equiv p^\alpha \frac{\partial f}{\partial x^\alpha} + (-\Gamma^\alpha_{\lambda\mu}p^\lambda p^\mu + \Phi^\alpha)\frac{\partial f}{\partial p^\alpha} = 0. \qquad \text{(X.3.6)}$$

Lemma X.3.2 $\mathcal{L}_Y\theta = 0$ *if $\partial\Phi^\alpha/\partial p^\alpha \equiv 0$.*

Exercise X.3.4   *Prove this result.*

Hint: Use the identity $\mathcal{L}_Y\theta \equiv d(i_Y\theta) + i_Y(d\theta)$.

If the particles have *electric charge e and move in an electromagnetic field F*, then their trajectories have tangent vectors in phase space

$$Y = (p^\alpha, G^\alpha + \Phi^\alpha), \quad \text{with} \quad \Phi^\alpha \equiv e|g(p,p)|^{-\frac{1}{2}}F^{\alpha\beta}p_\beta. \qquad \text{(X.3.7)}$$

We deduce from the expression for Φ and the antisymmetry of F that

$$\frac{\partial\Phi^\alpha}{\partial p^\alpha} \equiv 0.$$

Therefore, the volume form θ is invariant under the flow of Y, and particle number conservation gives the **EM–GR–Vlasov equation**

$$\mathcal{L}_Y f \equiv p^\alpha \frac{\partial f}{\partial x^\alpha} + (-\Gamma^\alpha_{\lambda\mu}p^\lambda p^\mu + e|g(p,p)|^{-\frac{1}{2}}F^{\alpha\beta}p_\beta)\frac{\partial f}{\partial p^\alpha} = 0. \quad \text{(X.3.8)}$$

The following lemma provides a coherent consideration of an EM–GR plasma where particles have both the same mass and the same charge.

Lemma X.3.3 *In an EM–GR–Vlasov plasma, the scalar $g(p,p)$ is constant along an orbit of Y in TV; that is,*

$$p^\alpha \frac{\partial g(p,p)}{\partial x^\alpha} + (G^\alpha + em^{-1}F^{\alpha\beta}p_\beta)\frac{\partial g(p,p)}{\partial p^\alpha} \equiv 0.$$

The proof is an easy consequence of the lemma proved in the purely gravitational case, because

$$\frac{\partial g(p,p)}{\partial p^\alpha} \equiv \frac{\partial(g_{\lambda\mu}p^\lambda p^\mu)}{\partial p^\alpha} = 2p_\alpha \quad \text{and} \quad F^{\alpha\beta}p_\beta p_\alpha = 0,$$

by the antisymmetry of F.

Exercise X.3.5 *Prove that if the particles all have the same mass m, then the distribution function f_m satisfies on P_m the following* **reduced EM–GR–Vlasov equation***:*

$$\frac{\partial f_m}{\partial x^0} + \frac{p^i}{p^0}\frac{\partial f_m}{\partial x^i} + \frac{G^i + \Phi^i}{p^0}\frac{\partial f_m}{\partial p^i} = 0.$$

The EM–GR–Vlasov equation can be generalized to particles with random charges (see Problem X.14.2).

X.3.3   Yang–Mills plasmas

Yang–Mills plasmas were not observed in laboratories until high-energy particle colliders were able to decompose baryons into their constitutive quarks and gluons. The Yang–Mills–Vlasov[10] equation is analogous to the EM–Vlasov equation, but the electromagnetic field is replaced by a Yang–Mills field taking its values in a Lie algebra $\mathcal{G}$ and the electric charge e, a constant, is replaced by a function q on the spacetime V with values in the Lie algebra $\mathcal{G}$. The phase space for the kinetic theory is now the product $\mathcal{P}_V \times \mathcal{G}$. Although the gravitational field does not seem to play a role at the scale of Yang–Mills plasmas, we write the general relativistic equations here. The trajectory of a particle in this phase space is a solution of the differential system

$$\frac{dx^\alpha}{ds} = p^\alpha, \quad \frac{dp^\alpha}{ds} = G^\alpha + q\cdot F^{\alpha\beta}p_\beta, \quad \frac{dq}{ds} = Q := -p^\alpha[A_\alpha, q], \quad \text{(X.3.9)}$$

where $[.,.]$ is the bracket in the Lie algebra $\mathcal{G}$, i.e. $[A_\alpha, q]^a := c^a_{bc}A^b_\alpha q^c$. The distribution function f is now a function of x, p, q. In the absence of collisions and other forces, f satisfies the Einstein–Yang–Mills–Vlasov equation

$$p^\alpha \frac{\partial f}{\partial x^\alpha} + (G^\alpha + q\cdot F^{\alpha\beta}p_\beta)\frac{\partial f}{\partial p^\alpha} + Q^a \frac{\partial f}{\partial q^\alpha} = 0. \qquad \text{(X.3.10)}$$

A realistic plasma may contain N charged particles of different kinds,[11] with rest masses m_I and electric charges e_I; their momenta lie

[10]Choquet-Bruhat and Noutcheguéme (1986). See also YCB-OUP2009, Chapter X, Section 3.3.

[11]A case treated in Marle (1969).

in mass hyperboloids $g_{\alpha\beta}p^\alpha p^\beta = -m_I^2$. If we take the p^i as coordinates on these mass hyperboloids and denote by $f_{m_I}(x,p)$ their distribution functions, then each f_{m_I} satisfies a reduced EM–GR–Vlasov equation on a phase space $\mathcal{P}_{m_I}$.

X.4 Solution of a Vlasov equation

X.4.1 Construction

Let (V,g) be a Lorentzian manifold. Assume given on the tangent spaces to V, $T_{M_0}V$, at points of an initial hypersurface $M_0 \subset V$, a function $\bar{f}$ that will be the initial value of a distribution function f solution of the Einstein–Vlasov equation, at least in a neighbourhood of $T_{M_0}V$.

The GR–Vlasov equation is a linear first-order partial differential equation

$$\mathcal{L}_X f := p^\alpha \frac{\partial f(x,p)}{\partial x^\alpha} + G^\alpha(x,p)\frac{\partial f(x,p)}{\partial p^\alpha} = 0.$$

As is classical for solving such an equation, we transform it into an ordinary differential equation by solving the differential system called the **characteristic system**:

$$\frac{dx^\alpha}{d\lambda} = p^\alpha, \quad \frac{dp^\alpha}{d\lambda} = G^\alpha \equiv -\Gamma^\alpha_{\lambda\mu}p^\lambda p^\mu. \qquad \text{(X.4.1)}$$

Let $x^0 = 0$ be the equation of M_0. Denote by ξ^i and π^α coordinates of a point of $\xi \in M_0$ and a point $\pi \in T_\xi V$, the tangent space to V at ξ. Assume given a function $\bar{f}(\xi,\pi)$; then the quasilinear first-order differential system (4.1) has, for λ small enough, one and only one solution,[12] a **geodesic flow** that takes, for $\lambda = 0$, the given values (ξ,π):

$$x^\alpha = \phi^\alpha(\lambda,\xi,\pi), \ p^\alpha = \psi^\alpha(\lambda,\xi,\pi), \quad \xi := (\xi^i), \pi := (\pi^\alpha),$$

with

$$\phi^0(0,\xi,\pi) = 0, \ \phi^i(0,\xi,\pi) = \xi^i, \ \psi^\alpha(0,\xi,\pi) = \pi^\alpha. \qquad \text{(X.4.2)}$$

Inserting these functions ϕ^α and ψ^α into the Vlasov equation gives the following ordinary differential equation along the trajectory issuing from (ξ,π):

$$\frac{df\{(\phi,\psi)(\lambda,\xi,\pi)\}}{d\lambda} = 0;$$

that is, f is constant along the trajectory, i.e.

$$f\{(\phi,\psi)(\lambda,\xi,\pi)\} = \bar{f}\{(\phi,\psi)(0,\xi,\pi)\}.$$

The mapping $\Lambda : (\lambda,\xi^i,\pi^\alpha) \mapsto (x^\alpha, p^\alpha)$ reduces to the identity for $\lambda = 0$; it is therefore invertible[13] (if C^1) for small enough λ, i.e. in

[12] A local C^1 solution exists, i.e. a mapping $(\xi,\pi) \mapsto (x(\lambda),p(\lambda))$ for λ small enough, if the coefficients of the differential system are C^1, i.e. if the spacetime metric is C^2. The solution is unique if the coefficients are Lipschitzian in x and p, and hence if the spacetime metric is $C^{1,1}$ (i.e. C^1 with Lipschitzian derivatives).

[13] In other words, the geodesic flow has no conjugate point.

a neighbourhood of M_0. The inverse mapping Λ^{-1} gives the following solution of the GR–Vlasov equation:

$$f(x, p) \equiv \bar{f}(\xi(x, p), \pi(x, p)). \qquad (X.4.3)$$

The same procedure gives the general solution of an EM–GR–Vlasov equation.

Exercise X.4.1   *Write the characteristic system of an EM–GR–Vlasov equation.*

X.4.2   Global existence theorem

[14]This global existence holds, in appropriate functional spaces, for linear hyperbolic equations, in particular wave equations in a Lorentzian metric whose null rays issuing from a point do not necessarily constitute (out of this point) a smooth hypersurface.

[15]Choquet-Bruhat (1971b).

The construction given in Section X.4.1 breaks down when the geodesic flow ceases to be regular. However, this breakdown does not imply that the Cauchy problem for the GR–Vlasov equation does not have a global solution.[14] This solution is not given by the formula (4.3), but its global existence can be proven by using energy-type estimates for first-order linear equations. We state a theorem,[15] leaving to the interested reader the formulation of relevant functional spaces.

Theorem X.4.1   *The Cauchy problem for the GR–Vlasov equation on the phase space $\mathcal{P}_V$ of a regularly sliced $(n + 1)$-dimensional spacetime (V, g), with data $\bar{f}$ on $\mathcal{P}_{\bar{M}}$, $\bar{M} := M \times \{0\}$, admits one and only one solution f on $\mathcal{P}_V$.*

It is probably possible to prove global existence on the tangent bundle of a globally hyperbolic manifold with appropriate hypotheses on the initial manifold and initial data.

X.4.3   Stress–energy tensor

The stress–energy tensor of a distribution f is its second moment:

$$T^{\alpha\beta}(x) := \int_{\mathcal{P}_x} f(x, p) p^\alpha p^\beta \theta_p. \qquad (X.4.4)$$

It is defined at $x \in V$ if $f p^\alpha p^\beta$ is integrable on the fibre $\mathcal{P}_x$ of the phase space, which requires a proper fall off of f in p at infinity in $\mathcal{P}_x$, satisfied in particular if f has compact support in p; this compactness results in a neighbourhood of $\bar{M}$ from the compactness of the support of the initial $\bar{f}$.

More generally, consider the case of interest for astrophysics:

$$0 < m^2 \leq -g(p, p) \leq M^2.$$

It holds that $p^\alpha p^\beta / (p^0)^2$ is uniformly bounded on $\mathcal{P}_V$, and hence the integrals (4.4) are defined if

$$\int_{\mathcal{P}_x} (p^0)^2 f(x, p) \theta_p$$

is defined.

X.5   The Einstein–Vlasov system

X.5.1   Equations

The Einstein–Vlasov system comprises the Einstein equations for a Lorentzian metric g with source the second moment of a distribution function f, coupled with the GR–Vlasov equation for f:

$$S_{\alpha\beta} = T_{\alpha\beta}, \quad \text{with} \quad S_{\alpha\beta} := R_{\alpha\beta} - \frac{1}{2}g_{\alpha\beta}R, \qquad \text{(X.5.1)}$$

$$T_{\alpha\beta}(x) := \int_{\mathcal{P}_x} f(x,p)p_\alpha p_\beta \theta_p, \quad \theta_p := |\text{Det } g|^{\frac{1}{2}} dp^0 dp^1 \ldots dp^n, \qquad \text{(X.5.2)}$$

coupled with

$$\mathcal{L}_X f \equiv p^\alpha \frac{\partial f}{\partial x^\alpha} + G^\alpha \frac{\partial f}{\partial p^\alpha} = 0, \quad G^\alpha := -\Gamma^\alpha_{\lambda\mu} p^\lambda p^\mu. \qquad \text{(X.5.3)}$$

X.5.2   Conservation law

The Einstein–Vlasov system is coherent if the stress–energy tensor (5.2) satisfies the conservation law

$$\nabla_\alpha T^{\alpha\beta} = 0.$$

We prove the following more general theorem:[16]

Theorem X.5.1   *If the distribution f satisfies the GR–Vlasov equation $\mathcal{L}_X f = 0$, then its moments satisfy the conservation laws*

$$\nabla_{\alpha_1} M^{\alpha_1 \alpha_2 \ldots \alpha_p} = 0, \qquad \text{(X.5.4)}$$

where ∇ is the covariant derivative in the spacetime metric g.

Proof. We choose at the point x coordinates such that the first derivatives $\partial_\alpha g_{\lambda\mu} := (\partial/\partial x^\alpha)g_{\lambda\mu}$, and hence the Christoffel symbols, vanish at x. At that point, in these coordinates, it holds that

$$\nabla_{\alpha_1} M^{\alpha_1 \alpha_2 \ldots \alpha_p} = \frac{\partial}{\partial x^{\alpha_1}} M^{\alpha_1 \alpha_2 \ldots \alpha_p}.$$

It also holds that $(\partial/\partial x^\alpha)\theta_p = \theta_p$, since $(\partial/\partial x^\alpha)(\text{Det } g)^{\frac{1}{2}} = 0$ when $\partial_\alpha g_{\lambda\mu} = 0$. Therefore,

$$\frac{\partial}{\partial x^{\alpha_1}} M^{\alpha_1 \alpha_2 \ldots \alpha_p} = \frac{\partial}{\partial x^{\alpha_1}} \int_{\mathcal{P}_x} f(x,p)p^{\alpha_1} \ldots p^{\alpha_p} \theta_p$$

$$= \int_{\mathcal{P}_x} \frac{\partial}{\partial x^{\alpha_1}} f(x,p)p^{\alpha_1} \ldots p^{\alpha_p} \theta_p,$$

but when the Christoffel symbols are zero,

$$\int_{\mathcal{P}_x} \frac{\partial}{\partial x^{\alpha_1}} f(x,p)p^{\alpha_1} \ldots p^{\alpha_p} \theta_p = \int_{\mathcal{P}_x} \mathcal{L}_X f(x,p)p^{\alpha_2} \ldots p^{\alpha_p} \theta_p. \qquad \text{(X.5.5)}$$

[16] For the first moment, this expresses the conservation law of the macroscopic momentum.

We have obtained a relation between tensors on V at an arbitrary point in particular coordinates; in arbitrary coordinates, it reads

$$\nabla_{\alpha_1} M^{\alpha_1 \alpha_2 \dots \alpha_p} \equiv \int_{\mathcal{P}_x} \mathcal{L}_X f(x,p) p^{\alpha_2} \dots p^{\alpha_p} \theta_p. \tag{X.5.6}$$

■

X.6   The Cauchy problem

In a Cauchy problem, one looks for solutions of a system of differential or integrodifferential equations taking given data on an initial hypersurface. A local-in-time existence theorem for a solution of the Cauchy problem for the coupled Einstein–Vlasov system was proved long ago;[17] here we will only give the idea of the proof and state the main results.

As in the general case of Einstein equations with source a stress–energy tensor satisfying the conservation laws, the solution of the Cauchy problem for the Einstein equations with a kinetic source splits into an elliptic system called the constraints for Cauchy data on the initial hypersurface and a hyperbolic evolution system for the metric in a suitably chosen gauge, for example a wave gauge. This evolution system is coupled with an evolution system for the source, here a Vlasov equation. The solution in wave gauge is proved, as always, to be a solution of the full Einstein equations by using the conservation law satisfied by the stress–energy tensor—here it is a distribution function solution of the GR–Vlasov equation.

X.6.1   Cauchy data and constraints

An *initial data set for the Einstein–Vlasov system* (see Chapter VIII) is a quadruplet $(M, \bar{g}, K, \bar{f})$, where M is a hypersurface of a differentiable manifold V, $\bar{g}$ and K are respectively a properly Riemannian and a symmetric 2-tensor on M, while $\bar{f}$ is a function on the bundle $\mathcal{P}_M := \cup_{x \in M} \mathcal{P}_x(V)$ corresponding to the kind of particles we consider.

As always, the data must satisfy on M the Hamiltonian constraint

$$R(\bar{g}) - |K|_{\bar{g}}^2 + (\mathrm{tr}_{\bar{g}} K)^2 = 2N^2 \bar{T}^{00} \tag{X.6.1}$$

and the momentum constraint

$$\bar{\nabla}.K - \bar{\nabla} \mathrm{tr}\, K = \bar{N} \bar{T}^{0i}, \tag{X.6.2}$$

$$\bar{T}^{\alpha\beta}(x) = \int_{\mathcal{P}_x} \bar{f}(x,p) p^\alpha p^\beta \theta_p, \quad x \in M, \quad p \in \mathcal{P}_x(V).$$

A general property of the Einstein equations is that the conformally formulated constraints decouple if the initial manifold has constant mean extrinsic curvature and the source has zero momentum, i.e. $\bar{T}^{0i} = 0$.

[17] Choquet-Bruhat (1971b), for particles with a given rest mass. For particles with rest mass in a range of positive numbers, see YCB-OUP2009, Chapter X.

Exercise X.6.1   *Show that $\bar{T}^{0i}(x) = 0$ if $\bar{f}(x, p^0, p^i) = \bar{f}(x, p^0, -p^i)$.*

In the **Maxwell–Einstein–Vlasov** case, the Cauchy data on M_0 are, in addition to $\bar{g}$, K, and $\bar{f}$, the data $\bar{F}_{\alpha\beta}$ of $F_{\alpha\beta}$ on M_0.

Exercise X.6.2   *Write the system of constraints satisfied by $\bar{g}, K, \bar{F}$.*

X.6.2   Evolution

As is often the case for a coupled system, a general method to solve the Einstein equations in wave gauge coupled with a GR–Vlasov equation is by iteration: find a solution f_1 of the Vlasov equation for a given Lorentzian metric g_1; then look for a solution g_2 of the reduced Einstein equations with source f_1; then iterate and study the convergence of the iteration, either numerically or mathematically. The mathematical method used by Choquet-Bruhat[18] to prove the existence of these iterates and their convergence is through classical energy estimates for wave equations for the metric and weighted-in-p^0 energy estimates for the Vlasov equation to be satisfied by the function f. It is probably possible to use instead constructive methods, relying on the parametrix in the Einstein case and the formula (4.3) for the Vlasov part.

[18]Choquet-Bruhat (1971b).

X.6.3   Local existence and uniqueness theorem

We state the theorem, leaving the formulation of relevant functional spaces to the interested reader:

Theorem X.6.1   *The Cauchy problem for the Einstein–Vlasov system with initial data $(\bar{g}, K)$ on M, $\bar{f}$, on $\mathcal{P}_M$, satisfying the constraints, admits a solution (g, f) on $V_\ell \times \mathcal{P}_V$, with $V_\ell := M \times [0, \ell])$, and ℓ small enough.*

X.6.4   Global theorems

The collisionless kinetic theory having no global problems of its own, one may hope to extend to the Einstein–Vlasov system global results obtained for the vacuum Einstein equations. Such results are already hard to prove. Global existence theorems, or proofs of the cosmic censorship conjectures, for the Einstein–Vlasov system have been obtained only in the presence of an isometry group. The first global existence theorem, for small initial data, was proved by Rein and Rendall[19] in the case of spherical symmetry. A paper by Dafermos and Rendall[20] proves cosmic censorship in the case of surface-symmetric compactly supported initial data. References to other global results concerning the Einstein–Vlasov system can be found in that paper and in a review by Rendall.[21]

[19]Rein and Rendall (1992).
[20]Dafermos and Rendall (2006).

[21]Rendall (2005).

X.7   The Maxwell–Einstein–Vlasov system

Charged particles in a kinetic model generate by their motion an average electric current J, and hence an electromagnetic field, a 2-form F.

X.7.1   Particles with given rest mass and charge

If all particles have the same charge e and rest mass m, the current is

$$J^\alpha(x) = m^{-1}e \int_{\mathcal{P}_{m,x}} f(x,p)p^\alpha \theta_{m,p}. \tag{X.7.1}$$

The electromagnetic field satisfies on V the Maxwell equations

$$dF = 0, \quad \delta F = J, \quad \text{i.e.} \quad \nabla_\alpha F^{\alpha\beta} = J^\beta. \tag{X.7.2}$$

[22] The Maxwell–Vlasov system in Special Relativity has been studied extensively.

The Maxwell–Einstein–Vlasov system[22] is the coupled system of these Maxwell equations with the EM–GR–Vlasov equation together with the Einstein equations with source a stress–energy tensor that is the sum of the second moment of the distribution function f and the stress–energy tensor τ of the electromagnetic field:

$$S^{\alpha\beta} = T^{\alpha\beta} \equiv \int_{\mathcal{P}_x \times R} f(x,p,e)p^\alpha p^\beta \theta_p \, de + \tau^{\alpha\beta},$$

$$\tau^{\alpha\beta} \equiv F^\alpha{}_\lambda F^{\beta\lambda} - \frac{1}{4}g^{\alpha\beta}F_{\lambda\mu}F^{\lambda\mu}.$$

[23] Recall that $\nabla_\alpha S^{\alpha\beta} \equiv 0$ and $\nabla_\alpha \nabla_\beta F^{\alpha\beta} \equiv 0$ for any metric g and 2-form F.

The following theorem makes the system coherent:[23]

Theorem X.7.1

1. *If the distribution function f satisfies the EM–GR–Vlasov equation, then the current J has zero divergence.*

2. *If, in addition, F satisfies the Maxwell equations, then the stress–energy tensor T is divergence-free.*

Proof. 1. The electric current is given by the product with $m^{-1}e$ of the first moment of f:

$$J^\alpha \equiv m^{-1}e \int_{\mathcal{P}_x} p^\alpha f(x,p)\theta_p. \tag{X.7.3}$$

Hence, by (5.6),

$$\nabla_\alpha J^\alpha = m^{-1}e \int_{\mathcal{P}_x} \mathcal{L}_X f(x,p)\theta_p, \tag{X.7.4}$$

which implies, if $\mathcal{L}_Y f(x,p) \equiv \mathcal{L}_X f(x,p) + \Phi^\alpha \partial f(x,p)/\partial p^\alpha = 0$,

$$\nabla_\alpha J^\alpha = -m^{-1}e \int_{\mathcal{P}_x} \int_{\mathcal{P}_x} \Phi^\alpha \frac{\partial f(x,p)}{\partial p^\alpha} \theta_p. \tag{X.7.5}$$

Using integration by parts and the equality $\partial \Phi^\alpha / \partial p^\alpha = 0$ gives

$$\int_{\mathcal{P}_x} \Phi^\alpha \frac{\partial f(x,p)}{\partial p^\alpha} \theta_p = -\int_{\mathcal{P}_x} f(x,p) \frac{\partial \Phi^\alpha}{\partial p^\alpha} \theta_p = 0, \qquad \text{(X.7.6)}$$

and hence

$$\nabla_\alpha J^\alpha = 0.$$

2. Analogous reasoning gives for the second moment when $\mathcal{L}_Y f = 0$

$$\nabla_\alpha M^{\alpha\beta} = \int_{\mathcal{P}_x} -\Phi^\alpha \frac{\partial f}{\partial p^\alpha} p^\beta \theta_p. \qquad \text{(X.7.7)}$$

The calculus identity

$$\Phi^\alpha \frac{\partial f}{\partial p^\alpha} p^\beta \equiv \frac{\partial}{\partial p^\alpha}(\Phi^\alpha f p^\beta) - \frac{\partial \Phi^\alpha}{\partial p^\alpha} f p^\beta - \Phi^\beta f, \qquad \text{(X.7.8)}$$

integration by parts, the property $\partial \Phi^\alpha / \partial p^\alpha = 0$, and antisymmetry of F show that

$$\nabla_\alpha M^{\alpha\beta} = -\int_{\mathcal{P}_x} \Phi^\beta f \theta_p \equiv -e \int_{\mathcal{P}_x} F_\lambda{}^\beta p^\lambda \theta_p. \qquad \text{(X.7.9)}$$

We have seen in Chapter III that the divergence of the Maxwell tensor is, if F satisfies the Maxwell equations,

$$\nabla_\alpha \tau^{\alpha\beta} = J^\lambda F_\lambda{}^\beta \equiv e \int_{\mathcal{P}_x} F_\lambda{}^\beta p^\lambda f \theta_p. \qquad \text{(X.7.10)}$$

We have proved that

$$\nabla_\alpha (M^{\alpha\beta} + \tau^{\alpha\beta}) = 0 \quad \text{if} \quad \mathcal{L}_Y f = 0. \qquad \blacksquare$$

The local existence theorem (Theorem X.6.1) extends to the Maxwell–Einstein–Vlasov system modulo hypotheses on the electromagnetic data (see the fluid case).

X.7.2   Particles with random masses and charges

If the particles have random masses and charges, then the electric current is given, at each point $x \in V$, by

$$J^\alpha(x) = \int_{\mathcal{P}_x \times R} f(x,p,e) p^\alpha \theta_p \, de. \qquad \text{(X.7.11)}$$

See Problem X.14.2 for the local existence theorem in this case.

X.8   Boltzmann equation. Definitions

When the particles undergo collisions, their trajectories in phase space are no longer connected integral curves of the vector field X: their momenta undergo jumps on crossing other trajectories. Consequently, the derivative of the distribution function f along X is no longer zero. In the Boltzmann model, this derivative is equal to the so-called **collision operator** $\mathcal{I}f$:

$$\mathcal{L}_X f = \mathcal{I}f. \tag{X.8.1}$$

$\mathcal{I}f$ is a quadratically nonlinear integral operator that is being interpreted as being linked with the probability that two particles of momenta respectively p' and q' collide at x and give, after the shock, two particles, one with momentum p and the other with momentum q. One says that the shock is elastic when the following law of conservation of momentum holds:[24]

$$p' + q' = p + q. \tag{X.8.2}$$

This equation defines a submanifold Σ in the fibre $(\times \mathcal{P}_x)^4$. For fixed p, q, one denotes by Σ_{pq} the submanifold of $(\times \mathcal{P}_x)^2$ defined by (8.2). The volume element ξ' (Leray form) in Σ_{pq} is such that

$$\xi' \wedge \left(\underset{\alpha}{\wedge}(d(p'^{\alpha} + q'^{\alpha})) \right) = \theta_{p'} \wedge \theta_{q'}. \tag{X.8.3}$$

The collision operator is

$$(\mathcal{I}f)(x,p) \equiv \int_{\mathcal{P}_x(q)} \int_{\Sigma_{pq}} [f(x,p')f(x,q') - f(x,p)f(x,q)]$$
$$\times A(x,p,q,p',q')\xi' \wedge \theta_q. \tag{X.8.4}$$

The function $A(x,p,q,p',q')$ is called the shock cross-section. It is a phenomenological quantity that depends on the physics of the shocks. No explicit expression is known for it in Relativity.[25] A generally admitted property is the reversibility of elastic shocks, namely

$$A(x,p,q,p',q') = A(x,p',q',p,q). \tag{X.8.5}$$

Lemma X.8.1   *When the particles have all the same, non-zero, proper mass m, the integral on Σ_{pq} can be written by using a formula of the type, with θ, φ canonical angular parameters on the sphere S^2,*

$$A(x,p,q,p',q')\xi' = S(x,p,q,\theta,\varphi)\sin\theta\, d\theta \wedge d\varphi.$$

Proof. Introduce at the point x, for a given pair p, q of timelike vectors, an orthonormal Lorentz frame with time axis e_0 in the direction of $p+q$. Set $p + q = 2\lambda e_0$, $2\lambda = [-g(p+q)]^{1/2}$. Then also $p' + q' = 2\lambda e_0$, and since p, q, p', q' are timelike vectors with length m,

$$p^{\alpha}q_{\alpha} = p'^{\alpha}q'_{\alpha} = m^2 - 2\lambda^2 \leq -m^2.$$

In such a frame, $p^0 + q^0 = p'^0 + q'^0 = 2\lambda$, while $p^i + q^i = p'^i + q'^i = 0$; hence $\Sigma(p^i)^2 = \Sigma(q^i)^2$, and therefore $p^0 = q^0 = \lambda$, while

$\Sigma(p^i)^2 = \lambda^2 - m^2 =: \alpha^2$. The same properties hold for the primed variables; in this frame, Σ_{pq} is represented by $\alpha 2$-sphere, $\Sigma(p'^i)^2 = \alpha^2$ of radius α in the plane $p'^0 = \lambda$. It holds that

$$\alpha = \frac{1}{2}(-p^\alpha q_\alpha - m^2)^{1/2} = \frac{1}{2}g(p-q, p-q)^{1/2}.$$

Take a vector parallel to $p - q$ (which is orthogonal to $p + q = 2\lambda e_0$) as axis for polar coordinates θ, φ on Σ_{pq}, denoting by θ the angle between the space vectors $p - q$ and $p' - q'$; the definition of ξ' then gives

$$\xi' = (2\lambda)^{-1}\alpha \sin\theta \, d\theta \wedge d\varphi.$$

The given relation holds with

$$S(p, q, \theta, \varphi) = A(x, p, q, p', q')(2\lambda)^{-1}\alpha, \qquad\qquad \text{(X.8.6)}$$

where on the sphere Σ_{pq} both p' and q' are given by the θ-, φ-dependent 4-vectors

$$p'_{\Sigma_{pq}} = q'_{\Sigma_{pq}}$$
$$= \left\{ -\frac{1}{2}g(p+q)^{1/2}, \frac{1}{2}g(p-q, p-q)^{1/2}(\cos\theta, \sin\theta\cos\varphi, \sin\theta\sin\varphi) \right\}.$$
$$\text{(X.8.7)}$$

∎

X.9   Moments and conservation laws

The moments of a distribution function f have been defined in Section X.2 by integrals over a fibre in phase space. The moment of order n is

$$T^{\alpha_1 \ldots \alpha_n}(x) =: \int_{\mathcal{P}_x} p^{\alpha_1} \ldots p^{\alpha_n} f(x, p)\theta_p. \qquad\qquad \text{(X.9.1)}$$

It satisfies the identity

$$\nabla_\alpha T^{\alpha\alpha_2 \ldots \alpha_n} \equiv \int_{\mathcal{P}_x} p^{\alpha_2} \ldots p^{\alpha_n} \mathcal{L}_X f(x, p)\theta_p. \qquad\qquad \text{(X.9.2)}$$

The right-hand side is zero if f satisfies the Vlasov equation.

We have interpreted the first and second moments as giving respectively the proper rest-mass–momentum vector and the stress–energy tensor of the macroscopic matter corresponding to the distribution function f. We prove a lemma, important for the coherence with the Einstein equations, when f satisfies a Boltzmann equation with reversible collisions:

Lemma X.9.1   *The first and second moments of a distribution f satisfying a Boltzmann equation on a spacetime (V, g) have zero divergence in the spacetime metric if the collisions are reversible.*

Proof. The first moment of f is the vector field on V, interpreted as proper rest-mass–energy–momentum density, given by

$$P^\alpha := \int_{\mathcal{P}_x} p^\alpha f(x,p)\theta_p. \tag{X.9.3}$$

The equation (9.2) reads in this case

$$\nabla_\alpha P^\alpha \equiv \int_{\mathcal{P}_x} \mathcal{L}_X f(x,p)\theta_p = \int_{\mathcal{P}_x} (\mathcal{I}f)(x,p)\theta_p. \tag{X.9.4}$$

The second moment of f, interpreted as the stress–energy tensor, is defined by

$$T^{\alpha\beta} := \int_{\mathcal{P}_x} p^\alpha p^\beta f(x,p)\theta_p. \tag{X.9.5}$$

For f satisfying a Boltzmann equation, we have

$$\nabla_\alpha T^{\alpha\beta} \equiv \int_{\mathcal{P}_x} p^\beta (\mathcal{I}f)(x,p)\omega_p. \tag{X.9.6}$$

Standard calculus, which we leave to the reader as an exercise, shows that, for reversible collisions, (9.4) and (9.6) have zero right-hand sides.

In the case of particles with non-zero rest mass, we have set $P^\alpha = ru^\alpha$, with u the unit flow vector of the macroscopic matter corresponding to the distribution function f. Equation (9.4) is identical to the matter conservation law found for fluids:

$$\nabla_\alpha(ru^\alpha) = 0. \tag{X.9.7}$$

Equation (9.6),

$$\nabla_\alpha T^{\alpha\beta} = 0, \tag{X.9.8}$$

is to be satisfied by all stress–energy tensors in General Relativity.   ∎

The higher moments do not satisfy conservation laws, but form an infinite hierarchy; see Section X.11.

X.10   Einstein–Boltzmann system

The Einstein equations with source the stress–energy tensor of a distribution function f satisfying a Boltzmann equation form a coherent system if the collision operator is such that the stress–energy tensor is conservative. We have said that this property holds if the collisions are reversible.

It has been proved[26] that the local Cauchy problem, with initial data set $(M, \bar{g}, K, \bar{f})$, is well posed for such a Einstein–Boltzmann system.

The same kinds of results hold for an Einstein–Maxwell–Boltzmann system obtained on replacing the GR–Vlasov operator by the EM–Vlasov operator.

[26]Bancel (1973) for the Boltzmann equation and Bancel and Choquet-Bruhat (1973) for the coupled system.

X.11   Thermodynamics

One of the main reasons for the interest in relativistic kinetic theory is the possibility of obtaining the laws of thermodynamics in a relativistic context. The problem is already difficult in Special Relativity, and new challenges arise in General Relativity because of the non-existence of an equilibrium distribution function in a non-stationary universe.

X.11.1   Entropy and the H theorem

The entropy density of a one-particle distribution f was computed by Boltzmann to be in phase space the function[27] $-k(f \log f)(x, p)$. The entropy flux in spacetime is the future-directed timelike vector[28]

$$H^\alpha(x) := -k \int_{\mathcal{P}_x} p^\alpha (f \, \log f)(x, p)\theta_p. \qquad (\text{X}.11.1)$$

The following theorem is the relativistic formulation of a theorem well known in non-relativistic thermodynamics:

Theorem X.11.1 *(H theorem) If collisions are reversible and satisfy the symmetry property (8.5), then the entropy flux H^α is such that*

$$\nabla_\alpha H^\alpha \geq 0. \qquad (\text{X}.11.2)$$

Proof. The divergence of H is found to be, by a computation similar to a previous one,

$$
\begin{aligned}
(\nabla_\alpha H^\alpha)(x) &\equiv -k \int_{\mathcal{P}_x} (\mathcal{L}_X[f \, \log f])(x, p)\theta_p \\
&\equiv -k \int_{\mathcal{P}_x} (\mathcal{L}_X f[\log f + 1])(x, p)\theta_p.
\end{aligned}
\qquad (\text{X}.11.3)
$$

Using the Boltzmann equation and the reversibility of collisions, which implies that

$$\int_{\mathcal{P}_x(q)} \int_{\Sigma_{pq}} [f(x, p')f(x, q') - f(x, p)f(x, q)]A(x, p, q, p', q')\xi' \wedge \theta_q \wedge \theta_p = 0, \qquad (\text{X}.11.4)$$

we find that

$$
\begin{aligned}
(\nabla_\alpha H^\alpha)(x) = -k \int_{\mathcal{P}_x(q)} \int_{\Sigma_{pq}} & [f(x, p')f(x, q') - f(x, p)f(x, q)] \\
& \times (\log f)(x, p)A(x, p, q, p', q')\xi' \wedge \theta_q \wedge \theta_p. \quad (\text{X}.11.5)
\end{aligned}
$$

Making, moreover, the natural assumption of symmetry,

$$A(x, p, q, p', q') = A(x, q, p, p', q'), \qquad (\text{X}.11.6)$$

---

[27] Boltzmann's constant k is the quotient of the perfect gas constant and Avogadro's number, with the value $1.3806568 \, \mathrm{J \, K^{-1}}$.

[28] From its physical definition, it holds that $0 \leq f \leq 1$; hence $-f \log f \geq 0$.

we rewrite the above divergence as follows (the dependence on x has been made implicit for brevity):

$$\nabla_\alpha H^\alpha = -\frac{k}{4} \int_{\mathcal{P}_x(q)} \int_{\Sigma_{pq}} [f(p')f(q') - f(p)f(q)]$$

$$\times \log \left[\frac{f(p)f(q)}{f(p')f(q')} \right] A(p,q,p',q') \xi' \wedge \omega_q \wedge \omega_p, \quad (X.11.7)$$

[29]Because $(a-b)(\log a - \log b) \geq 0$.

which is non-negative[29] when A is non-negative. ∎

The gas is said to be in thermal equilibrium if $\nabla_\alpha H^\alpha = 0$. A general relativistic gas undergoing collisions will not in general attain in a finite time thermal equilibrium (see Section X.11.2).

We can make a simple link between the microscopic and macroscopic properties of the gas when the distribution f is isotropic in phase space, namely when there exists a timelike vector v on spacetime such that f depends only on $v_\alpha p^\alpha$. We have seen (Section X.2) that the macroscopic gas is then a perfect fluid with momentum flow P collinear with V. The same type of proof gives the following lemma:

Lemma X.11.1 *Assume there exists on spacetime a timelike vector V such that the function f is symmetric in the spaces orthogonal to V, i.e. in each fibre,*

$$f(x,q) = f(x,p) \quad \text{if} \ \ V_\alpha p^\alpha = V_\alpha q^\alpha \ \ \text{and} \ \ \bar{p} = -\bar{q}, \quad (X.11.8)$$

where $\bar{p}$ and $\bar{q}$ denote respectively the projections of p and q on the subspace orthogonal to V. Then the first moment $P = rU$ as well as the entropy vector H are collinear with V.

Lemma X.11.2 *When the vectors H and P are collinear, one defines a positive specific scalar entropy S by setting*

$$H^\alpha = SP^\alpha. \quad (X.11.9)$$

This specific entropy S satisfies on spacetime the inequality

$$P^\alpha \partial_\alpha S \geq 0. \quad (X.11.10)$$

Proof. The lemma follows from inequality $\nabla_\alpha H^\alpha \geq 0$ and the conservation law $\nabla_\alpha P^\alpha = 0$. ∎

When f does not have the property (11.8), the *entropy and matter flux are not collinear.*

In all cases, integrating (11.2) on a spacelike slice V_T, with compact space or appropriate boundary conditions at spacelike infinity, we find that

$$\int_{M_T} H^0 N \mu_{\bar{g}} \geq \int_{M_0} H^0 N \mu_{\bar{g}}. \quad (X.11.11)$$

Exercise X.11.1 *Prove this inequality.*

The inequality (11.11) leads some cosmologists to think that it is the expansion of the universe that permits its ever-increasing organization from an initial anisotropy of f.

X.11.2   Maxwell–Jüttner equilibrium distribution

A gas is considered to be in thermal equilibrium if its entropy is conserved in the sense that

$$\nabla_\alpha H^\alpha = 0. \tag{X.11.12}$$

Equation (11.3) shows that a sufficient condition for this equality to hold is that the distribution function f be conserved along the trajectories of X in phase space, i.e. $\mathcal{L}_X f = 0$. This condition is, in a Boltzmann framework,

$$\mathcal{I}(f) = 0. \tag{X.11.13}$$

A sufficient condition for this equality to hold is that at each point x of spacetime, the function f is such that

$$f(p')f(q') - f(p)f(q) = 0, \quad \text{if} \quad p + q = p' + q'. \tag{X.11.14}$$

This condition is also necessary if $A(p, q, p', q')$ is strictly positive. It can also be proved to hold under weaker assumptions.[30] It can immediately be checked that a solution of the above functional equation is

$$f(x, p) = a(x) \exp[b_\alpha(x)p^\alpha], \tag{X.11.15}$$

with a a positive scalar function and b a covariant vector on spacetime. It can be proved that all continuous solutions of (11.14) are of this form. If a is non-negative, the same is true of f; if, moreover, b is future timelike, then $b_\alpha(x)p^\alpha < 0$ and $f(x, .)$ is integrable on $\mathcal{P}_x$. Such functions are called **Maxwell–Jüttner distributions**.

For such a Maxwell–Jüttner distribution f, we find by straightforward computation that

$$\mathcal{L}_X f \equiv e^{b_\lambda p^\lambda} \left[p^\alpha \frac{\partial a}{\partial x^\alpha} + \frac{1}{2} p^\alpha p^\beta (\nabla_\alpha b_\beta + \nabla_\beta b_\alpha) \right]. \tag{X.11.16}$$

A function f given by (11.15) will satisfy $\mathcal{L}_X f = 0$ on phase space if and only if on spacetime it holds that $\partial a / \partial x^\alpha = 0$ and $\nabla_\alpha b_\beta + \nabla_\beta b_\alpha = 0$, that is, iff a is a constant and b is a Killing vector field: the spacetime must therefore be stationary to admit a thermal equilibrium distribution function—an unsurprising result.

We deduce from Theorem X.2.1 that the rest-mass–momentum vector and the stress–energy tensor on spacetime associated with a Maxwell–Jüttner distribution with a timelike vector b are those of a perfect fluid with flow vector collinear to b. The same kind of proof shows that the entropy vector H is collinear to b, and hence to u. The integrals giving the moments can be expressed through modified Bessel functions of the second kind and index n.[31] The scalar $(-b^\alpha b_\alpha)^{\frac{1}{2}}$ is interpreted as the inverse of the product of the absolute temperature by Boltzmann's constant:

$$(-b^\alpha b_\alpha)^{\frac{1}{2}} = (kT)^{-1}. \tag{X.11.17}$$

[30] Marle (1969), p. 94.

[31] See Pichon (1967) and Marle (1969).

Remark X.11.1 *If the particles have zero mass (e.g. photons), one can satisfy the condition $\mathcal{L}_X f = 0$ only by imposing that b is a conformal Killing vector field, i.e. $\nabla_\alpha b_\beta + \nabla_\beta b_\alpha = \lambda g_{\alpha\beta}$. This is the reason why Robertson and Walker used the Planck distribution for photons in their expanding spacetimes.*

X.11.3   Dissipative fluids

In classical mechanics, the Navier–Stokes equations can be obtained from a distribution function that is a first-order perturbation of the Maxwell equilibrium distribution, either by the Chapman–Enskog method or by Grad's polynomial expansion method. Both methods have been extended by Marle, who found the stress–energy tensor corresponding to a first-order perturbation of the Maxwell–Jüttner distribution. The corresponding, very complex, system of equations have been shown to be of hyperbolic type only in some simplified cases. The fact that thermal equilibrium is not compatible with non-stationary spacetimes limits the validity of such equations in General Relativity.

X.12   Extended thermodynamics

[32]See, for a summary, YCB-OUP 2009, Chapter X, Section 11 (contributed by Tommaso Ruggeri) and, for a complete exposition, Müller and Ruggeri (1998).

The objective of extended thermodynamics[32] is to obtain equations for fluids in General Relativity that respect fundamental physical laws and also the relativistic causality principle. It uses for general relativistic fluids, in addition to the usual first moment $P^\alpha(x)$ and second moment $T^{\alpha\beta}(x)$, higher moments and relations between them that are physically and mathematically meaningful.

X.13   Solutions of selected exercises

Exercise X.3.1

Let $x^{\alpha'}$ be another coordinate system in V. Set

$$A_\alpha^{\alpha'} := \frac{\partial x^{\alpha'}}{\partial x^\alpha}.$$

The corresponding change of components of a vector in $T_x V$ is

$$p^{\alpha'} = \frac{\partial x^{\alpha'}}{\partial x^\alpha} p^\alpha = A_\alpha^{\alpha'} p^\alpha.$$

Therefore, the change of coordinates $(x^\alpha, p^\alpha) \to (x^{\alpha'}, p^{\alpha'})$ in TV implies

$$\frac{\partial p^{\alpha'}}{\partial p^\alpha} = A_\alpha^{\alpha'}, \quad \frac{\partial p^{\alpha'}}{\partial x^\beta} = \partial_\beta(A_\alpha^{\alpha'} p^\alpha). \tag{X.13.1}$$

Hence, for the components $(p^{\alpha'}, G^{\alpha'})$ of X in the new coordinates on TV,

$$p^{\alpha'} = A^{\alpha'}_\alpha p^\alpha, \qquad G^{\alpha'} = p^\beta \partial_\beta (A^{\alpha'}_\alpha p^\alpha) - A^{\alpha'}_\alpha \Gamma^\alpha_{\lambda\mu} p^\lambda p^\mu.$$

We know that (see Chapter I)

$$\Gamma^{\alpha'}_{\beta'\gamma'} = A^{\alpha'}_\alpha \partial_{\beta'} A^\alpha_{\gamma'} + A^{\alpha'}_\alpha A^\beta_{\beta'} A^\gamma_{\gamma'} \Gamma^\alpha_{\beta\gamma}. \tag{X.13.2}$$

Some straightforward manipulations shows that

$$G^{\alpha'} \equiv -\Gamma^{\alpha'}_{\beta'\gamma'} p^{\beta'} p^{\gamma'}. \tag{X.13.3}$$

Exercise X.3.2

$$\mathcal{L}_X g(p,p) \equiv p^\alpha p^\lambda p^\mu \frac{\partial g_{\lambda\mu}}{\partial x^\alpha} - 2g_{\alpha\rho} p^\rho \Gamma^\alpha_{\lambda\mu} p^\lambda p^\mu$$

$$\equiv p^\alpha p^\lambda p^\mu \frac{\partial g_{\lambda\mu}}{\partial x^\alpha} - g_{\alpha\rho} p^\rho g^{\alpha\nu} [\lambda\mu, \nu] p^\lambda p^\mu$$

$$\equiv p^\alpha p^\lambda p^\mu \left(\frac{\partial g_{\lambda\mu}}{\partial x^\alpha} - [\lambda\mu, \alpha] \right) \equiv 2p^\alpha p^\lambda p^\mu \left(\frac{\partial g_{\lambda\mu}}{\partial x^\alpha} - \frac{\partial g_{\lambda\alpha}}{\partial x^\mu} \right) \equiv 0.$$

Exercise X.3.3

$g_{\alpha\beta} p^\alpha p^\beta = \text{constant}$ implies

$$2g_{00} p^0 \frac{\partial p^0}{\partial p^i} + 2g_{0j} p^j \frac{\partial p^0}{\partial p^i} + 2g_{0i} p^0 + 2g_{ij} p^j \equiv 2p_0 \frac{\partial p^0}{\partial p^i} + 2p_i = 0;$$

hence

$$\frac{\partial p^0}{\partial p^i} = -\frac{p_i}{p_0}.$$

We have $f_m(x,p) := f(x, p^0, p^i)$, with $g_{\lambda\mu} p^\lambda p^\mu = -m^2$, and hence

$$\frac{\partial f_m}{\partial x^\alpha} = \frac{\partial f}{\partial x^\alpha} - \frac{p^\lambda p^\mu}{2p_0} \frac{\partial g_{\lambda\mu}}{\partial x^\alpha} \frac{\partial f}{\partial p^0} \quad \text{and} \quad \frac{\partial f_m}{\partial p^i} = \frac{\partial f}{\partial p^i} - \frac{p_i}{p_0} \frac{\partial f}{\partial p^0},$$

from which the result follows.

Exercise X.6.1

For an arbitrary function such that $\phi(t) = \phi(-t)$, it holds that

$$\int_{-\infty}^{+\infty} \phi(t) t \, dt = \int_{-\infty}^0 \phi(t) t \, dt + \int_0^{+\infty} \phi(t) t \, dt = 0,$$

because

$$\int_{-\infty}^0 \phi(t) t \, dt = \int_\infty^0 \phi(-t') t' \, dt' = -\int_0^\infty \phi(-t') t' \, dt'.$$

X.14   Problems

X.14.1   Liouville's theorem and generalization

We have shown in Problem I.14.1 in Chapter I that the volume form θ of the tangent bundle TV of a Lorentzian spacetime is invariant under the geodesic flow; that is, with $\mathcal{L}_X$ denoting the Lie derivative with respect to the vector $X = (p^\alpha, G^\alpha)$, it holds that $\mathcal{L}_X\theta = 0$.

Using the identity $\mathcal{L}_Y \equiv d(i_Y\theta) + i_Y d\theta$, show that the volume form θ is invariant under the flow of a vector $Y = (p^\alpha, G^\alpha + \Phi^\alpha)$ if Φ is orthogonal to p and has a vanishing divergence.

Solution

The volume form on TV is

$$\theta := |\mathrm{Det}\, g| dx^0 \wedge \ldots \wedge dx^n \wedge dp^0 \wedge \ldots \wedge dp^n.$$

Recall that $\mathcal{L}_Y\theta \equiv d(i_Y\theta) + i_Y d\theta$. Since the volume form is of maximum degree on TV, we have $d\theta \equiv 0$, and hence $\mathcal{L}_Y\theta = d i_Y\theta$. The linearity of the interior product gives

$$i_Y\theta = i_X\theta + i_{(0,\Phi)}\theta.$$

We already know that $d(i_X\theta) = 0$. We set $\theta = \theta_x + \theta_p$; then

$$i_{(0,\Phi)}\theta = \theta_x \wedge i_\Phi\theta_p, \tag{X.14.1}$$

with

$$(i_\Phi\theta_p)_{\alpha_1\ldots\alpha_n} = \Phi^\alpha(\theta_p)_{\alpha\alpha_1\ldots\alpha_n},$$

and hence, because of antisymmetry,

$$d(i_\Phi\theta_p) = \frac{\partial\Phi^\alpha}{\partial p^\alpha}\theta_p = 0 \quad \text{if} \quad \frac{\partial\Phi^\alpha}{\partial p^\alpha} = 0.$$

When $\Phi^\alpha = F^\alpha{}_\beta p^\beta$, it holds that

$$\frac{\partial\Phi^\alpha}{\partial p^\alpha} = F^\alpha{}_\alpha = 0.$$

X.14.2   Vlasov equation for particles with random charges

Assume that the considered 'particles' have, like momenta, random electric charges. The phase space $\mathcal{F}_V$ is the bundle over V whose fibre $\mathcal{F}_xV$ is the product $P_xV \times I$, with I an interval of R. The volume form in $\mathcal{F}_V$ is the product, with θ given by (2.2),

$$\theta_e := \theta \wedge de.$$

The particles create an *average electromagnetic field* F and are subjected to the Lorentz force

$$\Phi^\alpha := F^{\alpha\beta} j_\beta,$$

where j_β is the electric current associated to a particle of momentum p and charge e located at x, i.e. with u denoting its unit velocity,

$$j_\beta = e u_\beta \equiv e |g(p,p)|^{-\frac{1}{2}} p^\beta.$$

We denote by Y_e the vector field on $TV \times R$ with components

$$p^\alpha, G^\alpha + \Phi^\alpha, 1.$$

The volume form θ_e is clearly invariant under the flow of Y_e, as θ was under the flow of Y; that is, $\mathcal{L}_Y \theta_e = 0$. Particle number conservation gives for the distribution function f the **EM–GR–Vlasov equation with random charges**:

$$\mathcal{L}_Y f \equiv p^\alpha \frac{\partial f_e}{\partial x^\alpha} + (-\Gamma^\alpha_{\lambda\mu} p^\lambda p^\mu + F^\alpha{}_\beta j^\beta) \frac{\partial f_e}{\partial p^\alpha} + \frac{\partial f_e}{\partial e} = 0. \qquad \text{(X.14.2)}$$

X.14.3   Distribution function on a Robertson–Walker spacetime with Vlasov source

Consider a spacetime $M \times R$ with a Robertson–Walker metric.

$$g \equiv -dt^2 + R^2(t)\sigma^2, \quad \text{with} \quad \sigma^2 \equiv \gamma_{ij} dx^i dx^j.$$

σ is a given Riemannian metric on M.

1. Derive the Einstein–Vlasov equation for a general distribution function f.
2. Look for a solution f_m depending only on t and p^0 for particles of a given rest mass m. Show that it satisfies the equation

$$R \frac{\partial f_m}{\partial R} - [(p^0)^2 - m^2] \frac{1}{p^0} \frac{\partial f_m}{\partial p^0} = 0.$$

3. Show that f_m is an arbitrary function of the scalar $R^2 [(p^0)^2 - m^2]$.
4. Suppose that f_m vanishes at time t_0 for particles with momentum such that

$$(p^0)^2 \geq m^2 + k R^{-2}(t_0), \quad k \text{ some constant.}$$

Show that the maximum possible energy p^0 of particles with a given rest mass decreases with expansion, as foreseen physically.

Solution

1. Straightforward computation of $G^\alpha \equiv -\Gamma^\alpha_{\lambda\mu} p^\lambda p^\mu$ gives

$$\mathcal{L}_X f \equiv p^\alpha \frac{\partial f}{\partial x^\alpha} - RR' \gamma_{ij} p^i p^j \frac{\partial f}{\partial p^0} - 2R^{-1} R' p^0 p^i \frac{\partial f}{\partial p^i}.$$

For particles of a given rest mass m, i.e. such that

$$R^2 \gamma_{ij} p^i p^j = (p^0)^2 - m^2,$$

the equation for a distribution f_m depending only on t and p^0 reduces to

$$p^0 \frac{\partial f_m}{\partial t} - R^{-1} R' \left[(p^0)^2 - m^2 \right] \frac{\partial f_m}{\partial p^0} = 0.$$

Taking R instead of t as a variable, the equation reads as the following linear first-order partial differential equation:

$$R \frac{\partial f_m}{\partial R} - \left[(p^0)^2 - m^2 \right] \frac{1}{p^0} \frac{\partial f_m}{\partial p^0} = 0.$$

The general solution is constant along the rays (bicharacteristics) that satisfy the differential system

$$\frac{dR}{R} = -\frac{p^0 \, dp^0}{(p^0)^2 - m^2} = d\lambda.$$

These rays are such that

$$\log R + \frac{1}{2} \log \left[(p^0)^2 - m^2 \right] = \text{constant}, \quad \text{i.e.} \quad R^2 \left[(p^0)^2 - m^2 \right] = \text{constant}.$$

The distribution f_m is therefore an arbitrary function of the scalar $R^2 \left[(p^0)^2 - m^2 \right]$. Suppose, for instance, that f_m vanishes at time t_0 for particles with momentum such that

$$(p^0)^2 \geq m^2 + kR^{-2}(t_0).$$

Then the function f_m vanishes at time t for particles with momentum

$$(p^0)^2 \geq m^2 + kR^{-2}(t).$$

Hence the maximum possible energy p^0 of particles with a given rest mass decreases with t if R increases.

References

Alexakis, S., Ionescu, A. D., and Klainerman, S. (2013). arXiv:1304.0487v2 [gr-qc].

Anderson, A. and York, J. (1999). Phys. Rev. Lett. **82**, 4384–4387.

Andersson, L. and Moncrief, V. (2004). In *The Einstein Equations and the Large Scale Behaviour of the Gravitational Field: 50 Years of the Cauchy Problem in General Relativity*, ed. P. T. Chruściel and H. Friedrich, pp. 299–330. Birkhäuser, Basel.

Anile, A. M. (1989). *Relativistic Fluids and Magneto-fluids*. Cambridge University Press, Cambridge.

Anile, A. M. and Greco, A. (1978). Ann. Inst. Henri Poincaré A **29**, 257–272.

Arnowitt, R., Deser, S., and Misner, C. W. (1962). In *Gravitation: An Introduction to Current Research*, ed. L. Witten, pp. 227–265. Wiley, New York.

Bancel, D. (1973). Ann. Inst. Henri Poincaré A **18**, 263–284.

Bancel, D. and Choquet-Bruhat, Y. (1973). Commun. Math. Phys. **33**, 83–96.

Bel, L. (1958). C. R. Acad. Sci. Paris **247**, 1094–1096.

Bel, L. (1959). C. R. Acad. Sci. Paris **248**, 1297–1300.

Bieri, L. (2007). PhD thesis, ETH Zurich.

Blanchet, L. (2014). Living Rev. Relativity **17**, 2.

Blanchet, L. and Damour, T. (1989). Ann. Inst. Henri Poincaré A **50**, 377–408.

Bondi, H. (1947). Mon. Not. R. Astron. Soc. **107**, 410–425.

Bony, J.-M. (2001). *Cours d'Analyse: Théorie des distributions et analyse de Fourier*. Les Éditions de l'École polytechnique, Paris.

Bray, H. and Chruściel, P. (2004). In *The Einstein Equations and the Large Scale Behaviour of Gravitational Fields: 50 Years of the Cauchy Problem in General Relativity*, ed. P. T. Chruściel and H. Friedrich, pp. 59–70. Birkhäuser, Basel.

Bruhat, G. (1931). *Le soleil*. Alcan, Paris.

Bruhat, G. (1934). *Cours de mécanique*. Masson, Paris.

Bruhat (Choquet-Bruhat), Y. (1960). Astronautica Acta **6**, 354–365.

Bruhat (Choquet-Bruhat), Y. (1962). In *Gravitation: An Introduction to Current Research*, ed. L. Witten, pp. 130–168. Wiley, New York.

Brumberg, V. A. and Kopejkin, S. M. (1989). Nuovo Cim. B **103**, 63–98.

Buonanno, A. and Damour, T. (1999). Phys. Rev. D **59**, 084006.

Cantor, M. (1979). Compos. Math. **38**, 3–35.

Cantor, M. and Brill, D. (1981) Compos. Math. **43**, 317–330.

Cattaneo, C. (1959). Ann. Math. Pura Appl. (IV) **48**, 361–386.

Chaljub-Simon, A. and Choquet-Bruhat, Y. (1979). Ann. Fac. Sci. Toulouse **1**, 9–25.

Chandrasekhar, S. (1983). *The Mathematical Theory of Black Holes.* Oxford University Press, New York.

Cheng, T.-P. (2010). *Relativity, Gravitation and Cosmology: A Basic Introduction*, 2nd edn. Oxford University Press, Oxford.

Choquet, G. (2006). *Cours de mathématiques.* Ellipses, Paris.

Choquet-Bruhat, Y. (1965). C. R. Acad. Sci. Paris **261**, 354–356.

Choquet-Bruhat, Y. (1966). Commun. Math. Phys. **3**, 334–357.

Choquet-Bruhat, Y. (1968). In *Batelle Rencontres 1967*, ed. C. M. DeWitt and J. A. Wheeler, pp. 84–106. Benjamin, New York.

Choquet-Bruhat, Y. (1969a). Commun. Math. Phys. **12**, 16–35.

Choquet-Bruhat, Y. (1969b). J. Maths. Pures Appl. **48**, 117–152.

Choquet-Bruhat, Y. (1971a). Commun. Math. Phys. **21**, 211–218.

Choquet-Bruhat, Y. (1971b). Ann. Inst. Fourier **21**(3), 181–201.

Choquet-Bruhat, Y. (1974). Gen. Rel. Grav. **5**, 49–60.

Choquet-Bruhat, Y. (1983). In *Relativity, Cosmology, Topological Mass and Supergravity*, ed. C. Aragone C., pp. 108–135. World Scientific, Singapore.

Choquet-Bruhat, Y. (1989). Class. Quant. Grav. **6**, 1781–1789.

Choquet-Bruhat, Y. (1992a). J. Math. Phys. **33**, 1782–1785.

Choquet-Bruhat, Y. (1992b). C. R. Acad. Sci. Paris Ser. I **318**, 775–782.

Choquet-Bruhat, Y. (1993). In *Waves and Stability in Continuous Media*, ed. S. Rionero and T. Ruggeri, pp. 54–69. World Scientific, Singapore.

Choquet-Bruhat, Y. (1996). In *Gravity, Particles and Space–Time*, ed. P. Pronin and G. Sardanashvily. World Scientific, Singapore.

Choquet-Bruhat, Y. (2000). Ann. Phys. (Leipzig) **9**, 258–266.

Choquet-Bruhat, Y. (2004). In *The Einstein Equations and the Large Scale Behaviour of Gravitational Fields: 50 Years of the Cauchy Problem in General Relativity*, ed. P. T. Chruściel and H. Friedrich, pp. 251–298. Birkhäuser, Basel.

Choquet-Bruhat, Y. (2009). *General Relativity and the Einstein Equations.* Oxford University Press, Oxford. [Referred to in the notes as YCB-OUP2009]

Choquet-Bruhat, Y. and Christodoulou, D. (1981). Acta Math **146**, 129–150.

Choquet-Bruhat, Y., Chruściel, P., and Loiselet, J. (2006). Class. Quant. Grav. **23**, 7383–7394.

Choquet-Bruhat, Y. and DeWitt-Morette, C. (1982). *Analysis, Manifolds and Physics.* Part I: Basics. North-Holland, Amsterdam. [Referred to in the notes as CB-DMI]

Choquet-Bruhat, Y. and DeWitt-Morette, C. (2000). *Analysis, Manifolds and Physics.* Part II. North-Holland, Amsterdam. [Referred to in the notes as CB-DMII]

Choquet-Bruhat, Y. and Friedrich, H. (2006). Class. Quant. Grav. **23**, 5941–5949.

Choquet-Bruhat, Y. and Geroch, R. (1969). Commun. Math. Phys. **14**, 329–335.

Choquet-Bruhat, Y. and Greco, A. (1983). J. Math. Phys. **24**(2), 377.

Choquet-Bruhat, Y., Isenberg, J., and Pollack, D. (2007). Class. Quant. Grav. **24**(1), 800–29.

Choquet-Bruhat, Y., Isenberg, J., and York, J.W. (2000). Phys. Rev. D **3**(8), 83–105.

Choquet-Bruhat, Y. and Moncrief, V. (2001). Ann Inst. Henri Poincaré **2**, 1007–1064.

Choquet-Bruhat, Y. and Noutchegueme, N. (1986). C. R. Acad. Sci. Paris Ser. I **303**, 259–263.

Choquet-Bruhat, Y. and Novello, M. (1987). C. R. Acad. Sci. Paris Ser. II **305**, 155–160.

Choquet-Bruhat, Y. and Ruggeri, T. (1983). Commun. Math. Phys. **89**, 275–289.

Choquet-Bruhat, Y. and York, J.W. (1980). In *General Relativity and Gravitation: One Hundred Years After the Birth of Albert Einstein*, Vol. I, ed. A. Held. Plenum, New York.

Choquet-Bruhat, Y. and York, J.W. (1996). In *Gravitation, Electromagnetism and Geometric Structures*, ed. G. Ferrarese, pp. 55–74. Pythagora, Bologna.

Choquet-Bruhat, Y. and York, J.W. (2002). In *Cosmological Crossroads: An Advanced Course in Mathematical, Physical and String Cosmology*, ed. S. Cotsakis and E. Papantonopoulos, pp. 29–58. Springer-Verlag, Berlin.

Christodoulou, D. (1984). Commun. Math. Phys. **93**, 171–195.

Christodoulou, D. (2007). *The Formation of Shocks in Relativistic Fluids*. EMS, Zurich.

Christodoulou, D. (2009). *The Formation of Black Holes in General Relativity*. EMS, Zurich.

Christodoulou, D. and Klainerman, S. (1993). *The Global Nonlinear Stability of the Minkowski Space*. Princeton University Press, Princeton, NJ.

Chruściel, P. and Friedrich, H., eds. (2004). *The Einstein Equations and the Large Scale Behaviour of Gravitational Fields: 50 Years of the Cauchy Problem in General Relativity*. Birkhäuser, Basel.

Cotsakis, S. and Papantonopoulos, E., eds. (2002). *Cosmological Crossroads: An Advanced Course in Mathematical, Physical and String Cosmology*. Springer-Verlag, Berlin.

Dafermos, M., Holzegel, G., and Rodnianski, I. (2013). arXiv: 1306.5364v2 [gr-qc].

Dafermos, M. and Rendall, A. (2006). arXiv:gr-qc/0610075v1.

Damour, T. (1982). C. R. Acad. Sci. Paris Ser. II, **294**, 1355–1357.

Damour, T. (1983a). In *Gravitational Radiation*, ed. N. Deruelle and T. Piran, pp. 59–144. North-Holland, Amsterdam.

Damour, T. (1983b). Phys. Rev. Lett. **51**, 1019–1021.

Damour, T. (2006). *Once upon Einstein*. A. K. Peters, Wellesley, MA (originally published as *Si Einstein m'était conté*. Cherche Midi, Paris, 2005).

Damour, T. (2013a). In *Time: Poincaré Seminar 2010*, ed. B. Duplantier, pp. 1–17. Birkhäuser, Basel.

Damour, T. (2013b). *Experimental Tests of Gravitational Theories*. Particle Data Group. Available at http://pdg.lbl.gov/2013/reviews/rpp2013-rev-gravity-tests.pdf.

Damour, T. and Deruelle, N. (1981). Phys. Lett. A **87**, 81–84.

Damour, T. and Deruelle, N. (1985). Ann. Inst. Henri Poincaré A **43**, 107–132.

Damour, T. and Deruelle, N. (1986). Ann. Inst. Henri Poincaré A **44**, 263–292.

Damour, T. and Iyer, B. R. (1991). Phys. Rev. D **43**, 3259–3272.

Damour, T. and Mukhanov, V. F. (1998). Phys. Rev. Lett. **80**, 3440–3443.

Damour, T. and Nagar, A. (2011). In *Mass and Motion in General Relativity*, ed. L. Blanchet, A. Spallicci, and B. Whiting, pp. 211–252. Springer-Verlag, Berlin.

Damour, T., Nagar, A., and Bernuzzi, S. (2013). Phys. Rev. D **87**, 084035.

Damour, T., Soffel, M., and Xu, C. (1991). Phys. Rev. D **43**, 3273–3307.

Damour, T., Soffel, M., and Xu, C. (1992). Phys. Rev. D **45**, 1017–1014.

Darmois, G. (1927). Mem. Sci. Math. **25**, 1–48.

D'Eath, P. D. (1975). Phys. Rev. D **11**, 1387–1403.

DeWitt, C. M. and Wheeler, J. A., eds. (1968). *Batelle Rencontres 1967*. Benjamin, New York.

Dionne, P. (1962). J. Anal. Math. Jerusalem, **1**, 1–90.

Eardley, D. and Moncrief, V. (1981). Gen. Rel. Grav. **13**, 887–892.

Eardley, D. and Moncrief, V. (1982). Commun. Math. Phys. **83**, 171–212.

Eckart, C. (1940). Phys. Rev. **58**, 919–924.

Ferrarese, G. (1963). Rendic. Matem. **22**, 147–168.

Ferrarese, G. (2004). *Riferimenti generalizzati in relatività e appliazioni*. Pitagora, Bologna.

Ferrarese, G. and Bini, D. (2007). *Relativistic Mechanics of Continuous Media*. Springer-Verlag, Berlin.

Fischer, A. E. and Marsden, J. E. (1979) In *Isolated Gravitating Systems in General Relativity*, ed. J. Ehlers, pp. 322–395. North-Holland, Amsterdam.

Fourès (Choquet) Bruhat, Y. (1948). C. R. Acad. Sci. Paris **226**, 48–51.

Fourès (Choquet)-Bruhat, Y. (1952). Acta Math. **88**, 141–225.

Fourès (Choquet)-Bruhat, Y. (1953). Bull. Soc. Math. France **81**, 255–288.

Fourès (Choquet)-Bruhat, Y. (1956). J. Rat. Mech. Anal. **5**, 961–966.

Fourès (Choquet)-Bruhat, Y. (1958). Bull. Soc. Math. France **86**, 155–175.

Friedrich, H. (1986). Commun. Math. Phys. **107**, 387–609.

Friedrich, H. (1998). Phys. Rev. D **57**, 2317–2322.

Friedrichs, K. O. (1954) Commun. Pure Appl. Math. **7**, 345–392.

Gårding, L., Kotake, T., and Leray, J. (1966). Bull. Soc. Math. France **94**, 25–48.

Geroch, R. (1970). J. Math. Phys. **11**, 437–439.

Gu, C.-H. (1973). J. Fudan Univ. **1**, 73–78.

Gundlach, C., Calabrese, G., Hinder, I., and Martin-Garcia, J. (2005). Class. Quant. Grav. **22**, 3767–3774.

Hawking, S. and Ellis, G. (1973). *The Large Scale Structure of Space-Time.* Cambridge University Press, Cambridge.

Hinder, T. et al. (the NRAR Collaboration) (2014). Class. Quant. Grav. **31**, 025012.

Hoffman, F. and Teller, E. (1950). Phys. Rev. **80**, 692–703.

Hu, H.-S. (1974). J. Fudan Univ. **2**, 92–98.

Huisken, G. and Ilmanen, T. (2001). J. Diff. Geom. **59**, 353–437.

Isenberg, J. and Moncrief, V. (2002). Class. Quant. Grav. **19**, 5361–5386.

Kerner, R. (1987). C. R. Acad. Sci. Paris **304**, 621–624.

Klainerman, S., Luk, J., and Rodnianski, I. (2013). arXiv:1302.59512 [gr-qc].

Klainerman, S. and Nicolo, F. (2003). *The Evolution Problem in General Relativity.* Birkhäuser, Basel.

Klainerman, S., Rodnianski, I., and Szeftel, J. (2012a). arXiv:1204.1767v2 [math.AP].

Klainerman, S., Rodnianski, I., and Szeftel, J. (2012b). arXiv:1204.1772v2 [math.AP].

Landau, L.D. and Lifshitz, E. M. (1987). *Fluid Mechanics*, 2nd edn. Pergamon, Oxford.

Leray, J. (1953). *Hyperbolic Differential Equations.* Institute for Advanced Studies, Princeton (mimeographed notes).

Leray, J. and Ohya, Y. (1968). Math. Ann. **162**, 228–236.

Lichnerowicz, A. (1939). *Problèmes globaux en mécanique relativiste.* Hermann, Paris.

Lichnerowicz, A. (1944). J. Math. Pures Appl. **63**, 39–63.

Lichnerowicz, A. (1955). *Théories Relativistes de la gravitation et de l'électromagnétisme.* Masson, Paris.

Lichnerowicz, A. (1967). *Relativistic Hydrodynamics and Magnetohydrodynamics.* Benjamin, New York.

Lindblad, H. and Rodnianski, I. (2005). Commun. Math. Phys. **256**, 43–110.

Lindblom, L., Scheel, M. A., Kidder, L. E., Owen, R., and Rinne, O. (2006). Class. Quant. Grav. **23**, S447–S462.

Marle, C. M. (1969). Ann. Inst. Henri Poincaré A **10**, 67–194.

Moncrief, V. (1986). Ann. Phys. (NY) **167**, 118–142.

Moncrief, V. and Isenberg, J. (2008). Class. Quant. Grav. **25**, 195015.

Müller, I. and Ruggeri, T. (1998). *Rational Extended Thermodynamics*, 2nd edn. Springer-Verlag, Berlin.

Müller zum Hagen, H., Seifert, H. J., and Yodzis, P. (1973). Commun. Math. Phys. **34**, 135–148.

Newman, R. (1986). Class. Quant. Grav. **3**, 527–539.

Ohanian, H. and Ruffini, R. (2013). *Gravitation and Spacetime*, 3rd edn. Cambridge University Press, Cambridge.

Oppenheimer, J. R. and Snyder, H. (1939). Phys. Rev. **56**, 455–459.

Penrose, R. (1965). Phys. Rev. Lett. **14**, 57–59.

Penrose, R. (1968). In *Batelle Rencontres 1967*, ed. C. M. DeWitt and J. A. Wheeler, pp. 121–235. Benjamin, New York.

Penrose, R. (1979). In *General Relativity: An Einstein Centenary Survey*, ed. S. W. Hawking and W. Israel, pp. 581–638. Cambridge University Press, Cambridge.

Pichon, G. (1965). Ann. Inst. Henri Poincaré A **2**, 21–85.

Pichon, G. (1967). C. R. Acad. Sci. Paris Ser. A **264**, 544–547.

Pretorius, F. (2005a). Class. Quant. Grav. **22**, 425–452.

Pretorius, F. (2005b). Phys. Rev. Lett. **95**, 121101.

Pretorius, F. (2007). arXiv:0710:1338v1 [gr-qc].

Rein, D. and Rendall, A. (1992). Commun. Math. Phys. **150**, 561–583.

Rendall, A. (1992). J. Math. Phys. **33**, 1047–1053.

Rendall, A. (2005). Living Rev. Relativity **8**, 6.

Rezzolla, L. and Zanotti, O. (2013). *Relativistic Hydrodynamics*. Oxford University Press, Oxford.

Ringström, H. (2009). *The Cauchy Problem in General Relativity*. EMS, Zurich. See also <http://www.math.kth.se/~hansr/mghd.pdf>.

Rodnianski, I. and Speck, J. (2009). arXiv:0911.5501v2 [math-ph].

Salomon, C. (2013). In *Time: Poincaré Seminar 2010*, ed. B. Duplantier, pp. 173–186. Birkhäuser, Basel.

Sbierski, J. (2013). arXiv:1309.7591v2 [gr-gc].

Schneider, P., Ehlers, J., and Falco, E. E. (1992). *Gravitational Lenses*. Springer-Verlag, Berlin.

Schutz, B. (1971). Phys. Rev. D **2**, 2762–2773.

Segal, I. E. (1976). *Mathematical Cosmology and Extragalactic Astronomy*. Academic Press, New York.

Taub, A. H. (1954). Phys. Rev. **94**, 1408–1412.

Taub, A. H. (1959). Arch. Rat. Mech. Anal. **3**, 312–329.

Taub, A. H. (1969). Commun. Math. Phys. **15**, 235–240.

Tolman, R. C. (1934). Proc. Natl Acad. Sci. USA **20**, 169–176.

Vaillant-Simon, A. (1969). J. Math. Pures Appl. **48**, 1–90.

Villani, C. (2002). In *Handbook of Mathematical Fluid Dynamics*, Vol. I, ed. S. Friedlander and D. Serre, pp. 71–306. North-Holland, Amsterdam.

Wald, R. (1984). *General Relativity*. University of Chicago Press, Chicago.

Will, C. M. (2014). Living Rev. Relativity **17**, 4.

Witten, L., ed. (1962). *Gravitation: An Introduction to Current Research*. Wiley, New York.

Wolf, J. (2011). *Spaces of Constant Curvature*, 6th edn. AMS, Providence, RI.

York, J. W. (1972). Phys. Rev. Lett. **28**, 1082–1085.

York, J. W. (1974). Ann. Inst. Henri Poincaré A **21**, 319–332.

Index

Notes: references to marginal notes are indicated by the suffix 'n' followed by the note number, for example 188n33; page numbers in italics refer to figures.